THE

LIFE *of the*

COSMOS

Lee Smolin

THE

LIFE *of the*

COSMOS

Weidenfeld & Nicolson
LONDON

First published in Great Britain in 1997
by Weidenfeld & Nicolson

The Orion Publishing Group
Orion House
5 Upper Saint Martin's Lane
London WC2H 9EA

First published in the USA in 1997
by Oxford University Press

ISBN 0 297 81727 2

A catalogue record for this book is available
from the British Library

Printed in Great Britain by
Clays Ltd, St Ives plc

Dedicated to

Laura Kuckes

1961–1990

Physician, playwright, friend,

who inspired me to write this book.

Contents

CONTENTS

THE

LIFE *of the*

COSMOS

This interconnection (or accommodation)
of all created things to each other,
brings it about that each simple substance
has relations that express all the others,
and consequently,
that each simple substance
is a perpetual, living mirror of the universe.

Just as the same city
viewed from different directions
appears entirely different
and, as it were, multiplied perspectively,
in just the same way
it happens that,
because of the infinite multitude of simple substances,
there are, as it were, just as many different universes,
which are, nevertheless,
only perspectives on a single one, . . .

And this is the way of obtaining as much variety as possible,
but with the greatest order possible,
that is, it is the way of obtaining as much perfection as possible.

—G. W. Leibniz, The Monadology, 56–58, 1714

Prologue/Revolutions

As the story is told, Nicolaus Copernicus received the first copy of his first and only book as he lay dying in the tower of the castle in northeastern Germany where he had lived and served as Deacon for the last half of his life. The book was titled *Revolutions of the Spheres* and it was, in time, to inspire such a radical shift of the world view of both the erudite and the ordinary that its title has since become our word for radical political and philosophical change. In this book, Copernicus expounded the astonishing idea that the Sun, and not the Earth, was at the center of the Universe, and that much of the mystery of the motions of the Sun, stars and planets could be explained by this simple hypothesis.

The legend of Copernicus, and the revolution that followed, has served—more than any other episode—to define and explain science, its power and its role in European, and now world, civilization. Like most legends, it contains shades of truth, and shades of distortion and misinformation. For example, while it is true that Copernicus did delay publication of his book as long as he could, it is not true that he did this because of real, or imagined, fear of prosecution by the Catholic Church. The church, which was his sole employer during his life, had supported him in his work; even a Pope had expressed interest in these new ideas.

It is more likely that this reluctant revolutionary was simply afraid of provoking controversy and public ridicule for advocating a view that went against both common sense and established science.

Whatever the truth might be, what we can be certain that Copernicus would have been even more reluctant had he known how different a universe from the one he inhabited would arise from the revolutions of his book. For, apart from exchanging the role of the Sun and the Earth, most of the steps we associate with the shift in world view, which we call the Copernican Revolution, were not even mentioned in his book. For Copernicus, and even for Kepler and Galileo, the Universe was finite and spherical. It was only large enough to contain the orbits of the six known planets, and the distant stars were lights affixed to the outer sphere that was its final boundary. There is no evidence that any of these, the main protagonists of the revolution, ever doubted that the universe had been created six thousand years earlier by God, who waited and watched just outside of that stellar sphere.

The more radical notions that led to our modern world view came after Copernicus. It was the mysterious monk turned mystic, Giordano Bruno, who proclaimed that space is infinite, that the stars are other suns, and that around them are other planets on which live other peoples. It was for these and other heresies, much more threatening to the authority of the church than the rather minor question of whether to reinterpret scripture to allow for the Earth's motion, that he was burned alive in the Campo dei Fiori in Rome in 1600. Copernicus also had no notion of the central ideas that, over the next century and a half, would shape a new view of the universe: that everything in the world, on the Earth and in the heavens, is made of atoms; and that all motion is governed by simple and universal laws. Indeed there is much more distance between the new picture of the universe as we find it finally synthesized in Newton's great book, the *Principia*, and Copernicus's work published almost one hundred and fifty years earlier, than there was between Copernicus and his contemporaries. Copernicus started a revolution, but it is doubtful whether he would have approved of the result.

A second great revolution in physics and cosmology was begun in the opening years of this century. It was launched with the introduction of relativity and quantum theory, each of which breaks decisively with the world view of Newtonian physics. These two theories were put in final form in 1916 and 1926, respectively. However, in spite of the fact that the basic formulations of relativity and quantum theory have not been modified for more than sixty years, the revolution in world view which they make necessary is not yet over. The reason for this is easy to see. The Aristotelian world view, which the Copernican revolution overthrew, described a single, unified theory of nature which could account for everything that happened, or might happen, in the universe then known to human beings. It explained what space and time are, what the shape of the cosmos is, what it contains, and how and why change takes place. The Newtonian world view, which was the result of the Copernican revolution, was also a comprehensive, unified theory, which applied to everything in the cosmos as it was then conceived. However, if the Newtonian world view has been overthrown in the 20th century, what has replaced it is not one new theory, but two. These two theories have been extremely successful in explaining both old and new phenomena, but neither of them can claim to be universal. The reason is, in each case, the existence of the other. Quantum theory has yet to successfully encompass the phenomena of gravitation, and Einstein's theory of relativity can only explain gravitation by ignoring quantum theory and treating matter as if Newton's world view still held.

Thus, the revolution which began with the creation of quantum theory and relativity theory can only be finished with their unification into a single theory that can give us a single, comprehensive, picture of nature. At the present time the construction of such a theory is the main goal of much work being done by theoretical physicists, and a number of different ideas are being actively pursued. In the last ten years there have been some remarkable developments that have brought us closer to this goal; still, it remains true that no one has been able to construct a theory which is completely satisfactory as a unification of quantum theory and relativity. It is still not even clear whether this can be accomplished without a radical change in the basic principles of either, or both, theories.

A successful unification of quantum theory and relativity would, for reasons I will explain in detail later, necessarily be a theory of the universe as a whole. It would tell us, as Aristotle and Newton did before, what space and time are, what the cosmos is, what things are made of, and what kinds of laws those things obey. Such a theory will bring about a radical shift—a revolution—in our understanding of what nature is. It must also have wide repercussions, and will likely bring about, or contribute to, a shift in our understanding of ourselves and our relationship to the rest of the universe.

What kind of universe will we live in after this revolution is over? At present, when the basic shape of the theory that will finish the revolution is unknown, there is very little we can say with assurance about this question. At the same time, it is possible, indeed, I would argue, necessary, to speculate. Because we cannot invent what we cannot conceive, the construction of a new theory must involve, or perhaps be preceded by, attempts to imagine the outcome. This book is one such attempt.

Because this book is frankly speculative, I want to be very clear so that the reader is not deceived as to its purpose or foundations. The person who is writing this book is a theoretical physicist who has been involved for many years in the attempt to construct a theory that unifies quantum theory and relativity. As a scientist he has had some success, but on the whole has been as unable to construct a theory that achieves that goal as anyone else who has been engaged in that struggle. Thus, what the reader holds is not, strictly speaking, a work of science. In order to make the argument clearer, a number of things from particle physics, cosmology, relativity and quantum theory will be explained. Many of these are now part of established knowledge, but others are new ideas or hypotheses that have not yet been sufficiently tested to be considered part of science. I will try to be very clear about which is which.

Instead, what this book does do is to attempt to sketch out a vision that has been taking shape in my mind during the almost twenty years that I have been studying and working on the problem of the unification of quantum mechanics and relativity theory. This vision has not so far been realized in a complete theory; were I able to do that I would be writing a very different kind of a book. It has provided the motivation for much of my work as a physicist, but I must confess at the beginning that the success I have had in constructing a mathematical framework to realize the ideas I will be describing here has been, compared to the aims of the work, modest.

Thus, what I am presenting in this book is a frank speculation, if you will, a fantasy. This fantasy is inspired by diverse sources and issues , some physical and mathematical, some biological, and others philosophical. Similarly, this book is addressed to several different audiences, which include interested lay people, philosophers, biologists and astronomers, and my own colleagues in theoretical physics. Indeed, it is necessary that I write for a wide audience, as the number of people who are acquainted with both the scientific ideas and the philosophical sources I will use in these arguments is rather small. Thus, if this book is to be worth writing, I have to explain the philosophical background to the scientists and the scientific problems to the philosophers. After this, only a little more is needed to make the book accessible to anyone with a general curiosity about the world, and so that is what I have done.

As the point of view I will presenting is personal, I hope the reader will not mind if I introduce from time to time something of my own history and evolution. I do this at the risk of seeming egotistical, especially to that part of my audience that are scientists. To recount the story of our own thought is not something that we do in scientific discourse. There are good reasons for this and, while I reject the objective, detached view of what a scientist is, I do think that there is a purpose to the pretense of objectivity and impersonality that is normal scientific practice. It expresses a kind of modesty, a conviction that if our thought is to be ultimately worth something it will be agreed to by any scientist, in spite of what ever personal evolution they have come through.

However, I will take this risk, for two reasons. First, I think that it will make the book more interesting for the nonscientists, who—not sharing the conventions of formal scientific monologue—can relate more easily to a personal story than to a ritualized display of impersonality. Second, I think it is a good thing if those who are not scientists see us more closely, as people. It demystifies who we are and what we do and, given the authority that science has come to have in our society, I think that this is not a bad thing. However, let me stress also that there are as many different ways people come to be in science as there are ideas and points of view about science. There is no such thing as a typical scientist.

In my case, my engagement with physics started when I was a high school student in the early 1970's, infatuated with rock and roll, what I understood to be revolutionary politics, and my girlfriend. I also was studying mathematics, not because I really liked it, but because my high school had decided that I was not bright enough to take the advanced track in mathematics, and to prove them wrong I had done the three years of advanced high school mathematics in one. It was, to begin with, another thing to do to demonstrate the incorrectness of their ideas about education, and if it was not as subversive as rock and roll and publishing an underground newspaper, I was slowly discovering it was almost as much fun.

In eleventh grade I arranged for Buckminster Fuller to speak at my high school, and after listening to him wanted passionately to study architecture. In the next year my room filled up with straw, pipe cleaner and thread fantasies of domes, suspension buildings and other exotic structures. I was especially fond of warped and stretched geodesic domes; that is, buildings constructed with all kinds of curved surfaces. An overdeveloped sense of responsibility made me wonder how one calculated the stresses on these curved surfaces, in order to insure that they did not fall down on some hapless future clients, and going to the library, I found I needed something called "tensor calculus". It was my good luck that my spitefulness had taken me through enough mathematics that I could read the books on this subject. It was also luck (the kind of luck that one's life is made out of) that tensor calculus is exactly the mathematics with which Einstein built his theory of space and time, so that every book on the subject ended with a chapter on relativity theory.

To learn more about relativity I went back to the library, and there found a book of essays dedicated to Einstein, edited by Paul Schilpp. It was prefaced by Einstein's only attempt at autobiography, titled "Autobiographical Notes." Although he quickly slips into scientific autobiography, there are a few pages at the beginning in which he describes his growing up and reasons for going into science. Now, at that moment I was a very disillusioned young man, my band had broken up, my girl friend had left me, and as for our beautiful revolution . . . need I say anything? Einstein reported a similar melancholy (I'm sure now, with much better reason) and described science as a transcendent calling, whereby one could rise about the pettiness, impermanence, and struggle of human life and visit a bit with the beauty and truth that underlies the real, permanent, nature of the world. This was strong stuff for a failed sixteen-year-old rock-and-roll star, and it occurred to me then and there that if I could do nothing else with my life, perhaps I could do that. I resolved at that moment to become a theoretical physicist.

In the rest of the essay Einstein decries the sorry state of theoretical physics as he saw it in 1950. His general theory of relativity had revolutionized our understanding of what space and time are, but it had failed to produce a picture of the world that accounted for most known phenomena. In particular, it had nothing to say about atoms, the particles that make up the atom, and their interactions with light, electromagnetic fields, and the other forces that hold the atom together. As the great physics of the twentieth century, at least from an experimental point of view, was atomic and nuclear physics, this meant that for most of physics general relativity was, at best, irrelevant. But, what for Einstein was worse was that the theory that had proved successful in the atomic and nuclear realm made little sense to him.

This theory, quantum mechanics, owed its birth to Einstein as much as to any other single person. But the final form of the theory, as it had been arrived at in 1926 did not seem to him to fit the criteria of a fundamental physical theory. While he had to concede that it worked, he could not accept it, and the growing success of quantum mechanics meant Einstein himself was more and more irrelevant—marginalized, in the modern lingo. Indeed, Einstein spent the last twenty years of his life in a vain attempt to invent a theory, the so called unified field theory, that would supersede quantum mechanics. Einstein was well aware of his failure to achieve this and what he so forcefully communicated in that essay was that until there is some theory that brings together the successes of relativity and quantum theory, we will not have a physical theory that can be relied upon to give us a complete picture of what the world is. And, as the ultimate task of physics is to provide that picture then, without it, despite all of the experimental successes of relativity and the quantum theory, we do not have, in the deepest sense, any physical theory at all.

Thus, two days after taking the Shilpp book out of the library, I not only had a profession, I had a mission. I would learn physics, become a theoretical physicist, and then take on the problem of the relationship between quantum theory and relativity. If there was not going to be a revolution in society, at least maybe there was some kind of revolution to which I could contribute.

Before starting college I took a job for a few months as a sheet metal apprentice in Los Angeles. It was a hot and uncertain summer, I was sure of my intention to be a scientist, but I had failed chemistry, been refused admission to the physics class, and had finally left high school to study on my own. I did the job all right, notwithstanding the time I backed the truck into the boss's Cadillac, but business was slow and we were often idle. I don't imagine I fit in, but they were nice to me and no one objected when I spent the empty afternoons sitting in the sun reading books with titles like *Lectures on Quantum Mechanics* and *What is Life?* It was during one such sultry day that I wrote down three questions that I hoped I could devote myself to answering.

1. What is the universe?
2. Within the context of the answer to the first question, what is a living thing? What is life?
3. Within the context of the answers to the first two questions, what is a human being? Who are we?

I have tried to put in this book what I have learned about the first two of these questions, in the time since then.

Introduction

What is the universe? Is it infinite, or finite? Is it eternal, or did time begin at some first moment? If it began, what began it? Everyone has wondered about these questions, and every culture has told a story about them. As human knowledge and culture has expanded, this story has changed and grown, in a succession of stages, as the known world expanded and the realm of mythology and speculation was pushed back.

Now at the end of the twentieth century we stand at the verge of another step in our growing understanding of the universe. We live in the middle of one of the great revolutionary periods in our understanding of nature, as we attempt to meld into one framework what we have learned about relativity, the quantum and the expanding universe. This framework must provide answers to the questions I began with, as well as other questions, which the developments of this century have made urgent, such as: Why is there life in the universe? Why is the universe full of such a variety of beautiful structures? Are the laws of physics eternal truths, or were they somehow created, with the world? Is it possible to conceive of, and understand, the universe as a whole system, as something more than the sum of its parts?

How will we see the universe when we have completed this revolution, when we have succeeded in unifying the different developments of twentieth century science? Of course, we must begin by admitting that we don't know, the story is not complete, the great synthesis that will close the revolution, as Newton closed the revolution begun by Copernicus, has not yet been invented. I believe, however, that a picture is emerging of what this new universe will look like. The purpose of this book is to describe this new picture and to explain why I have come to believe in it.

The evidence for this new picture of cosmology comes from several different directions and sources. It comes first of all from the new information that observational astronomers and cosmologists have recently gathered about the organization of the universe. Beyond the simple fact that the universe is evolving and changing, we are finding we live in a world that is much more dynamical, much more intricately structured, much more interesting than we had previously imagined. Moreover, we are now confronted with observations that, if current theory is to be believed, extend our view of the universe out to appreciable fractions of its apparent size and back in time almost to its beginning. Observational astronomy is beginning to do what previous generations of scientists could only have dreamed about, which is to give us a view of the entire universe.

A different kind of evidence comes from the attempts to combine quantum theory and relativity into a unified theory of the whole universe. This problem is not yet completely solved, but at the same time definite progress has been made in the last decade. In order to do this we have had to learn to give up certain ideas, which seemed previously basic to our understanding of nature, and adopt new ones. This is an exciting story, not yet finished, but far enough along for us to learn something from it about what the new universe must look like.

Further evidence for this new view comes from philosophy. Philosophy cannot settle scientific questions, but it has a role to play. A bit of philosophical thought may prevent us from getting hung up on a bad idea, and the record of people who have struggled with the deep questions we face, such as they meaning of time and space, may suggest new hypotheses for us to play with.

There is one philosophical question that faces us urgently, given the tremendous progress in observation and theory of the last years: What might it mean to extend science to encompass the whole universe? Is it possible to describe the whole of the universe in scientific terms? And, if it is possible, how must we modify our current theories in order to be able to do this?

I have come to believe that this is the central issue that we must confront if we are to solve many of the key open problems in theoretical physics. How we think about the universe as a whole affects such apparently diverse questions as the problem of unifying quantum theory with general relativity, the problem of understanding the origin of the properties of the elementary particles, the problems of the interpretation of the quantum theory, the problem of what "caused" the Big Bang, and the question of why the universe is hospitable to life. These are all problems we have so far failed to solve, in spite of having made partial progress on some of them. In my view, part of the reason for this is that we have not paid enough attention to the ways in which a theory that could be sensibly applied to the whole universe must differ from our present theories.

Before wasting too much effort, we might ask whether it is even possible to have a scientific theory that takes as its subject the whole universe. Certainly there are good reasons why this might seem presumptuous. By definition, the universe contains all that exists. How then could we imagine that we could ever come to know more than a tiny fraction of it?

What I mean by asking whether science can be extended to encompass the whole universe is something rather different from asking whether we can know about everything that exists. The question is instead whether we can *understand* the whole of the universe as comprising a single, interrelated system. To make clear what I mean let me begin with the assertion that what science has been doing for the last 400 or so years is giving a description, not of the universe as a whole, but of small parts of it. For example, we have theories of atoms, nuclei, living things, stars, galaxies, and so on. In each case we are studying something that can be contained in a region of the universe that is very small compared to the whole.

We have made, in these last three centuries, wonderful progress in our understanding of parts of the universe. However, it must be appreciated that to describe a particular part of the universe is a very different thing than to give a theory about the whole. There is a simple reason for this, that is easy to explain. This is that in practice we never really describe a portion of the world as if it were completely detached from the rest. Even if the object of our theory is to describe only a single atom, aspects of the rest of the world still enter the description, at least implicitly. Even the simple act of describing where something is, or when something happened, involves implicit reference to the rest of the world. Because of this, all the theories that describe parts of the world actually need the rest of the world in order to make complete sense.

Let us call what is left out, when we study a particular part of the universe, the *background*. As it is not the subject of the theory, the background is always assumed to be fixed, and this provides in each case a necessary reference or framework, with respect to which the properties of the system we are studying can be defined. For example, when we do an experiment on an atom the experimental equipment we use to "see" that atom is part of the background. When people give a theory of the human personality, there are often assumptions about the nature of society that are kept fixed, and are hence part of the background. Or, when we give a theory of motion, there are certain fixed assumptions we make about the properties of space and time, which form aspects of the background.

It is no shame to make such use of a background when we are studying a part of the universe. For one thing, it is often very helpful to be able to describe the properties of things with respect to reference points that we may take as being fixed. The problem arises when we attempt to make a theory of the whole universe, for in that case there cannot be anything outside of the system. There is then no fixed background, with respect to which the properties of things in the world can be defined. One cannot speak of where the universe is, or when it happened. We can only speak of where and when things happened inside of it. And without a fixed background, the only reference points available to describe where or when something happens are other events.

The problem of how to make a theory of a whole universe is thus the problem of how to construct a theory without making any reference to anything that exists, or anything that we might have imagined happened, outside of the system we are describing. We must discover a way to speak of the properties of the things in the universe, the particles, atoms, fields, stars, galaxies, as well as space and time themselves, without ever once referring to something that is not one of those things, and which hence lies outside of the universe.

This problem is not new. As long as people have been doing science, there were those who pointed out that a theory of the whole world must be different in certain respects from the theories people were constructing. Newton's physics was a great achievement, but it relied heavily on the use of a fixed background and, for this reason, it could never have stood as a theory of a whole universe. Even at the moment that Newton wrote down his laws of physics, towards the end of the seventeenth century, there were those who criticized them on this basis. Among these critics was the great philosopher and mathematician Leibniz, about whom I will have occasion to speak more than once in the following pages. The reader who knows something of the history of philosophy will easily recognize that a great deal of what I have to say here is not new.

However, our situation is now very different from that of Newton's seventeenth century critics. We live in the ruins left by the overthrow of Newtonian science, trying to make sense of the many new discoveries that have grown up suddenly like a lush forest among the scattered stones of an ancient temple. In the confusion and promise of this situation our current theories are temporary dwellings, built with what is at hand, which includes both the stones of the old science and the trunks and branches of the new. Our task is to merge these together into something as coherent and lasting and, we hope, as beautiful, as the old Newtonian edifice.

Underlying the view of the world that grew up out of Newton's science have been certain assumptions about what matter is, what space and time are, what the universe as a whole is. Since they have led to such success, these assumptions have not often been questioned, at least in scientific circles. But all of the evidence I have mentioned tells us that nevertheless, we must question them. The fundamental principles on which the Newtonian picture of the world was built included the idea that the universe is eternal, that everything is made out of particles which obey absolute and unchanging laws, and that everything in the world can be reduced ultimately to the action of these absolute laws. If this view is correct, then the only truly fundamental science must be the study of what these particles are, and how they move and interact with each other. Everything else, whether it is biology or astronomy, is to be understood ultimately in terms of these fundamental particles and the laws they obey. The whole living world around us must then be understood as being, in a sense, accidental and historical; all that is truly necessary or essential in nature is the fundamental particles and laws.

This point of view has often been opposed because it seems to cheapen life, to make our existence meaningless, to make beauty irrelevant. Of course, these kinds of arguments have not mattered much to science, because this view has seemed a necessary underpinning of its momentous progress. But now, at this moment of crisis, it seems that it is exactly this view of things that must be challenged. It is not just a question of ethics, or what makes us feel comfortable, this view is no longer working as science.

When I began my study of physics I imagined that the reality behind the world we see around us is formed by some beautiful mathematical law, which exists eternally, transcending the short and petty existence of living beings like myself. This was the picture I learned from my adolescent reading of Einstein. As I grew up and slowly became a physicist, I learned that I was far from the first to have been seduced by this vision. Platonism, the search for the eternal and abstract behind the transient and perceived world, has driven the searchings of physicists and mathematicians from ancient times to the present. And who can blame us, for certainly the mathematical beauty of relativity and quantum theory stand as the strongest confirmation of this vision. But yet, one thing I want to communicate in this book is how the difficulties that we face in extending those theories to a complete description of the universe have caused me to doubt whether the foundations of the world are indeed to be grasped *solely* by the discovery of a perfect and eternal mathematical law. In its place, I believe that we are beginning to see evidence of an alternative view. In this new view it becomes possible to imagine that a great deal of the order and regularity we find in the physical world might have arisen just as the beauty of the living world came to be: through a process of self-organization, by means of which the world has evolved over time to become intricately structured.

The idea that the world must be understood to be the result of processes of self-organization, and not just a reflection of fixed and eternal natural law, may be a difficult one for many readers to accept. I myself rejected such ideas when I first encountered them. But, through a chain of reasoning I will do my best to present here, I have come to believe that the transition from a notion of law as absolute and eternal to a notion of a universe whose regularities have evolved through processes of self-organization is a natural and necessary consequence of the shift from a science of parts of the world to a science of the whole universe.

Another aspect of this shift is the realization that how the world is organized is as fundamental a question as what it is made of. From the point of view of the old, Newtonian-style physics, the structure of the world is accidental. The law of increasing entropy tells us that the natural state of a world described by nineteenth-century physics is a dead equilibrium. But a common conclusion of the different arguments of this book will be that, from the point of the view of the new physics, complexity must be an essential aspect of the organization of the world. Indeed, it is not only that a world with life must necessarily be complex. As I will explain when we come to it, in the twentieth century our very understanding of space and time, of what it means to say where something is or when something happened, requires a complex world.

This means that the picture of the universe in which life, variety and structure are improbable accidents must be an outmoded relic of nineteenth-century science. Twentieth-century physics must lead instead towards the understanding that the universe is hospitable to life because, if the world is to exist at all, then it must be full of structure and variety.

Just as the problem of how to construct a theory of the whole universe is not new, neither is the idea that this must involve a change in how we understand what laws of nature are. Over the last century or so, several deep thinkers, confronted with the new discoveries in physics, astronomy and biology, have argued that the Platonic conception of law as mathematical and eternal must give way to a view in which the laws are themselves formed as a result of process of evolution or self-organization. The great theorist and teacher John Wheeler long ago came to this view, and I have found aspects of it in the writings of a number of theorists, such as Per Bak, Paul Davies, Stuart Kauffman, Andrei Linde, Yochiro Nambu, Holgar Nielsen, Sam Schweber and Walter Thirring. Moreover, it seems that quite soon after Darwin more than one philosopher reached the view that the idea of evolution must be significant for the problem of constructing a theory of cosmology. One finds in the writings of the late nineteenth century American pragmatist Charles Pierce the idea that the laws of physics must be understood to be the result of a process of evolution. At the same time, the French philosopher Henri Bergson suggested that the description of the universe as a whole must be closer to the description of a living organism than it is to the description of a simple physical system isolated in a small part of the universe.

These philosophers had good reasons for their views. But a good measure of the progress of twentieth-century science is how much better a position we are in now to actually try to implement such ideas. Part of the reason we can now contemplate a shift in our understanding of what the laws of physics might represent is the very success we have recently achieved towards what has traditionally been taken as the purpose of physics: to discover what the laws of nature are. For we now have in our possession laws that can describe correctly every experiment we have been able to invent. We are certainly not finished with this task, as the existence of open questions such as the ones I have mentioned indicates. But we are far enough along that it now becomes possible to ask as a serious question: not, What are the laws of nature? but, Why are the laws of nature as we find them to be, and not otherwise?

There is yet another reason why we are in a much better position to try to answer such a question than Pierce and Bergson were a century ago. In the meantime, we have made a monumental discovery that completely changes the context in which we ask it. This is that the universe we see around us is not eternal. Nor is it static. It was born a finite time ago, and it evolved over time to reach its present state. As far as we can tell, it began in a very different state, the so-called "Big Bang," when the densities and temperatures were at least as high as any now to be found in the universe. Furthermore, the universe left this state not really very long ago, when we think in terms of an appropriate time scale, for the universe seems barely older than the stars and galaxies it contains. Nor is it very much older than the history of life on Earth.

In recent years we have been able to gather many details about how our universe evolved to its present state. We see back to the time of the quasars, when the galaxies were most likely being formed, before most of the elements that make us up had been forged in stars. We see further back to the time of the cosmic blackbody radiation, when the whole universe resembled the conditions at the surface of a star. And we have been able to assemble reasonable arguments that enable us to discuss critically what happened in the first minutes and seconds of its life, when we surmise the universe was still denser and hotter. Thus, rather than living in an eternal cosmos, we live in a young world, the story of whose maturation we see spread out before us as we look out with our telescopes and antennas. This makes it possible to ask, as scientific questions, not only how was the world we see around us made, but what existed before *this* world? It is not too much of an exaggeration to say that the question of what happened during, and perhaps even before, the Big Bang is slowly coming into focus in the last years of this century in the same way that the question of what happened before the origin of our species came into focus during the last.

The fact that our universe is young and evolving puts the question of the origin of the laws of nature in a quite different light. If the universe is eternal, there are only two possible answers for the question of why the laws of nature are as we find them to be: religion or Platonism. Either God (who is, in most tellings, eternal) made the laws of nature as he made the world; or they are as they are because there is a mathematical form for the laws that is somehow fixed by some abstract principle. But although deism and Platonism seem, at first, poles apart, in a certain sense these two kinds of explanation are not really very different. Mathematical truth is supposed to be eternal, as is a god. Mathematical truth is supposed to be something that holds irrespective of what is in the world, or indeed whether the world exists at all. A world made by mathematical law, like a world made by a god, is a world constructed by something that exists eternally and outside of the world it creates.

But if our world is not eternal, then new possibilities open up. It seems, all of a sudden, a bit too much to postulate eternal laws for a world whose origin we can, almost literally, see. If our world could have been made a short time ago, could not the laws that govern it also have been made? If it is possible to imagine natural processes that created the world, can we not also imagine processes that could have created or selected the laws that the universe would obey?

In the nineteenth century, biologists learned to give up the idea that the species are eternal categories, and replace it with a dynamical view in which the living world constructs itself through the process of natural selection. They gained from this a much more rational framework for biology, for the properties of the living things now have reasons, and these can be traced if one knows their histories. Although it may not seem so to a platonist, who confuses rationality with the invention of an imaginary world of eternal ideas, biologists and geologists have learned a priceless lesson: by bringing natural phenomena into time, by making them dynamical and contingent, we make possible a more complete and more rational understanding of them.

What I am proposing is that a similar shift must take place in our understanding of physics and cosmology. The laws of nature themselves, like the biological species, may not be eternal categories, but rather the creations of natural processes occurring in time. There will be reasons why the laws of physics are what they are, but these reasons may be partly historical and contingent, as in the case of biology.

Let me emphasize that the question is not whether there are some general principles that limit how the world is formed. Certainly there are, and it is likely that we know some of them. I also personally believe that recent developments in string theory and quantum gravity have uncovered evidence for one or two more. The question is, instead, whether these general principles will suffice to uniquely determine all the properties of our universe and the particles that live in it. It is this much more naive and radical ambition that I would like to suggest is wrong. That it is likely to be wrong can, I will argue, be seen from general philosophical arguments about such issues as how space and time are described, and how the intrinsic conflict between atomism and the search for unifying principles may be resolved. It may also be seen in recent developments in string theory and quantum gravity, which I will describe.

In the following pages I hope to convince the reader that the desire to understand the world in terms of a naive and radical atomism in which elementary particles carry forever fixed properties, independent of the history or shape of the universe, perpetuates a now archaic view of the world. It suggests a kind of nostalgia for the absolute point of view, a way of seeing the world that was lost when the Newtonian conception of space and time was overthrown.

I will argue that this view cannot be maintained because it is inconsistent with quantum theory and general relativity, as well as the new theories which underlie our modern understanding of the elementary particles. Instead, in different ways these theories take steps from the Newtonian view of absolute properties to another view, which may be called the relational view. This view is not new, it was championed by Leibniz, in a series of attacks on Newton's physics. It was his answer to the question of how to construct a theory of a whole universe. In a relational world, the properties of things are not fixed absolutely, with respect to some unchanging background. They arise instead from the interactions and relationships among the things in the world. As I will explain, twentieth century physics represents a partial triumph of this relational view over the older Newtonian conception of nature.

But what I will offer is not only philosophical argument for the possibility that the laws of nature were constructed through natural processes of self-organization. It is possible to construct an example of a theory in which exactly this happens. In the second part of the book I will present this theory. I hope the reader will agree that it represents a rational and reasonable alternative, and that—given everything that we know about the world—this theory (or something like it) might be true. The particular theory I will present is, in fact, testable, and it makes predictions that have so far stood up to comparison with the world. But whether or not it ultimately succeeds, it shows that the kind of theory the different arguments of the book point to can be concretely realized.

Another aspect of the nostalgia for the absolute Newtonian universe is the desire to be able to see the universe from the outside, as a disembodied observer. This also, I will argue, is a remnant of the old physics and is inconsistent with relativity and quantum theory. Instead, we have to confront the problem of how to construct a rational and complete understanding of the world that allows the observer to be in the world. But observers are not simple things, and any universe that naturally gives rise to, and is hospitable to, an observer must be complex. Thus, a theory of a whole universe, if it is to be consistent with what we know of quantum theory and relativity, must be a theory of a complex, self-organized universe.

In such a universe, the familiar divisions and hierarchies between phenomena that are considered fundamental and emergent, organized and simple, kinematic and dynamic, and perhaps even between what is considered biological and what is physical, are redrawn and redefined. These divisions, which for the nineteenth century were absolute may, after this century of transition, come to be understood as dependent on what question is being asked, on what scale of phenomena is being probed. If we can construct a science that does this, we open up the possibility of describing the universe as a coherent whole, in relationship only to itself, without need of anything outside itself to give it law, meaning or order.

As there are several different arguments that lead towards this conclusion, this book is divided into five parts. Each of these is organized around a simple question which any complete theory of the universe must be able to answer. In the order in which we will encounter them, these questions are:

1. Why is the universe hospitable to life? Why is it full of stars?
2. Is there a unique fundamental theory that determines the properties of the elementary particles? Or might the laws of nature themselves have evolved?
3. Is it accidental or necessary that the universe have such a large variety of structure? Why is the universe so interesting?
4. What are space and time?
5. How can we, who live in the world, construct a complete and objective description of the universe as a whole?

Each of these is easy to state. But we will see that the attempts to answer them, separately and together, open up a window on a new world much different from that inhabited by our predecessors.

The book that follows is written for anyone interested in the questions it asks. There are no equations, and it assumes only a level of acquaintance with the situation in modern physics and astronomy that most readers of popular science books and science sections of newspapers will have. In the different parts of the book, I have tried to strike a balance between presenting so little detail that the reader will remain unconvinced, and too much that he or she may be in danger of losing the thread of the argument. There may, of course, be readers who desire less information about some topics than is presented here, they are invited to simply skip over the details and proceed to the end of a given chapter, where the results of the discussion are usually summarized. Readers who, on the other hand, want more detail about a particular argument or topic may find it in the notes at the end of the book or in the references they point to.

PART ONE

THE CRISIS *in*

FUNDAMENTAL

PHYSICS

Why is the universe hospitable to life? Why is it full of stars?

ONE

LIGHT *and* LIFE

S cience is, above everything else, a search for an understanding of our relation-ship with the rest of the universe. We may begin it with the simplest, most basic fact about ourselves: Each of us is a living thing. As such, the most obvious and fundamental medium of our connection to the universe is light. For we, living things, live in a universe of light. We all see; even the simplest fungus or protozoa has receptors that respond to the presence of light. And is it not true that the most feared thing about imprisonment, or even death, is the loss of contact with the light? The dependence of life on light underlies so many metaphors and so much of the imagery of our culture (think of the fear of the dark) that to even quote examples is to risk cliché, but let me mention one, that to understand something is to attain an in*sight*.

But, of course, light is the ultimate source of life. Without the light coming from the sun, there would be no life here on earth. Light is not only our medium of contact with the world; in a very real sense, it is the basis of our existence. If the difference between us and dead matter is organization, it is sunlight that provides the energy and the impetus for the self-organization of matter into life, on every scale, from the individual cell to the life of the whole planet and from my morning awakening to the whole history of evolution.

We will never know completely who we are until we understand why the universe is constructed in such a way that it contains living things. To comprehend that, the first thing we need to know is why we live in a universe filled with light. Thus, the problem of our relationship with the rest of the world rests partly on at least one question that science ought to be able to answer: Why is it that the universe is filled with stars?

But before we approach this question, there is another kind of relationship that I must comment on; that between author and reader. This is, as my literary friends have been telling me for some time, a rather problematic relationship. After all, the dominant literary theory taught by my colleagues these days is that books must be read as if, in a certain sense, the author does not exist. But, beyond my natural protest against such an assertion (in my present situation), there are special problems that arise when the author is a scientist and the reader is not. Anyone who sets out to teach ideas from physics to those who are not specialists, whether as a teacher in a lecture room or through a book such as this one, faces a curiously paradoxical situation. To begin with, there is no doubt that a great many people have a deep interest in physics and cosmology. Who has not looked up at the stars, or gazed at a tree or a kitten and wondered what the universe is and what our place in it might be? And, what culture has not had a story about how the universe was created?

It is a cliché to say that in the twentieth century science has replaced religion as the dominant cosmological authority. While this does not seem to have actually done much to decrease the popularity of religion, it is true that at the present time, for many of the cultures of the planet, we physicists are the official makers and keepers of the story of the cosmos. This, perhaps more than anything else, accounts for the peculiar combination of interest and distance that many people seem to bring to a meeting with a physicist. At the same time, it is unlikely that there is any subject in high school or university that is more disliked than physics. If a great many people want to know about what we think the universe is, almost no one seems much interested in the tools with which we acquire and construct this knowledge.

I have been teaching physics to non-science students for much of my career. While I am considered a good teacher, what has most impressed me is how unsuccessful, on the whole, I have been at imparting my love of physics. Thus, at some point during the last few years I began to ask my students directly why they don't like physics. Of course, a number found the sustained attention required to learn how to think in new ways disagreeable. Others are understandably put off by the unfortunate connection between physics and weapons of mass destruction. But from the most interesting students, the artists, the philosophers, the hot shot literary theory types who can sail through Derrida and Christeva but cannot penetrate textbooks developed and marketed at great expense especially for nonscientists, I began to hear another kind of answer. They find physics difficult because they don't like and don't believe the picture of nature embodied by the science they are being taught.

There is at least one good reason not to believe the physics that is taught in most courses for nonscientists. It isn't true. For a reason that after many years of university teaching remains opaque to me, physics is the only subject in the university curriculum in which the first year's study rarely gets beyond what was known in 1900. Now, Newtonian physics is a beautiful subject, as are the plays of Shakespeare. But no one tries to teach first year students to think about Shakespeare the way critics thought in the nineteenth century. A good literature teacher will teach the classic books in the context of the current debates about the nature of texts. Almost no one teaches Newtonian physics to beginning students in the context of the current debates about the nature of space and time.

Newtonian physics is useful, even if it is not true, as an approximation that helps us to understand many different phenomena. But it is completely discredited as an answer to any fundamental question about what the world is. It has a great deal of historical and philosophical interest, but this is rarely mentioned in beginning courses. Thus, it is not surprising if students find the subject uninspiring.

But, beyond the fact that they are given little reason to believe in it, I find that students simply are not drawn to the description of the world offered by Newtonian physics.

Once I suspected this I began to ask myself what exactly is it that they don't like about the Newtonian view of the cosmos?

I believe that the answer is that there is no place for life in the Newtonian universe. On the basis of the physics that was known in the nineteenth century, it is impossible to perceive a connection between ourselves as living things and the rest of the universe.

But physics must provide a way to understand what life is and why we are here. It is the "science of everything" whose task is to uncover those facts and laws that apply universally. Physics must underlie and explain biology because living creatures, like all things in the universe, are made out of atoms which obey the same laws as do every other atom in the world. An approach to physics that does not make the existence of life comprehensible must eventually give way to one that does.

One might have expected that before the twentieth century people would have been concerned about the fact that life did not fit easily into the Newtonian cosmos. If few scientists worried about this, it may in part be due to a philosophy called vitalism that was popular in the last century. According to it, there is no reason to expect that physics should illuminate the processes of life because living and non-living matter may obey different laws. Imagine how disappointing it would be were vitalism true, it would mean that there is no essential connection between us and what we see when we look around us. Still, there is no denying the attraction such a view holds for many people. The idea that life is not reducible to physics seems a remnant of the Greek and Christian cosmologies in which earth and sky are made from different essences. Behind it one can sense the ancient desire to escape nature and partake of heaven.

But in any case, before Einstein, people had little choice. Had Newtonian physics turned out to be correct, vitalism would have been necessary. It is only with the physics of the twentieth century that we have been able to understand how living things are constructed from the same ordinary atoms that make up rocks and stars. Thus, part of the movement from the Newtonian world to the modern one is a transition from a universe in which life is impossible to one in which life has a place. It is partly for this reason that the question of the existence of life becomes central to the twentieth century revolution in physics. Quantum physics, for all its intrinsic weirdness, gives us for the first time an opportunity to comprehend our relationship to the rest of the universe in a way that avoids both the Aristotelian fiction of our absolute centrality and the Newtonian fiction of our absolute alienation.

To appreciate the meaning of this change, we must first understand why it is that we would not expect to find anything like life in a universe governed by Newton's laws. Let us begin with an image that comes to mind when one asks the question of what our place is in the universe. This is the image of a warm, living earth, lost in the depth of an infinite, cold and dead cosmos. This image, which embodies one of those basic ideas that are so obvious as to seem almost beyond examination, hides, in my opinion, an absurdity. To see why, we may start by asking what must be true about the universe in order that it contain living things.

The first thing required for life is a variety of different atoms that can combine to form a very large number of molecules, which differ greatly in their sizes, shapes and chemical properties. It is often stressed that carbon is required because it is the only element that forms a sufficient variety of stable molecular structures. All of the living things on earth are made out of carbon compounds that are built with copious amounts of carbon, hydrogen, oxygen and nitrogen, as well as traces of many other atoms. But beyond the specifics of carbon chemistry, life would be impossible were there not a sufficient variety of atoms. A universe containing only one kind of atom would almost certainly be dead.

The problem with Newtonian physics is that it does not allow the existence of many distinct kinds of atoms. A Newtonian atom would be something like a solar system, but held together by the electrical attraction of the nuclei and electrons rather than by gravity. However, there is a problem with this, because when the electrons move in circles they radiate light waves, which carry energy away from the atom. The result is that the electrons lose energy and spiral into the nucleus.

If the world suddenly became Newtonian it would take only a fraction of a second for most of the electrons to fall into the nuclei. This was, in fact, the direct motivation for the introduction of the quantum mechanical picture of the atom. The fact that atoms are like solar systems, with most of the mass in the nuclei and most of the space taken up by the electrons, was discovered by Ernest Rutherford, in his laboratory in Cambridge in 1911. Within months his young protégé, Niels Bohr, had invented the first quantum mechanical theory of the atom. Whatever else one may say about the quantum theory, its central success is that it explains the stability of atoms.

However, it is not enough that the laws of physics allow the existence of a variety of stable atoms. The universe must, during its history, produce these atoms in copious quantities so that they may be available for the development of living things.

Thus, we must ask what is required of a universe so that large amounts of carbon, oxygen and the other ingredients of life are plentifully produced. This question has a simple answer: the universe must contain stars. All but the lightest elements were forged in stars. Thus, it is not a coincidence that when we look up we see stars, just as it is not a coincidence that when we look around we see plants and trees. Just as the plants produce the oxygen we breath, it is the stars that produced all the chemical elements out of which we, and the plants, are made.

This, at least in outline, settles the question of where the ingredients for life come from. But there is another, deeper question we must ask. Given the ingredients, what are the conditions that make the universe hospitable to life? What must be true about the world so that some of its atoms will spontaneously invent the astoundingly intricate dance which makes them living? That life arose from a simpler world seems the ultimate miracle. But, if we are to understand our place in the universe, we must come to understand it.

In its capacity to create organization and complexity where none existed before life seems to run contrary to the laws of physics. This was, in any case, what was thought by many in the nineteenth century, who worried that the law of increasing entropy (or, as it is also called, the second law of thermodynamics) contradicted the observed record of biological evolution.

Most people have an intuitive idea of the meaning of the law of increasing entropy. A hot cup of tea cools down until it is the same temperature as the air in the room. Snow melts on a warm day. These examples illustrate the tendency for differences between the temperatures in different parts of a system to be erased. The configuration in which all parts of a system have the same temperature, density and chemical composition is called thermodynamic equilibrium. The law of increasing entropy says that if I have a closed system, which is isolated from the rest of the universe, it is overwhelmingly probable that it will come to, and remain in, a state of thermodynamic equilibrium.

Living things, of course, do not behave like this. My body stays at about the same temperature, no matter what the temperature of my environment might be, at least as long as I am healthy. My cat also maintains a constant body temperature, which is different from mine. If I sleep with my clothes on, then when I wake up they have the same temperature as me. If I sleep with my cat, he wakes up (for the few minutes he condescends to enter that state) with his own unique temperature.

Another, related meaning of entropy is that it is a measure of disorganization. The atoms in a gas are disordered to the extent that there is no way to tell one from another. In equilibrium there is maximal disorder, because every atom moves randomly, with the same average energy as any other atom. A living system, on the contrary, continually creates an enormous number of different kind of molecules, each of which generally perform a unique function. The entropy of a living thing is consequently much lower, atom for atom, than anything else in the world.

It is an interesting historical fact that the laws of thermodynamics were put in their modern form during the second half of the nineteenth century, more or less at the same time that the theory of natural selection was introduced by Darwin and Wallace. Since that time, many people, both inside and outside of science, have made a great deal of the apparent contradiction between these two developments. The fossil record tells us that the biosphere has become more organized and more varied over time. The laws of thermodynamics say that there is a tendency for systems to become less organized and less varied over time. Thus, one argument that was often made for vitalism during the last century is that the matter living things are made of must be excluded from the strictures of the laws of thermodynamics.

In fact, the case of thermodynamics is different from that of Newtonian physics. The laws of thermodynamics are not in contradiction with the existence or the evolution of life. Not only is the existence of life compatible with thermodynamics, the two subjects are actually so intimately related that the clearest characterization of life I know of is one given in thermodynamic terms. This is because once we understand what it means for a system to be in thermodynamic equilibrium, we can understand its opposite: what is required for a system to be out of equilibrium, as all living things are, for arbitrarily long periods of time.

Nothing can live in an environment in thermal equilibrium. If life is to exist there must then be regions of the universe that are kept far from thermodynamic equilibrium for the billions of years it takes for life to evolve. We then want to ask, What is required of the universe so that it contains such regions? The answer to this question is easy. There must be things in the universe that are much hotter than the rest of it, and are able to maintain themselves as constant sources of light and heat for enormous periods of time.

What kinds of things can do this? The answer to this question is the same as the answer to the other questions we raised in this chapter: There must be stars.

We can thus begin to see what is wrong with the picture of a warm living earth inside a cold dead cosmos. If the universe really were cold and dead, if it contained no stars, there would be no living planets. The existence of stars is thus the key to the problem of why the cosmos is hospitable to life.

I would like to inject one note of caution before proceeding. If we were interested only in feeling better about ourselves, we might be happy to jump from vitalism to a kind of pantheism according to which life exists because the universe itself is alive. But our goal should be more than inventing a story that explains what we are doing in the universe. In the end what is wrong with the Newtonian theory of the universe is its essential irrationality, as it leaves unexplained too many aspects of the world that we may hope to comprehend. What is needed is a deeper understanding of what both life and the universe are that allows us to comprehend why it is natural to find one inhabited by the other.

The scientific revolution did not take off when Copernicus simply switched the places of the earth and sun in the Aristotelian cosmos. To put the earth on one of the crystal spheres was logically absurd, as it contradicted the basic assumptions behind the Aristotelian cosmos such as the immutability of the heavens. Any intelligent sixteenth-century person could explain why what Copernicus had done didn't really make sense. The revolution began in earnest when Kepler abolished the crystal spheres and cast the planets adrift in empty space. Then he had to ask a new question: How does a planet in the midst of empty space know where to move? It was this and other new questions that drove the revolution.

Similarly, to assert simply that the universe is alive is absurd. Instead, I would like to suggest that the time has come for us to knock our understanding of what the laws of physics represent off a kind of philosophical mooring that has become as outdated as Aristotle's crystal spheres were in the seventeenth century. Set adrift, we have now to ask new questions about how the regularities we refer to as the laws of physics came to be and whether, and how, they can change. The search for answers to these questions may then lead us to reconsider our familiar understandings about the relationships between the fundamental and the emergent and between physics and biology. To put it another way, one of the questions we will be seeking to answer in the following chapters is whether it is purely an accident, or whether it is to some extent necessary, that this, or any, cosmos is a universe of light and life.

You must become an ignorant man again
And see the sun again with an ignorant eye
And see it clearly in the idea of it.
                    —*Wallace Stevens,*
        "Notes Toward a Supreme Fiction"

T W O

THE LOGIC *of* ATOMISM

"*We are stardust*" Joni Mitchell sings, and it rings so true that we have to pinch ourselves to remember that it is less than seventy years since we learned that everything we are made of, except hydrogen, was fused in stars. What is, on the other hand, very old is the idea that the world is made of atoms. The philosophy of atomism goes back at least to the Greek philosophers Democritus and Leucippus in the 6th century BC. According to them, the universe consists of a large number of fundamental particles, moving in empty space. As obvious as this idea may seem to us now, it was rejected by Aristotle, and was only revived many centuries later at the start of the scientific revolution. But atomism triumphed only in this century, as quantum physics opened up the atom to our understanding. In quick succession we descended through several levels of structure, so that we now study the quarks: the things within the things within the atom.

The triumph of atomism is by now so complete that any challenge to it seems at first to point outside of the boundaries of science. The Greeks could only dream of a science in which the properties of anything in the world could be explained by decomposing it into its atoms. We have this science, it is the foundation of everything we understand from immunology to transistors to nuclear physics. And if the atoms and the nuclei turned out to be divisible, we have now reasonable candidates for truly elementary particles in the electrons, neutrinos and quarks.

But if it is hard to conceive of it being wrong, there are still questions that the atomistic philosophy cannot help us answer. Some of these have to do with the elementary particles themselves: The electron is lighter than the proton, but not as light as the neutrino. Why? Why is the neutron just a bit heavier than the proton? Why doesn't the neutrino have any electric charge?

We cannot understand the elementary particles, as we do everything else, by breaking them into parts. First of all, there is no evidence that they are made of still smaller things. But even if there were, eventually the game must stop, we must at some point arrive at some truly elementary particles. In a whimsical mood we may entertain the idea that there is an infinite regress, but this seems unlikely. For one thing, there is good reason to believe there really is a smallest ultimate size to things, which I will explain later, in Part Five. Whatever the elementary particles are, we are going to have to understand them, and we are going to have to do this in terms different than those we use to understand everything made from them.

According to the Greek philosophers, the elementary particles are eternal, never created or destroyed. This seemed to them the only alternative, for if they were created they would have to be put together out of some parts. Then they would no longer be the smallest things. Making the elementary particles eternal puts the questions as to their properties in the realm of the absolute: They are like that because they always were and always will be. As a consequence, each of the elementary particles exist independently of all of the others. Neither the history of the universe nor its present configuration can have any effect on the properties of any single elementary particle. It is completely conceivable that the universe might have but one neutron in it. And, according to this philosophy, that neutron would be exactly the same as one found in an atom of my cat's whisker.

Modern elementary particle physics does allow the elementary particles to be created and destroyed. But their properties are determined by laws, which endow each particle, when created, with certain properties, completely independent of whatever else may exist in the universe. These laws are presumed to be absolute and to hold for all time. Thus, the idea of the absolute plays an essential role for us, as it did for the Greeks. It has just been abstracted, from eternal atoms to eternal laws. That the laws of physics might be created or modified seems to us as nonsensical as it would have seemed to Democritus to build a machine that creates elementary particles.

The idea that there is an absolute law of nature, which fixes once and for all the properties of the elementary particles, has been so successful it is difficult to imagine a scientific approach to understanding nature that does not begin there. But, in fact, there are very good reasons to believe that in the end this idea cannot be right. Some of these reasons come from the logic of atomism itself. I will argue as we go along that the reductionist philosophy that underlies atomism is necessarily incomplete. A philosophy that tells us to explain things by breaking them into parts will not help us when we confront the question of understanding the things that have no parts. At that point we must turn to some different strategies if science is to progress.

For most of the last century, elementary particle physics moved at a rapid pace, with a new discovery appearing at least once a decade. During this time we have come to see it as the route to answering all the most fundamental questions about nature. When I was trained as an elementary particle theorist, I believed myself to be joining the exalted ranks of those whose task is to discover the fundamental reality behind our perceptions of nature. I always felt a bit sorry for scientists who were not elementary particle theorists, for I could never understand how they could find complete satisfaction in investigating nature at any other than its most fundamental level. Nor was I terribly interested in the "higher order sciences", such as biology or astronomy, because nothing that they could learn could have any bearing on the fundamental questions, which were about the elementary particles.

Unfortunately, for the last twenty years elementary particle physics has not moved at the pace it once did. In the middle-1970's, there was a great triumph, in which the theory that we call the standard model of elementary particle physics was constructed. This theory puts us in the position to predict the results of virtually any experiment that could be done with present technology, with one significant exception, which encompasses anything having to do with gravity. But it leaves open a large number of questions, and these past twenty years have been a very frustrating period because almost none of these questions have been answered.

The most important of these questions is how to include gravity, and this cannot be done until we know how to unify general relativity with quantum theory. But there are also other questions that the standard model does not answer, which have remained mysteries. Many of these have to do with the properties of the elementary particles, such as: Why do they have particular masses and charges?

The persistence of these problems does not imply that important work has not been done. On the theoretical side especially, new ideas have been invented that are likely to help explain some of the questions left over by the standard model. But on the experimental side, nothing has been discovered that could not be explained in terms of the standard model. At the same time, not one of the theoretical ideas intended as answers to the questions left open by the standard model has been confirmed experimentally. Perhaps if elementary particle physics had not been so successful, this situation would not be so worrying. But one has to look back more than a century to find a comparable twenty-year period without definitive progress in our understanding of the basic laws of nature.

There are several reasons for this, one of which is certainly the great difficulty and expense of making new experiments that probe layers of structure smaller than those described by the standard model. But I believe that part of the present crisis is inevitable, and is due to our having reached the limits of what we can learn solely by breaking things into their parts. The very success of the reductionist philosophy may have brought us to the moment when we have in our hands at least some of the truly elementary particles. If so, it should not surprise us if methods that have been so successful up to this point seem to be failing.

In science, detective movies, love or any other area of life, when one is confronted with a situation in which the old assumptions are no longer working as they used to, it is perhaps time to look for new questions to ask. But how does one search, not for new answers, but for new questions? Perhaps the first thing to do is to try to look around us with fresh eyes and examine the evidence that is close at hand. Sometimes a crucial piece of evidence lies right in front of us that has up till now lacked any significance. Looked at in a new way, our familiar world can all of a sudden reveal new meanings.

It is exactly such a new look that I would like to propose we take to the problems of elementary particle physics. The important question, if we are to try to begin again, is which assumptions we should keep and which we should throw away. To begin with, there can be nothing wrong with atomism, as long as we take that to mean only the simple idea that most things in the world are made of elementary particles, which are not themselves composed of anything smaller. But we may question the more radical assumption that the properties of these elementary particles themselves are fixed eternally in terms of absolute laws. To distinguish this idea from commonsense notions of atomism and reductionism, I will give it a name. I will call it *radical atomism*. Similarly, there is no need to question the idea that there are laws of nature. But we can question the idea that if we knew only those laws, and nothing else about the history or organization of the universe, we could deduce the properties of a quark or electron.

One reason to question radical atomism is that it must eventually lead either to infinite regress or to a brick wall. If the elementary particles have no parts, then no explanation of any of their properties can be found by looking inside them. The only alternative may be to look outside them. This means we must try to determine if the properties of the elementary particles might be somehow influenced by their relationships with the things that are around them. If elementary particles are so influenced, then perhaps those properties are not absolute and eternal. Instead, to understand a quark or an electron, we may have to know something about the history or organization of the universe.

I must confess that it is still not completely easy, even after the years I have spent thinking about it, to write these last sentences. The weight of all the philosophy that lay behind my training as a physicist tells me that this is the wrong thing to try to do. There is, of course, absolutely no evidence that the elementary particles are affected by the environments in which we find them. Observations of the light from distant stars affirms that the protons they are made of are exactly the same as those I am breathing now. But this does not mean that there can be no effect by which an elementary particle is influenced by its environment. It only means that to find such effects, we probably have no alternative but to look at scales much larger than stars and galaxies.

Another problem with the philosophy of radical atomism is that it gives us little ground to understand why the universe is as organized as it seems to be. If the universe is nothing but atoms moving in a void, then it is hard to understand why it isn't far simpler than it is. From a fundamental point of view, a universe filled with a gas of atoms in thermal equilibrium is as plausible as a world full of a variety of structures. Indeed, it is much more than plausible, for according to the law of increasing entropy, it is much more probable that the world be disorganized, be merely a gas in thermal equilibrium.

Why is the universe so dynamical? Why is it not closer to thermal equilibrium, as nineteenth century cosmologists expected? As I suggested in the last chapter, the answer to these questions is that there are stars. For they are the primary sites for transformations of energy and matter in the universe. In each star, as the elements are forged, gravitational and nuclear energy are converted into light and radiation and sent out into the universe. Indeed, just as our life is embedded in the ecological cycles of the biosphere, our whole planet exists as a part of a much older cycle of material and energy that forms the galaxy.

Another thing that must strike us when we look around at the universe is that it seems to be structured hierarchically. Imagine that we have stepped for the first time, not into a universe, but into a library. To use it we need to know how it is organized. We find first that the library is divided into sections, each of which is divided into a large number of books. Most of the books are further divided. For example, this book is divided into parts, each of which is further divided into chapters. The meaning of the chapters is conveyed in paragraphs, which are composed of sentences. A sentence is made of words, in certain orders, each of which is made out of letters. Finally, each letter is a combination of a small number of basic shapes, lines, circles, and arcs.

Our universe has at least as many levels of organization as a library. The elementary particles, small in number, are something like the basic shapes; the atoms something like the letters. In each case there are several dozen of each. The atoms are organized into an enormous number of different molecules, just as the letters spell out an enormous number of words. As the order of the letters on the page is relevant for the meaning of the word, the arrangement of the atoms in three dimensional space is crucial for the properties of the molecule. Molecules can be organized many different ways, as solids, crystals, liquids, gases, just as there are many kinds of texts. The arrangement of the elementary particles in the world is much more interesting than the ancient atomists pictured it to be, for the atoms are not just dancing about. Instead their organization is structural, containing a great complexity, on which depend the enormously diverse chemical and physical properties of the molecules.

But when we raise our eyes from the molecules, and look at the universe on a large scale, we also see a hierarchical structure. One of the great discoveries of the present period is that the galaxies are not distributed randomly in space. Instead, we find structure in the arrangements of the galaxies on every scale up to the largest that has so far been surveyed. The largest structures that have so far been mapped are great systems of galaxies, each of which contains many clusters, each of which contains dozens to thousands of galaxies. An example of such a great system is the "Great Wall," which is a sheet of clusters of galaxies spread over a large part of our sky, at a distance of about thirty million light years from us.

Seen from the largest scales, the galaxies are the basic structural units of the organization of the universe. And what then are the galaxies? We will later devote a whole chapter to this question, but the simple answer is that galaxies are great systems for making stars.

The hierarchies of structures that we see in the sky are not random, they are created and maintained by processes that go on in stars and galaxies. To comprehend them requires more than just knowing how to break everything into its parts; we must understand how it is that such a complex hierarchy of structures and processes arose as the universe evolved. The question of the origin of the structure in the universe is then not unlike the question of the origin of life. We need to know if, given the laws of physics, it was probable that such structures and processes spontaneously form.

As long as we do not comprehend why it was probable that living things formed spontaneously as soon as conditions in the earth's oceans allowed, our understanding of biology must be considered incomplete. Similarly, any philosophy according to which the existence of stars and galaxies appears to be very unlikely, or rests on unexplained coincidence, cannot be satisfactory. In the next chapter, I will explain that the radical atomist philosophy is in great danger in this regard. We shall see that, in spite of all that we have learned, given the basic principles and laws of nature as we understand them now, it is extraordinarily improbable that the universe be full of stars.

What I am really interested in is whether God
had any choice in the creation of the world.
—*Albert Einstein*

THREE

THE MIRACLE
of the STARS

There is one way to resolve all of the questions I've raised in the last chapters. Suppose that there was only one possible theory that could describe a world like our own. This might be the case if, for example, it were so difficult to construct any theory that the requirement of mathematical consistency was enough to rule out all candidates except one. Suppose that we had this theory. Suppose also that when we worked with it we found it gave only right answers to the questions we posed about the elementary particles. In this case all of the worries I've raised would be moot. There would be only one logically possible world, and it would be ours.

For better or worse, no such theory has ever been found. Nor is there any reason, besides faith, to hope that a consistent theory that was able to describe something like our world should be unique. It then seems at least prudent to wonder what we are to do if there turn out to be many different theories that might describe a possible universe, all equally consistent. In this circumstance we would have to wonder whether the world was to some extent the result of a choice, made somehow at some time in the past.

The theory that we do have, the standard model of particle physics, is very far from being unique. In spite of the fact that it represents our deepest knowledge about what the world is made of, it leaves open many questions about the properties of the elementary particles. These open questions have to do with the values of certain numbers that characterize the particles. These numbers measure things like the masses of the different particles and the strengths of their electrical charges. According to our best present understanding, these numbers are free to vary within wide ranges. They are then *parameters*, whose values may be set arbitrarily. Physicists set the values of the parameters so as to make the theory agree with observation. By doing so, for example, we make the electron, proton, neutron and neutrino all have the right masses. But as far as we can tell, the universe might have been created so that exactly the same laws are satisfied, except that the values of these parameters are tuned to different numbers.

There are about twenty of these parameters in the standard model of particle physics. The question about why the universe has stars can then be posed in the following way: We may imagine that God has a control panel on which there is a dial for each parameter. One dial sets the mass of the proton, another the electron's charge and so on. God closes his eyes and turns the dials randomly. The result is a world governed by the laws we know, but with random values of these parameters. What is the probability that the world so created would contain stars?

The answer is that the probability is incredibly small. This is such an important conclusion that I will take a few pages to explain why it is true. In fact, the existence of stars rests on several delicate balances between the different forces in nature. These require that the parameters that govern how strongly these forces act be tuned just so. In many cases, a small turn of the dial in one direction or another results in a world not only without stars, but with much less structure than our universe.

Although many different kinds of elementary particles have been discovered, almost all the matter in the universe is made of four kinds: protons, neutrons, electrons and neutrinos. These interact via four basic forces: gravity, electromagnetism and the strong and weak nuclear forces. Each of these forces is characterized by a few numbers. Each has a *range*, which tells us the distance over which the force can be felt. Then, for each kind of particle and each force there is a number which tells us the strength by which that particle participates in interactions governed by that force. These are called the coupling constants. One of these is the electrical charge, which tells how strongly a particle may attract, or be attracted by, other charged particles. The parameters of the standard model consist primarily of the masses of the particles and these numbers that characterize the four forces.

In order to understand why the existence of stars is so improbable, it helps to know some basic facts about the four different interactions. We may start with gravity, which is the only universal interaction. Every particle, every form of energy, feels its pull. Its range is infinite, which means that although the gravitational force between two bodies falls off with distance, it is never zero, no matter how far apart the two bodies may be. Gravity has another distinguishing feature, which is that it is always attractive. Any two particles in the universe attract each other through the gravitational interaction.

The strength by which any particle is affected by gravity is proportional to its mass. The actual force between two bodies is given by multiplying the two masses together, and then multiplying the result times a universal constant. This constant is called Newton's gravitational constant, it is one of the parameters of the standard model. The most important thing to know about it is that it is a fantastically small number. Its actual value depends on the units we use, as is the case with many physical constants. For elementary particle physics it is natural to take units in which mass is measured by the proton mass. In these units you or I have a mass of about 10^{28}, for that is how many protons and neutrons it takes to make a human body.* By contrast, in these units the gravitational constant is about 10^{-38}. This tiny number measures the strength of the gravitational force between two protons.

The incredible smallness of the gravitational constant is one of the mysteries associated with the parameters of particle physics. Suppose we had a theory that explained the basic forces in the universe. That theory would have to produce, out of some calculation, this ridiculous number, 10^{-38}. How is it that nature is so constructed that one of the key quantities that govern how it works at the fundamental level is so close to zero, but still not zero? This question is one of the most important unsolved mysteries in all of physics.

It may seem strange that a force as weak as gravity plays such an important role on earth and in all the phenomena of astronomy and cosmology. The reason is that, in most circumstances, none of the other forces can act over large dis-

---

*In physics and cosmology we often must refer to very large numbers, so the *exponential* notation I use here is very convenient. (This is the only mathematics that the reader must know to read this book.) Thus, 10^3 is a shorthand for a 1 with three zeros after it, which is 1,000, while 10^{28} stands for a one which twenty-eight zeros after it. When there is a minus sign, we mean the inverse of the quantity, thus 10^{-1} stands for 1/10, which is the same as .1, while 10^{-5} stands for 1/105 , which is .00001. When numbers like this are used, they are meant very approximately; thus, here I do not care exactly how many protons I contain, what is relevant is only which power of ten comes closest.

tances. For example, in the case of the electrical force, one almost always finds equal numbers of protons and electrons bound together, so that the total charge is zero. This is the reason that most objects, while being composed of enormous numbers of charges, do not attract each other electrically.

Gravity is the only force that is always attractive, which means that it is the only force whose effects must always add, rather than cancel, when one considers aggregates of matter. Thus, when one comes to bodies composed of enormous numbers of particles, such as planets or stars, the tiny gravitational attractions of each of the particles add up and dominate the situation.

The incredible weakness of the gravitational constant turns out to be necessary for the existence of stars. Roughly speaking, this is because the weaker gravity is, the more protons must be piled on top of each other before the pressure in the center is strong enough that the nuclear reactions ignite. As a result, the number of atoms necessary to make a star turns out to grow as the gravitational constant decreases. Stars are so huge exactly because the gravitational constant is so tiny.

It is fortunate for us that stars are so enormous, because this allows them to burn for billions of years. The more fuel a star contains the longer it can produce energy through nuclear fusion. As a result, a typical star lives for a long time, about ten billion years.

Were the gravitational force somewhat stronger than it actually is, stars would still exist, but they would be much smaller, and they would burn out very much faster. The effect is quite dramatic. If the gravitational force were stronger by only a factor of ten, the lifetime of a typical star would decrease from about ten billion years to the order of ten million years. If its strength were increased by still another factor of ten, making the gravitational force between two protons still an effect of order of one part in 10^{36}, the lifetime of a star would shrink to ten thousand years.

But the existence of stars requires not only that the gravitational force be incredibly weak. Stars burn through nuclear reactions that fuse protons and neutrons into a succession of more and more massive nuclei. For these processes to take place, protons and neutrons must be able to stick together, creating a large number of different kinds of atomic nuclei. For this to happen, it turns out that the actual values of the masses of the elementary particles must be chosen very delicately. Other parameters, such as those that determine the strengths of the different forces, must also be carefully tuned.

Let us think of the three most familiar particles: the proton, neutron, and electron. The neutron, it turns out, has almost the same mass as the proton, it is in fact just slightly heavier, by about two parts in a thousand. In contrast, the electron is much lighter than either, it is about eighteen hundred times lighter than the proton.

In the masses of these three particles there are as many mysteries. Why are the neutron and proton so close in mass? Why is the electron so much lighter than the other two particles? But what is most mysterious is that the two small numbers in this problem, the electron mass and the tiny amount by which a neutron is just slightly more massive than a proton, are quite comparable to each other. The neutron outweighs the proton by only about three electron masses.

We are so used to the idea that protons and neutrons stick together to make hundreds of different stable nuclei, that it is difficult to think of this as an unusual circumstance. But in fact it is. Were the electron's mass not about the same size as the amount that the neutron outweighs the proton, and were each of these not much smaller than the proton's mass, it would be impossible for nuclei to stick together to form stable nuclei. These are then facts of great importance for the world as we know it, for without the many different stable nuclei, there would be no nuclear or atomic physics, no stars and no chemistry. Such a world would be dramatically uninteresting.

According to the standard model of elementary particle physics, the masses of the proton, neutron and electron are set by completely independent parameters. There is a dial on the control panel by which each of the masses may be tuned. To visualize this we may think of a graph in a plane, in which one axis is labeled by the value of the electron mass and the other by the neutron mass. Each may be measured in units of the proton mass. A point in this space then corresponds to a choice of these masses. Each point then denotes a possible universe, in which the parameters have been chosen differently.

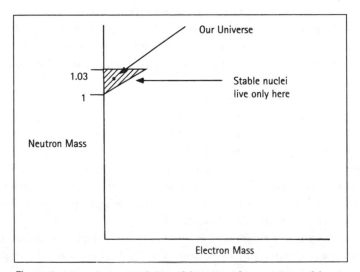

Figure 1   A two dimensional slice of the space of paramenters of the standard model of particle physics, labeled by the value of the neutron and electron masses. The small region correcsponds to values of the parameters in which there are stable nuclei.

One way to ask how probable it be that the world have atomic nuclei is to ask how large of a region in this space would correspond to a world that had stable atomic nuclei. The answer is that nuclei could only be stable if the parameters are chosen in a small corner of the graph (See Figure 1). Since stars cannot burn if there are not stable nuclei, only those possible universes that lie within that small region may have stars.

These are not the only parameters that must be tuned carefully if there are to be stars. For example, there is the mass of the neutrino. Here we face an embarrassing situation: we still do not know whether the neutrino has any mass at all. The experimental evidence is inconclusive, but we can assert that if it does have a mass, it is no more than one hundred thousandth that of the electron. But in spite of our ignorance as to its actual value, we do know that the mass of the neutrino cannot be too large if the nuclear reactions that energize the stars are to happen.

While we are discussing physical constants that must be finely tuned for the universe to contain stars, we may consider another kind of question. Why is the universe big enough that there is room for stars? Why is it not much smaller, perhaps even smaller than an atom? And why does the universe live for billions of years, which is long enough for stars to form? Why should it not instead live just a few seconds? These may seem silly questions, but they are not, because the fact that the universe can become very big and very old depends on a particular parameter of the standard model being extremely tiny. This parameter is called the cosmological constant.

The cosmological constant can be understood as measuring a certain intrinsic density of mass or energy, associated with empty space. That a volume of empty space might itself have mass is a possibility allowed by Einstein general theory of relativity. If this were sizable, it would be felt by matter, and this would effect the evolution of the universe as a whole. For example, were there enough of it, the whole universe would quickly pull together and collapse gravitationally, as a dead star collapses to a black hole. In order that this not happen, the mass associated with the cosmological constant must be much smaller than any of the masses we have so far mentioned. In units of the proton mass, it can be no larger than about 10^{-40}. If this were not the case, the universe would not live long enough to produce stars.

It seems that physics is full of ridiculously tiny numbers. For example, we may wonder what might be the most massive elementary particle that could be imagined. This is one that would be so massive that it would be overwhelmed by its own gravitational force and collapse instantly to a black hole. There is an actual mass, above which this must happen. It is called the Planck mass, after Max Planck, the founder of quantum mechanics. The Planck mass is enormous compared to the scale of the elementary particles. In units of the proton mass it would be about 10^{19}. In ordinary units it is about 10^{-5} of a gram—about the size of a living cell. To turn it around, this means that in units of the largest possible mass, the proton's mass is 10^{-19} , the electron's is 10^{-22} and the cosmological constant is no larger than 10^{-60} .

To an elementary particle theorist, there is no greater mystery than the values of the different masses of which we have been speaking. Mystery number one is why the proton mass is so tiny compared to the Planck mass. Mystery number two is why the cosmological constant is so much tinier still. Between the scale of the cosmological constant and the Planck mass is a ratio of 10^{60}. It is extraordinary that such a huge ratio should come into fundamental physics. But this is not all. Taking these values into account, it turns out, apparently coincidentally, that the lifetime of a typical star is about the same as the lifetime of the universe, measured as best we can by the speed of its expansion.

Why should the expansion rate of the universe have been set to the scale of the lifetime of stars, if the first stars formed millions of years after the big bang? What kind of physical mechanism could account for this? It is in mysteries like this that we see most clearly the limitations of the philosophy of radical atomism, according to which properties of the elementary particles (such as the mass of the proton or the strength of the gravitational force) should have nothing to do with the history of the universe.

Perhaps the reader is still not convinced that there is something incredible to be understood here. Let me then go on, we have only discussed gravity, there are three more interactions to consider. These forces are described by still additional parameters. The story for many of these is the same.

We may consider next the force which is most evident in our lives, the one which was the theme of the first chapter, electromagnetism and light.

The importance of electromagnetism for our modern picture of nature cannot be overstated, as almost all of the phenomena of everyday life which are not due to gravity are manifestations of it. For example, all chemistry is an aspect of electromagnetism. This is because chemical reactions involve rearrangements of electrons in their orbits around atomic nuclei, and it is the electrical force that holds the electrons in those orbits. Light is also an aspect of electromagnetism, for it is a wave traveling through the fields that convey the electric and magnetic forces.

Electromagnetism differs in two important respects from gravity. The first is that the electrical force between two fundamental particles is much stronger than their gravitational attraction. The strength of the electrical interaction is measured by a number, which was called alpha by the physicists of the last century, because it is a number of the first importance for science. Alpha, which is essentially a measure of the strength of the electric force between two protons or electrons, has a value of approximately 1/137. Physicists have been wondering about why alpha has this value, without resolution, for the whole of the twentieth century.

The second way in which electricity differs from gravity is that its effect is not only attractive: two electrical charges may attract or repel each other, depending on whether they are like or unlike.

As we did for gravity, we may ask how important the existence of a force with these properties is for the existence of stars. Light does, indeed, do something essential for stars. For it must be possible for the energy produced in stars to be carried away to great distances. Otherwise, stars could not radiate, and being unable to get rid of the energy they produce, they would simply explode. Light is precisely the medium by which the energy produced in stars is conveyed to the rest of the universe.

However, the existence of electrical forces makes another problem for stars. Like charges repel, and the nucleus of most atoms contain a number of protons, all of like charge, which are packed closely together. What keeps the nuclei from being blown apart by the repulsion of all the protons in them?

There is no way either electricity or gravity could save the situation. What is needed if nuclei are to exist is another force with certain properties. It must act attractively among protons and neutrons, in order to hold the atomic nuclei together. It must be strong enough to counteract the repulsions of all the protons. But it cannot be too strong, otherwise it would be too difficult to break the nuclei apart, and chain reactions of nuclear reactions could not take place inside of stars.

This force must also be short-ranged, otherwise there would be danger of its pulling all the protons and neutrons in the world together into one big nucleus. For the same reason, it cannot act on electrons, otherwise it would pull them into the nuclei, making molecules and chemistry impossible.

It turns that there is a force with exactly these required properties. It is called the strong nuclear force, and it acts, as it should, only over a range which is more or less equal to the size of an atomic nucleus.

Remarkably, the existence of more than a hundred kinds of stable nuclei is due to the fact that the strength of the attractive nuclear force balances quite well the electrical repulsion of the protons. To see this, it is necessary only to ask how much we have to increase the strength of the electrical force, or decrease the strength of the nuclear force, before no nuclei are stable. The answer is not much. If the strong interaction were only 50% weaker, the electrical repulsion is no longer overcome, and most nuclei become unstable. Going a bit further, perhaps to 25%, all nuclei fall apart. The same effect can also be achieved by holding the strong interaction unchanged and increasing the strength of the electrical repulsions by no more than a factor of about ten.

Thus we see that the simple existence of many species of nuclei, and hence the possibility of a world with the complexity of ours, with many different types of molecules each with distinct chemical properties, is ultimately the result of a rather delicate balance between two of the basic interactions, the electromagnetic and strong nuclear force.

There is, finally, one more basic interaction, which is called the weak nuclear interaction. It is called a nuclear interaction because the scale over which it can act is also about the size of the atomic nucleus. But it is much weaker than the strong nuclear force. It is too weak to play any role binding things together, but it does play an important role in transforming particles into each other. It is this weak interaction that governs the basic nuclear reaction on which the physics of stars is based, by means of which an electron and a proton are transformed into a neutron and a neutrino.

The reader to whom these things are new might pause and ponder the characteristics of these four basic forces, for it is they that give our world its basic shape. With their different properties, they work together to allow a world that is both complex and harmonious. Eliminate any one, or change its range or strength, and the universe around us will evaporate instantly and a vastly different world will come into being.

Would any of these other worlds contain stars? How many could contain life? The answer to both of these questions, as we have seen, is not many.

Physicists are constantly talking about how simple nature is. Indeed, the laws of nature are very simple, and as we come to understand them better they are getting simpler. But, in fact, nature is not simple. To see this, all we need to do is to compare our actual universe to an imagined one that really is simple. Imagine, for example, a homogeneous gas of neutrons, filling the universe at some constant temperature and density. That would be simple. Compared to that possibility, our universe is extraordinarily complex and varied!

Now, what is really interesting about this situation is that while the laws of nature are simple, there is a clear sense in which we can say that these laws are also characterized by a lot of variety. There are only four fundamental forces, but they differ dramatically in their ranges and interaction strengths. Most things in the world are made of only four stable particles: protons, neutrons, electrons and neutrinos; but they have a very large range of masses, and each interacts with a different mix of the four forces.

The simple observation we have made here is that the variety we see in the universe around us is to a great extent a consequence of this variety in the fundamental forces and particles. That is to say, the mystery of why there is such variety in the laws of physics, is essentially tied to the question of why the laws of physics allow such a variety of structures in the universe.

If we are to genuinely understand our universe, these relations, between the structures on large scales and the elementary particles, must be understood as being something other than coincidence. We must understand how it came to be that the parameters that govern the elementary particles and their interactions are tuned and balanced in such a way that a universe of such variety and complexity arises.

Of course, it is always possible that this is just coincidence. Perhaps before going further we should ask just how probable is it that a universe created by randomly choosing the parameters will contain stars. Given what we have already said, it is simple to estimate this probability. For those readers who are interested, the arithmetic is in the notes. The answer, in round numbers, comes to about one chance in 10^{229}.

To illustrate how truly ridiculous this number is, we might note that the part of the universe we can see from earth contains about 10^{22} stars which together contain about 10^{80} protons and neutrons. These numbers are gigantic, but they are infinitesimal compared to 10^{229}. In my opinion, a probability this tiny is not something we can let go unexplained. Luck will certainly not do here; we need some rational explanation of how something this unlikely turned out to be the case.

I know of three directions in which we might search for the reason why the parameters are tuned to such unlikely values. The first is towards some version of the *anthropic principle*. One may say that one believes that there is a god who created the world in this way, so there would arise rational creatures who would love him. We may even imagine that he prefers our love of him to be a rational choice made after we understand how unlikely our own existence is. While there is little I can say against religious faith, one must recognize that this is mysticism, in the sense that it makes the answers to scientific questions dependent on a faith about something outside the domain of rationality.

A different form of the anthropic principle begins with the hypothesis that there are a very large number of universes. In each the parameters are chosen randomly. If there are at least 10^{229} of them then it becomes probable that at least one of them will by chance contain stars. The problem with this is that it makes it possible to explain almost anything, for among the universes one can find most of the other equally unlikely possibilities. To argue this way is not to reason, it is simply to give up looking for a rational explanation. Had this kind of reasoning been applied to biology, the principle of natural selection would never have been found.

A second approach to explaining the parameters is the hypothesis that there is only a single unique mathematically consistent theory of the whole universe. If that theory were found, we would simply have no choice but to accept it as the explanation. But imagine what sense we could then make of our existence in the world. It strains credulity to imagine that mathematical consistency could be the sole reason for the parameters to have the extraordinarily unlikely values that result in a world with stars and life. If in the end mathematics alone wins us our one chance in 10^{229} we would have little choice but to become mystics. This would be an even purer mysticism than the anthropic principle because then even God would have had no choice in the creation of the world.

The only other possibility is much more mundane than these. It is that the parameters may actually change in time, according to some unknown physical processes. The values they take may then be the result of real physical processes that happened sometime in our past. This would take us outside the boundaries of the platonist philosophy, but it seems nevertheless to be our best hope for a completely rational understanding of the universe, one that doesn't rely on faith or mysticism.

In Part Two, I will describe one approach to such a theory. However, before coming to this I will turn, in the next two chapters, to some of the theories that have been proposed during the last twenty years to explain why the elementary particles are as we find them to be. I do this for two reasons. First, because there are some important lessons to be learned from these theories. Both their successes and their failures must be our landmarks as we venture into this difficult landscape. But, more importantly, the solution I will propose is speculative, and may seem even desperate. To judge it the reader will want to know what alternatives have been proposed, and how well they have done towards solving the same problems.

Whatever may have been the case in years gone by, the true use for the imaginative faculty of modern times is to give ultimate vivification to facts, to science and to common lives, endowing them with the glows and glories and final illustriousness which belong to every real thing, and to real things only.

          —Walt Whitman, *"A Backward Glance Over Troubled Waters"*

FOUR

THE DREAM *of* UNIFICATION

A ny theory interesting enough to have a hope of explaining our universe must confront the issue of the relationship between unity and variety. This is certainly true in politics, as the progress of democracy at the present time requires us to understand society as a network of very different cultures and individuals whose many and varied interactions work to weave together a common life. It is no less true in science. We have seen in the last chapters how the tremendous variety and complexity of the universe is built from a simple and common set of elements; four forces governing the lives of the four stable particles. But we have also seen how the incredible variety of phenomena in nature is a manifestation of a diversity of the properties of the elementary particles and their interactions. And we have seen that the large variations in the properties of the elementary particles and forces is necessary if our world is to filled with stars, which by forging the elements and filling the world continually with light and energy make it a home for life.

However, just as all the fantastic variety of phenomena we see in nature hides a common construction from only four basic forces and four stable particles, we want to ask whether the diversity of the elementary particles and forces hides as well a common origin. While we appreciate the significance of their diversity, we still want to know if there is some commonality that ties them all together. Is it possible that the four fundamental forces are all just manifestations of one basic force? Is it possible that underneath, protons, neutrons, electrons, neutrinos and all the others are all built from a common element?

This desire to discover unity among the diversity of the fundamental particles and their interactions has been the source of much of the progress in elementary particle physics for more than a century, and it is to this dream that we must now turn as we continue our search for the roots of the present crisis in theoretical physics and cosmology. In this chapter we will trace the progress of this dream and how it led to its greatest triumph, the standard model of elementary particle physics. Then, in the following chapter we will try to understand what has happened to this dream in the twenty years since that theory was invented.

Unification—the discovery that two phenomena which hitherto seemed completely separate have in fact a common origin—is what theoretical physicists, at least most of us, dream of. The discovery of a unification represents a great step in our understanding of nature. Each time it happens, it reassures us that our hubris that nature is understandable continues to be answered with comprehension. Furthermore, when some beautiful idea is found to be at the core of the unification, we see again at work the mysterious power we human beings seem to have to imagine what is behind the appearances of nature.

The first great unification in physics came in the middle of the last century, when a Scottish physicist named James Clerk Maxwell discovered that electricity and magnetism were really different manifestations of a single phenomenon, which he called electromagnetism. His reward was to have one of the greatest insights in the entire history of science. When he first tried to put the equations that describe the electric and magnetic fields into one system, he found a certain disturbing asymmetry in the forms of the equations. For purely esthetic reasons he changed the equations in order to make them more symmetric. He then discovered that his new equations predicted that waves should travel through the electric and magnetic fields. He was able to compute their speed, and this led to a great discovery: He found that their speed was equal to the speed of light!

I have often tried to imagine his feelings at that moment. He had discovered something no one before him had known but that everyone after him would take for granted—that light is a wave through the fields that carry the forces between electric charges and magnets. Surely, one such moment in a life of science can make all of the hard work and disappointment worthwhile; even one such moment in many thousands of working lives suffices.

In this wonderful century there have been several more such moments of discovery. As a result we now understand that Maxwell's great discovery was only the first step of a development that has led to our understanding that not only electricity and magnetism, but also the nuclear forces, are different manifestations of a single principle.

However, the desire for unification is not the only theme to have energized twentieth century physics. The imperatives of atomism and reductionism have been no less important. But what makes the story really interesting is that there is in the end a conflict between the logic of atomism and the desire for unification. Although it is often said that the goal of physics is to discover a completely unified theory of fundamental particles, there is a hidden tension between the notion that the elementary particles have absolute properties, independent of each other and the history of the universe, and the idea of complete unification, according to which all the elementary particles and forces are manifestations of a single fundamental entity. As we shall see, this conflict is the key to understanding the relationship between unity and variety in our understanding of the physical universe.

The conflict arises because, if the world is to have any variety of phenomena in it, it cannot be composed of only one type of fundamental particle or governed by only one force. As we have seen in the last chapter, an interesting universe requires that there be forces with different properties that act to balance each other. It requires also that the masses of the different elementary particles differ by large ratios. A universe built from only one kind of particle or one force would be crushingly boring.

The standard model of particle physics succeeds, at least partly, in unifying the fundamental particles and forces, because it is able to explain why the elementary particles and forces differ from each other. How this is understood—how diversity arises in the context of a unified theory of the fundamental particles and their interactions—is one of the important lessons we can learn from the story of the progress of elementary particle physics in this century.

The modern science of elementary particle physics began in the 1930s, with the discovery that the several hundred different kinds of atomic nuclei are all composed of protons and neutrons. The simplification this brought didn't last long. As larger particle accelerators were built over the next twenty years, many fundamental particles were found, each apparently as elementary as the proton and neutron. By the late 1950s their count was in the hundreds.

Something had to be done. In the nineteen-sixties it was proposed that most of these particles, including the proton and neutron, are, like the atoms, not elementary. Rather, each is composed of a number of more fundamental entities, called quarks.

By the time I entered graduate school in 1975, the quark theory, having been elaborated into the standard model of elementary particle physics, had triumphed. The atmosphere that a young student interested in elementary particle theory encountered at that time could not have been more dramatic and challenging. The triumph of the quark theory was part of a revolution that had taken place in the early 1970s in our understanding of the forces that are at work inside the atomic nucleus. The result of that revolution was the standard model, within which the strong and weak nuclear are, to some extent, unified with the electromagnetic forces into a single theory.

The standard model actually consists of two closely related theories. The first is an extension of Maxwell's theory that incorporates the weak nuclear interaction. This is called the Weinberg Salam theory. The second is a theory of how quarks bind together to make protons, neutrons and many other particles. This is called *quantum chromodynamics*, for a reason that will become apparent shortly. In the wake of these developments, no field of science could match elementary particle theory for excitement and glamour. Those of us beginning our work in physics in the shadow of this triumph went to bed, irrespective of our personal levels of ambition or modesty, with dreams of how we might somehow, if our mathematics was imaginative enough and we worked hard enough, be lucky enough to stumble upon the next level of unification.

The standard model made comprehensible a great many different pieces of experimental data that had been accumulating for several decades. It also predicted new phenomena. During the middle and late 1970s, many experiments were done to see if the theory could hold up to detailed test. This period was thus marked by a great deal of interaction and collaboration between theoretical and experimental physicists. Many seminars were devoted to the question of testing the standard model. Feelings ran high, and occasionally these seminars ended with people yelling at each other. This is rather disconcerting to observe if you are a new graduate student looking forward to your first opportunity to present your work in a seminar. At the same time, the passion was a sign, if one was needed, that something important was happening.

The standard model does have one great weakness, which I have already described: its dependence on a large number of free parameters. Thus, while a great triumph, it was immediately clear to everyone involved that it could not be taken as a fundamental theory. No theory with twenty parameters that can be freely chosen can be considered to be a fundamental theory of anything. What is clearly missing are some additional principles that set the values of these parameters. Any proposal for such principles must confront the puzzles associated with the improbabilities of their actual values, as we have discussed.

But the first thing that must be understood about the standard model is how it resolves the tension between unity and diversity. How is it that phenomena as different as electromagnetism and the weak and strong nuclear interactions can be encompassed within a single theory? The theory succeeds in doing this because it is based on two simple ideas, which are called the *gauge principle* and *spontaneous symmetry breaking*. It will be useful for us to know something about each of these.

The gauge principle is based on a simple philosophical idea, which is an answer to the question of how the elementary particles can have distinct properties if they are not made of any parts. To understand it we may begin by asking whether it really makes sense, as the philosophy of atomism supposes, that a single particle (say a neutron) would be exactly the same were it the only particle in the universe? While it is easy to imagine a world with one particle in it, we never actually observe anything in isolation. Just the act of observing something means that it is interacting with something else—light, us, or the measuring instrument. It makes sense then to ask whether the properties of an elementary particle like an electron are intrinsic to it, or are in part a manifestation of the interactions between it and the other things in the world.

The debate about whether the properties of an elementary particle are absolute or arise only from the relationships and interactions that tie it to the rest of the universe is very old. It goes back at least to the debates between Leibniz and Newton in the seventeenth century. But this controversy has turned out to be of more than philosophical interest, for it is central to all of the important developments of twentieth century theoretical physics. Relativity, quantum theory and the gauge principle which underlies the standard model can each be understood to have evolved from attempts to answer this question. Furthermore, they all come down on the same side of the issue, as they are all based in one way or another on the point of view that the properties of things arise from relationships.

To understand the gauge principle we may consider the simple case of electrical charge. We all learn in school that the electron has a negative charge and the proton has a positive charge. But, really, can there be any meaning to which is the positive one and which the negative? Certainly what is important is only the relationships between the charges, which ones are the same and which opposite.

This may seem a trivial question, for how could it be other than that the notion of which is positive and which negative is just a question of language, of convention. And as I have asked it here there is not much more to it. But, in the 1920's a very prescient mathematician and physicist named Herman Weyl realized that this question could be asked in a way that makes it much more interesting.

What Herman Weyl noticed is that most experiments involve looking at things in a small region of space. Thus, if all there is to charge is relationships, all that matters is the relationships between the charges I am playing with in a particular experiment. If I am playing with some atoms in my kitchen and you are doing the same in yours, can it matter if you and I use the same convention about which charge is negative and which positive?

It might seem that you and I are free to use different conventions, but what then are we to do when we meet? Suppose I carry an electron from my kitchen to yours. Can each of us continue to call each positive or negative, according to our own convention? Or must we agree to choose the same convention when our two electrons interact? It seems we must come to an agreement over conventions, otherwise, how will we decide if our respective electrons attract or repel each other? But how should we choose, as there can be no absolute reason to favor one convention over another? It seems that we may be in need of some arbitrary authority to decide how to label our charges. But, Herman Weyl made an interesting move here, which has had profound implications for twentieth century physics. Rather than accepting the possibility of an arbitrary labeling he insisted that there be some way that each of us can remain free to call which particles we like positive and which negative, no matter whether they interact with each other or not.

In insisting on this, Weyl might be said to have followed a principle of Leibniz's philosophy, which is called the *principle of sufficient reason*. This requires that in the description of the world we not be forced to make any choice unless there is a rational reason for making it one way or the other. According to this principle, either there must be a rational, objective, reason to call certain charges positive and certain negative or we must remain free to make these choices however we like.

Weyl found that there was a way to preserve our freedom to label charges as we would like. It requires that the force between the charges not be communicated directly. Instead the electrical force must be mediated by a field. A field is something that exists at each point of space. The force is carried by the field in the sense that each charge interacts only with the field in its immediate vicinity. The presence of a charge causes a change in the field nearby, and that change is then communicated through the entire field. Each charge feels the other only through the effect it has had on the field.

Because all that matters is the relationship between each charge and the field around it, Weyl discovered that it is possible to arrange the law by which the fields and the particles interact so that we keep the freedom to choose as we like which charges are negative and which positive. The field carries information about the presence of a charge in a form that does not depend on our conventions. As a result we can choose differently in different places, and we can change our minds about our choice at any time. But we can only do so if the field satisfies certain equations. Weyl wrote them down, and immediately saw he had made a great discovery; his equations were the same as those satisfied by the electromagnetic field! This meant that the story I have been telling is not just an imaginary adventure in culinary philosophy, it is about nature. And the field whose existence is necessary to preserve our freedom to call charges positive or negative as we like is real: it is the electromagnetic field.

Thus, the idea that charges be defined completely in terms of their relationships is more than philosophy. Through the chain of reasoning I have described, this idea can lead to the prediction of the existence of new fields that carry forces between particles. In this form it has become a physical principle, which is called the gauge principle.

This principle can be extended from electric charge to more complicated circumstances. In doing so a new class of theories was invented that extend in a beautiful way the physics of the electric and magnetic fields. It is these new theories that underlie the standard model of elementary particle physics.

To understand how this can happen, let us consider a more complicated kind of electron, which can have not one, but three kinds of charge. Let us name these charges after the primary colors, so now we can have red charges, yellow charges and blue charges. Following Weyl's ideas, several people then asked what would happen if we were each able to change our minds about which color was which, freely, in different places and at different times. They found that this could be accomplished if there was a field that interacted with the colored particles. This new field is a fancier object than the electromagnetic fields; it is something like *eight* different electromagnetic fields, which interact not only with the colored particles but with each other. The new theories were called Yang-Mills theories after C. N. Yang and Richard Mills, who together were among those who proposed them, in 1954.

It took another twenty years to understand that the Yang-Mills theories describe the strong and weak interactions. The reason is that it was not so simple to understand how to describe these theories in the language of quantum mechanics. This task was only accomplished in 1971, primarily by a Dutch graduate student named Gerard 't Hooft.

Shortly after this, several people realized that, when combined with quantum mechanics, Yang-Mills fields can have some remarkable properties. Everyone is familiar with the fact that in electricity opposite charges attract. But when there is more than one kind of charge, as in our example with colors, this tendency can be realized in a way that is much more drastic. Opposite colors not only attract—they cannot be separated from each other. No combination of colored particles can be separated from others unless all the color averages out completely. This property is called the *confinement of colors*. It means that one can never observe a colored particle in nature. One can only see combinations of particles in which the colors cancel each other out.

As soon as people understood this property of confinement, the application to physics was obvious. Physicists already had good reason to believe that protons and neutrons are each composed of three particles, which had been called quarks. Moreover, every one of the many strongly interacting particles that had been seen in experiments could be interpreted as containing an equal mix of the three colors. The result was that one can understand all the phenomena of the strong interactions, including all of nuclear physics, by supposing that each of the quarks comes in three colors, and that the forces between them are the result of their interactions with the Yang-Mills field.

This new theory, called *quantum chromodynamics*, or QCD, for short, must be considered to be one of the triumphs of twentieth century science. Once one understands it an enormous amount of experimental data: all of nuclear physics and a great deal of elementary particle physics, is revealed to be a manifestation of a single phenomenon. What is also very beautiful is how all of these phenomena can be understood to be the direct manifestation of the general principle that all properties of objects are based on relationships between real things and have no *absolute* meaning.

The other half of the standard model is Weinberg and Salam's unification of the weak and electromagnetic interactions. This is also described by a Yang-Mills theory. Here the theory acts in a still more interesting way, because it succeeds in unifying things that are, apparently, quite different. The range of the electromagnetic force is infinite, while the range of the weak nuclear force is no more than the diameter of an atomic nucleus. How can these be seen as different aspects of one single force? Furthermore, in this theory the electron and the neutrino turn out to be each manifestations of a single type of particle. But how can this be? One has an electrical charge, and the other has none. One also has a mass much larger than the other.

It is here, in the joining of these diverse particles and forces, that the quest for unification collides with the logic of atomism. The philosophy I have called radical atomism holds that the properties of fundamental particles are intrinsic, and owe nothing to their history or environment. The electron and neutrino are both fundamental particles. If they have different properties, aren't they intrinsically different? But if this is so, there can be no further unification.

The second key idea behind the standard model, spontaneous symmetry breaking, can be understood as an escape from this dilemma. The central point is that to make further progress in the search for unification some part of the radical atomist philosophy has to be given up. If we are to understand how fundamental particles, such as the electron and neutrino, are to be unified while maintaining their difference, one has to look to something that is not intrinsic to the particles. What can this be? There are not many answers we could give to this question. There must be some effect coming from their interaction with the environment that plays a role in the explanation of why they are different.

This is exactly what happened in the Weinberg-Salam model. In that theory, the mass of the electron is not intrinsic; it comes instead from its interaction with certain other particles, which are called Higgs particles. If there were no Higgs particles, the electron would have no mass. It would move at the speed of light, like a photon. But if it finds itself surrounded by a gas of Higgs particles, an electron is not able to move so quickly. The electron seems to gain mass because it is moving, not through empty space, but through a muck of Higgs particles. It becomes heavier because when one pushes it, one also pushes all the Higgs particles around it.

In fact, there is good reason to believe that the world is filled with a gas of Higgs particles, which are responsible for giving the electron its mass. But that's not the whole story. Adding the Higgs particles doesn't really remove the distinction between electrons and neutrinos: they are now different because the electron interacts with the Higgs particles, while the neutrino does not.

To remove the distinction between electron and neutrino, the theory must be written in terms that are completely symmetric. As far as the theory is concerned, electrons and neutrinos must be identical. This can be done. It means that we must add a second set of Higgs particles that interact with the neutrino. There are then two kinds of Higgs particles, which we may call electron Higgs and neutrino Higgs. When no Higgs particles are present, the electron and neutrino will be the same. If Higgs are present then either electrons or neutrinos become massive, depending on which kind of Higgs are around.

This scheme achieves the goal of having the distinction between electrons and neutrinos depend on the environment. The electron is different than the neutrino, according to this theory, because it happens that the world is filled only with the electron Higgs. Although I will not describe it here, it turns out that this is also the explanation of why the weak interactions are different from electromagnetism.

One question remains: Why the world is filled with only one kind of Higgs particle? Could we not have equal amounts of each, in which case the electron and neutrino would be still the same? In order to prevent this, the laws which govern these Higgs particles must be arranged so that such a symmetric configuration would be unstable. Instead, the only stable configurations are those in which the world is filled only with one kind of Higgs particle. To preserve the symmetry of the theory, it cannot matter which kind fills up the world—the theory stipulates only that it be one or the other.

This kind of situation is rather common in physics. There are many situations in which the laws of nature are symmetric in some way, but the only stable configurations are asymmetric. Because the theory is symmetric, it cannot tell which stable configuration is chosen. Instead, the choice must be made by the system itself. When this happens we say that *a symmetry of the laws has been spontaneously broken*.

Imagine a pencil balanced on its point. It cannot stay that way long for it is unstable, a little push to one side or the other and it will fall. If it is perfectly balanced, the law of gravity cannot tell us which way it will fall; any way is as good as another. But any small disturbance will break the symmetry, leading to a choice of a more stable, but less symmetric configuration, in which the pencil is lying on its side.

At the risk of seeming frivolous, perhaps an analogy drawn from life can also illustrate this idea. Think of the young people in a town or a city, as they grow up and begin to look for love in relation to someone else. For some, only one person is right for them, while others may date a number of people before settling down with their chosen mate. There may then be a number of different ways in which these people might organize themselves into couples. We might say that there is a symmetry among these possibilities, in each of which there is more or less the same amount of happiness. However, this being the case, it is not true that the most stable or happiest situation is the symmetric one in which everyone spends time with everyone else. Instead, as a result of seemingly random and serendipitous happenings, people meet and fall in love, and the result (at least ideally) is the establishment of a stable social life based on particular choices. This is an example of spontaneous symmetry breaking.

Furthermore, to the extent that each of our social identities are defined by our intimate and family relationships, one can say that each of us has as we grow up a large number of different potential identities, only one of which may be realized in a stable community. In the same sense, each elementary particle has a set of different potential properties, only one of which can be realized in a stable universe.

The dynamics of the Higgs particles are arranged so that they are just like these examples. As the Higgs particles were an invention of Weinberg and Salam, they were free to posit whatever forces they liked between the different Higgs particles. It was not hard to invent forces so that any symmetric configuration of the Higgs particles, in which there are the same numbers of each type, is unstable. This includes empty space because if there are no particles, then the numbers of the two types are still equal, because both are zero. The only stable configurations are those in which the world is filled with a gas of only one type of Higgs particle.

Even if the Higgs particles have yet to be seen, all of the tests that have been done which confirm the standard model tell us that Weinberg and Salam were right about this. Spontaneous symmetry breaking is the way that nature contrives to resolve the dilemma of unification and variety. As far as the laws of nature are concerned the electron and neutrino are identical. They are different only because the environment in which they move distinguishes them. The electron is different than the neutrino only because the world happens to be in the state in which it is filled with a gas of electron-Higgs.

This means that the theory does not itself determine all the properties of the elementary particles. Whether the electron has mass or not depends on whether the world is filled with Higgs particles. In turn, whether the world is filled by a gas of Higgs particles can depend on the overall conditions, such as the temperature. For if the temperature is high enough even a very unstable configuration can be maintained by the thermal energy. This makes it possible to speak of the universe having a number of different possible phases, which are analogous to the different phases of matter. The atoms that make up water can be organized into a solid, a liquid, or a gas. Analogously, the Higgs particles can exist in different phases. In some of these phases, the world is full of one kind or another of them, but in others there may be equal numbers of the two kinds, depending on the overall temperature and density of matter.

The laws of elementary particle physics do not choose among these configurations, they merely allow them as different possibilities. This means that the properties of the elementary particles are in the end influenced by the history and state of the whole universe. The dream of a connection between the microscopic and the cosmological is no longer a fantasy of philosophers—it is concretely realized in the standard model of elementary particle physics.

FIVE

THE LESSONS *of* STRING THEORY

The standard model of elementary particle physics stands as a monument to more than a century of continual discovery. In it is captured all of our knowledge of the very small, accumulated from the invention of the electric and magnetic fields in the 1840's to the discovery of the W and Z particles in the 1980s. If it has been difficult to do improve on it, this is because it represents such a total triumph. In its imperfections, however, the standard model also forces us to confront all that was not understood during this long period.

The main challenge which has faced theorists of the elementary particles since the mid-1970s is how to improve on the standard model. Is there some extension of the gauge principle that will reveal all the different interactions in nature to be manifestations of one basic force, and all the particles to be different realizations of a single fundamental entity? And, if so, can such a theory overcome the limitations of the standard model and explain why all the particles have the properties they do, without resorting to setting by hand a large number of parameters? Or is some new principle needed to explain to us how nature chooses the values of these parameters?

I'm sure that, to an outsider, theoretical physics seems a difficult and mysterious business. But it is actually not very complicated. Physics students have a lot less information to absorb than those in most other disciplines. The education of a physicist consists instead in mastering the crafts which are useful to approaching certain kinds of questions about the world. Among these are a set of powerful tools of the trade that we use to simplify problems to the point where we can think about them intuitively. One of the most powerful of these tools is the ability to think in terms of scales.

The idea behind this is that most physical phenomena take place at characteristic scales of length, time, energy and mass. For example, all atoms are around the same size, about 10^{-8} centimeters. All atomic nuclei are about a hundred thousand times smaller. Most stars are within a factor of ten or so of the mass of the sun, while most galaxies contain roughly the same number of stars as our Milky Way. A first step towards understanding any of these objects is to appreciate the reasons why they come in typical sizes. To do this we look for simple arguments that let us estimate, to within a factor of ten or so, what the typical scale of the object will be. This is usually more illuminating than the complicated calculations that are required for a more exact description.

I can illustrate this with a simple example. If we want to do fundamental physics we may ask what the scale of the simplest and most basic phenomena in nature should be. These should be phenomena that involve only those aspects of physics which are universal. We know of three universal phenomena. Everything that moves is described by the principles of relativity, and everything that exists seems to be described by quantum theory. Among the forces, only gravity applies universally to everything. Thus, we can ask: On what scale will we find the simplest possible processes that involve only gravity? These would be processes that can be described purely in terms of gravity and quantum theory.

This is easy to do because associated with the relativity and quantum theories, there are three universal physical constants. These are Newton's gravitational constant, G, which measures the strength of the gravitational force, Planck's constant, h, which measures the scale of quantum phenomena, and the speed of light, c. By putting these constants together, one can construct a set of units which describe the scales on which elementary quantum gravitational processes will take place. These are called the Planck units. We have already met the Planck mass, which is the scale of the most massive possible elementary particle. In terms of the elementary constants it is given by a simple expression: (The non-mathematical reader shouldn't worry; the definitions of these units are the only formulas in the book.)

$$\text{The Planck mass} = \sqrt{hc/G} = 10^{19} \text{ proton masses} = 10^{-5} \text{ grams}$$

Similarly, we can find a simple unit for length which is:

$$\text{The Planck length} = \sqrt{hG/c^3} = 10^{-33} \text{ centimeters}$$

This length tells us the scale of any simple processes that involve only effects of gravity and quantum theory. To see why, suppose that we have a quantum theory of gravity, and that it predicts the size of something. That prediction must result in a mathematical expression which describes a length. Any such expression will look like some simple combination of numbers times the Planck length. It must be so because the Planck length is the only way to get a length out of the three constants G, h, and c, and these are the only constants that go into the theory. Any simple collection of numbers will involve factors such as 2,3, 5, p, and so on. The product of a few such simple numbers cannot be too big or too small. Thus, any length predicted by the theory may be expected to be around the size of the Planck length.

This may seem like a bit of a trick, but in physics arguments like this are usually reliable. In fact, several different approaches to quantum gravity do predict that there is a smallest length and that it is about equal to the Planck length. (We will discuss this in more detail in Chapter 22.)

The most striking thing about the Planck units is how far they are from the scales of atomic and nuclear physics. Protons and neutrons are about 10^{-13} centimeters in diameter. With current accelerators we can probe down to about 10^{-15} centimeters. The Planck length is 10^{-33} centimeters—*eighteen powers of ten smaller.* This is very disconcerting. It means that, basic as they are to the construction of our world, quarks and electrons are still absolutely enormous compared to what we expect should be the scale of the truly elementary things in the world.

What makes it even worse is that the Planck scale is far removed from phenomena we can explore with any technology we can conceive of, currently. The distance we have yet to go to reach it is roughly the ratio of the orbit of the moon to the size of the atom. If we think about how much our picture of nature changed from 300 AD, when the Egyptian astronomer Ptolemy developed the first accurate theory of the orbit of the moon, to 1911 when Niels Bohr wrote down the first formula for the size of the atom, we can get a sense of how much it may still change before we have a fundamental theory which can tell us how to think about phenomena at the Planck scale. This is certainly the greatest obstacle facing any attempt to extend the gauge principle beyond the standard model.

A second obstacle arises from the theory's reliance on the idea of spontaneous symmetry breaking to explain why each of the elementary particles we see in the world has different properties. While it is a beautiful idea, there is a certain ad hoc quality to how it is realized. To this date, no one has so far observed a Higgs particle and we have only a very imprecise idea of their actual properties. As a consequence, the Higgs particles are described in terms of models in which a large set of free parameters remain to be specified. If there is to be any hope of reducing the number of free parameters in our physical theories, we must understand them in some way that makes their properties necessary consequences of some more fundamental theory.

This question of the nature of the Higgs particles is closely connected to the problem of the remoteness of the Planck scale. In spite of our ignorance, we can estimate roughly the amount of energy that would be necessary to create a Higgs particle. If there is to be a complete unification of all the interactions, there must be Higgs particles that play the role of distinguishing the strong interactions of the quarks from the other interactions. The scale for these is very high, it must be around 10^{15} times the mass of the proton. This is much closer to the Planck scale than to the scale of the physics we understand. Thus, any attempt to further unify the different interactions involves thinking about scales more than ten orders of magnitude removed from conceivable experiments. There is then a kind of conspiracy between these obstacles that makes the problem of going beyond the standard model such a hard nut to crack. Looking back, we should not be too surprised that two decades has not sufficed to discover the more fundamental theory.

Of course, this is not how it looked in the late 1970s. Physicists are no different from other people and, fresh from the dramatic successes of the standard model, it seemed the only thing to do was to find an extension of the ideas that had worked so well in that case. The combination of the Yang-Mills theory with the idea of spontaneous symmetry breaking seemed powerful enough to unify all the interactions into a Grand Unified Theory.

In 1975 it was clear how such a theory might be constructed. One must put all the particles together in one large family and posit that all the distinctions among them arise only from interactions. Following Weyl and Yang and Mills one would then demand that the labels among these different particles be purely conventional, and that those conventions be freely specifiable by different observers. The result would be a single unified interaction in which all the particles participate. The differences between the particles and interactions would then be introduced through the trick of spontaneous symmetry breaking. All the laws of physics, except perhaps gravity, would be derivable from this simple unified theory.

It was a good idea, indeed, it was a wonderful idea. In 1975 it seemed certain that someone with a little luck would hit upon the right way to do this, and all of particle physics would then be unified into a Grand Unified Theory. In the late

1970s and 1980s many different variations of this idea were invented. They had names like SU(5), SO(10), E6, E8, technicolor, and supersymmetry, which were the names of the mathematical structures on which they were built. Indeed, many of them worked, in the sense that they reproduced all the known physics of the standard model. And this was, ultimately, the source of the problem. Without any opportunity to distinguish these different theories experimentally, there are too many possibilities, too much arbitrariness.

But even more than this, it rather soon became clear that any attempt to copy the standard model too closely perpetuated as well its main weaknesses. Many of the attempts to extend the gauge principle to a Grand Unified theory ended up with the same two problems we've just discussed. The huge hierarchy of scales and a set of Higgs particles with just the right properties often had to be put in by hand. Furthermore, none of these theories eliminated the need to tune parameters delicately to fit the data, although in some cases their number was reduced. By the early 1980s it began to be clear to many people that if a truly unified theory were to be found, some new ideas would be needed.

Perhaps the cathartic experience that liberated physicists from the idea that progress could come from simply copying the standard model was a particularly dramatic failure of the simplest grand unified theories that took place in the early 1980s. There had been one possibility for an experimental test of the grand unified theories which, had it worked, would have meant that the further unification of physics could proceed without leaving behind the traditional collaboration of theorists and experimentalists that has characterized physics for the last centuries. This was the possibility that, as a result of the additional interactions that such theories necessarily predict, protons would be unstable to radioactive decay.

The reason goes to the heart of the basic philosophical idea that is behind the gauge theories. This is the principle that the properties of the elementary particles should be the consequence of their interactions with each other. For the same physical processes that distinguish the particles from each other can also act to transform them into each other. These processes may be extremely rare, but they must be there at some level. Among these there must be processes that transform quarks into electrons and neutrinos. If this happened to a quark inside of a proton, the result would be the disintegration of the proton into electrons and neutrinos.

Of course, if we are not to wake up tomorrow morning to a much plainer world, such processes must be very rare. Even so, this is a wonderful situation because proton decay, were it to occur, could only be a sign that a grand unified theory were true. Furthermore, even if the event is very rare—so that any single proton decays with a half-life of many millions of billions of times the age of the universe—any large collections of protons (say the water in a large swimming pool) contains so many that, by the laws of probability, more than one such event a year should take place. As such an event would liberate a huge amount of energy, to see protons decay it is only necessary to line the swimming pool with detectors, isolate it deep in a mine to avoid the false triggering of the detectors from cosmic rays, and sit back and wait.

A lot of ingenuity was put into such experiments in the early 1980s, but unfortunately the result was negative. No protons were seen to decay. This does not kill the idea of grand unification, it only means that proton decay, if it occurs, is too rare an event to be seen in these experiments. But what it did kill was the one realistic hope of a direct experimental test of the grand unified theories. As a result, those of us who were chasing the dream of unification had to face a rather unpleasant fact: if we want to proceed with the dream of unification of all physics, we must do so without any realistic expectation of soon receiving guidance from experimental physics.

Certainly we may hope that if something wonderful emerges from the program of unification, sooner of later experimental physics will catch up. A new theory might even reveal new kinds of experiments, which would not have been thought of in its absence. The history of science is full of examples in which exactly that has happened. Before Maxwell's theory of electromagnetism, it would not have made any sense to look for the transmission of radio waves. Proton decay is yet another example: no one would have thought it reasonable to look for it before the development of the grand unified theories. Thus, it is not so crazy to proceed without the guidance of experiment for a while. We can only hope that we will not be in this situation for too long.

As a result of the failure of the proton decay experiments, physicists were freed to make more radical proposals for going beyond the standard model. The most important of these was undoubtedly string theory, which began to be studied intensively in 1984. The story of string theory is important for the arguments of this book for two reasons. First, because it is an attempt at an extension of the gauge principle that holds unique promise in the search for a unification of all the fundamental particles and forces. Second, because it arose historically from a line of thought that was deeply critical of the naive atomism of the early quark theories.

The origins of string theory actually go back to the 1960s, before the invention of the standard model. It arose first as an attempt to understand the vast proliferation of apparently equally elementary particles that were discovered experimentally in the 1950s. One response to this situation had been the quark theory that says that each of these elementary particles is actually made up of smaller, more fundamental units. But a part of the community of theorists dissented from taking this step, which they perceived to be a kind of simple-minded atomism. Led by Geoffrey Chew, they believed that the tendency to explain the properties of little things at one level by breaking them into still littler things must somewhere come to an end.

Instead, they opposed to this reductionist strategy a principle that they called "nuclear democracy," according to which all particles in nature are equally fundamental. Moreover, the properties of each particle were to be understood as arising from their potential interactions with all the others. But, they not only embraced the Leibnizian philosophy that all properties arise from relations, they postulated that this idea, together with the principles of relativity and quantum theory, should be sufficient to explain all of the properties of the many elementary particles.

According to this view, if the properties of any one particle are determined by its interactions with all the others, while that particle itself participates equally in the determination of the properties of those others, then the laws of physics are a kind of system in which the influence of any one particle on the others feeds back to effect its own properties. The laws of physics then cannot be postulated a priori, one must find a self-consistent set of properties and interactions such that each particle in the system both contributes to and is determined by the network of interactions. The task of the elementary particle theorist then must be to discover a set of particles and interactions that determine each other in this self-consistent way.

Geoffrey Chew and his colleagues were able to write down a system of equations that express this idea, which were called the bootstrap equations. They conjectured that there might be a unique solution to these equations, subject only to the constraint that the theory agree with the principles of relativity and quantum theory. Thus, by solving these bootstrap equations, the whole of nature would emerge, with no more input than basic principles and mathematical consistency.

On the whole, this bootstrap program, as it came to be called, failed. The bootstrap equations were never solved and, in any case, the evidence that the proton and the neutron are each constructed out of three smaller particles is by now pretty substantial. But the program did, in a few cases, succeed. In these cases the conditions of consistency could be expressed in a way that could be solved, and they led to a correct description of certain experiments involving collisions of certain subatomic particles called mesons.

What was even more significant is that in the cases where the program worked, physicists—beginning with Yochiro Nambu, Holgar Nielsen and Leonard Susskind —realized that the solutions they found did not correspond to the traditional conception that a fundamental particle is a point that has no extension or dimension. Rather than behaving like mathematical points, they behaved more like stretched, one dimensional objects, something like rubber bands.

This led to the idea that perhaps atomism is right, because there are fundamental things in the world. Only these things are not to be visualized as point particles; they are instead one- dimensional. These fundamental one dimensional objects are what we call now strings. Just as a point has no size, these also take up no space, as their diameters are zero. But they do have length.

Because it postulated that there are basic things that the world is made of, string theory did not really satisfy the anti-atomist philosophy of the original bootstrap theorists. It represented instead a kind of accommodation of atomism to the problem of constructing a theory that unifies the many different particles and forces in the world. Indeed, looking back from a distance, one can see that string theory arose as a response to the conflict between the desire for unification and the logic of atomism. If we follow the philosophy of radical atomism, we must believe that the elementary particles are points—they must take up no space. If they did take up space they would have parts or regions. But then they would be, at least in principle, divisible; thus they would not be elementary.

At the same time, the idea of unification requires that the different kinds of elementary particles, such as quarks, electrons or neutrinos, arise as different manifestations of one single, most elementary particle. This requires that this most elementary particle has the possibility of existing in different, distinguishable states.

Now, if the elementary particle was something of a certain size, there would be no difficulty imagining it to exist in different states. It might be, for example, that the particle could take on different shapes. But it is hard to imagine how something that is just a point, that has no shape and takes up no space, could exist in different states or configurations. But if the elementary particles have no parts, we must imagine them as points.

String theory resolves this paradox, because it says that the end of the process of reductionism is that the most fundamental entities are one dimensional strings and not points. As these are the most fundamental things they cannot be further decomposed-there are no point particles into which a string might be decomposed. At the same time, it is easy to imagine the string existing in different configurations. Just like an ordinary guitar string, a fundamental string can vibrate in different modes. And it is these different modes of vibration of the string that are understood in string theory as being the different elementary particles.

As beautiful as this idea is, it did not succeed when it was applied to the interactions of the elementary particles in the 1960s. One reason was that when it was elaborated in detail, it was discovered that the idea of fundamental particles as one-dimensional objects could only be consistent with quantum mechanics and relativity theory if space had 25 dimensions. Perhaps this was sufficient, but what really killed interest in string theory was the success of the standard model as it was developed during the 1970s.

In his book *Against Method,* the philosopher of science Paul Feyerabend advises us that no theory can be so discredited that it might not have something to offer theorists in the future. This certainly turned out to be the case with string theory. For when most people began working on Yang-Mills theory in 1971, a small group—really just two or three individuals—continued to work on string theory. Having failed as a theory of the strong interactions they decided to up the ante and read string theory as a potential unified theory of all the interactions.

It was clear that in order to succeed at this, string theory would have to incorporate those achievements of the standard model, whose truth seems unassailable—especially the realization that the forces in nature can be understood as arising from the gauge principle. But, as both that principle and string theory originated from the philosophical idea that the properties of any one thing arise from its relations with the other things in the world, it is not surprising that this was not difficult to achieve. Indeed, as string theory was understood better, it became clear that the gauge interactions naturally emerged from it. But even more than this, during their period of exile from the mainstream, the string theorists realized that their theory naturally gave rise to an interaction that had all of the hallmarks of the gravitational force. In order to get the force to come out with the right strength, all they had to do was fix the length of the string to be about the Planck length. Thus, string theory had the potential to unify all of physics in a simple framework, in which all phenomena arise from the motion and vibrations of fundamental one-dimensional strings.

Beyond this striking discovery, by the early 1980s string theorists had found that a very beautiful way to extend the gauge principle. Since at least the time of Newton, those who tried to conceive nature's basic workings have been speaking of two distinct things: particles and forces. There are the things that make up the world, and then there are the interactions between them. However, Leibniz, among others, was suspicious of this distinction, for how can things that are truly the most fundamental and simple somehow have information about the others with which they interact? To have this information something must be changeable, or have parts. In either case, the thing is question is not the simplest possible thing.

Even without insisting on the purity of Leibniz's conception of the simple, it is clear that the idea that particles and forces are two separate kinds of things confines the ambition of the dream of unification. So, in the 1970s physicists invented an extension of the gauge principle that bridged this gulf between particles and forces, so that both were different manifestations of the same fundamental entity. This unification was called supersymmetry. When introduced into string theory, it was found to work miracles, so that it was possible to show, at least to a certain approximation, that string theory could give consistent predictions, something that had not been possible before. There remained the problem that string theory could not apparently be consistent with quantum theory if the world had three spatial dimensions, but at least supersymmetry brought the required number of dimensions down from 25 to nine.

Since the 1960s, particle theory had been split into two groups: those following the atomism of the quark theory and those who had followed the anti-atomism that had led from the bootstrap program to the string theory. What happened in 1984 was that it was realized that string theory could combine and satisfy the aspirations of both approaches to fundamental physics. Thus, the community of gauge theorists, driven by the failure of the proton decay experiments to search for new ideas that could unify physics, all of a sudden encountered their old friends, the string theorists, in the middle of what might be called a desert of disappointed expectations. By this time both groups had confronted the fact that further pursuit of the idea of unification would require them to work at scales far from those that could be experimentally probed, thus the desert in which they met was parched from a drought of experimental support, usually crucial for the flowering of science. There were not many flowers blooming in that desert, and this one—string theory—promised to satisfy at once the imperative to understand the world as constructed from some smallest fundamental entity and the desire to understand the laws of nature as arising solely from the postulation of a self-consistent and interrelated world.

It is not hard to understand why so many people have gotten excited about string theory. There is evidence that it is a consistent quantum theory of all the interactions, including gravity. This is even more than the grand unified theories had promised. Further, string theory seems to fulfill the dream of the original bootstrap theorists that there might be only one unique consistent theory that incorporates all the particles and forces in nature.

Since 1985 it has been clear that only a few open problems stand in the way of a triumph of string theory as the next great step forward in physics. One of these is to get the number of dimensions of space down from nine to three. This may seem daunting, but there is an idea about how to do it, which comes from some earlier attempts to unify gravity and electromagnetism. This is to introduce a new cosmological effect, in which the physics of the very small would depend on an aspect of the configuration of the whole universe.

The idea is to postulate that our world does have nine dimensions, but that six of them are rolled up, so that the diameter of the universe in these directions is not much more than a Planck length. There would then be no way for any of the elementary particles, such as the protons—which are twenty orders of magnitude larger than the diameter of these curled up dimensions—to know about anything other than the three remaining dimensions. We may then expect that once this curling up is understood, it might be just a technical matter to calculate what physics on ordinary scales looked like—what the masses of the particles are, what the different forces are, and what the ratios of their strengths are.

To recall now the great excitement about the discovery of this one, unique, theory of everything is a little like recalling 1968 and the great revolution we were about to make. People, among them the brightest and most creative young people working in theoretical physics, were saying things like, "If you want to be involved in this final stage of physics you should get in quickly, as things will certainly be all wrapped up in the next twelve to eighteen months." More than one person was saying that string theory was the most important development in physics in the last hundred years. Some even proclaimed the dawning of a new, postmodern, age of physics, in which mathematics would now play the driving role that had, since the time of Galileo, been played by experiment.

It was, indeed, a time of polemics, and the seminars that began with fifteen-minute speeches about our unique opportunity to make the final theory of physics were answered with seminars that began with thirty-minute speeches about how the string theorists were abandoning physics and threatening the integrity of science with their wild claims. However, for all the wisdom of experience, the first true theory of everything was apparently at hand. Could anyone be blamed for throwing themselves into this great adventure?

Now, more than ten years later, one can say that string theory did not quickly lead to a new unification of physics. But , on the other hand, it has been shown to be neither inconsistent nor wrong. The dream of a final unification has inspired a great deal of very good work, as a result of which our understanding of both string theory itself and the promise and difficulties of the general project of unification is now much deeper. String theory remains, for many who work in fundamental physics, the most promising hope for a unified theory yet proposed.

Further, whatever its future as a unified theory of physics, string theory has led to the discovery of beautiful mathematical ideas that have revolutionized the study of several branches of mathematics. Some of its proponents like to say that string theory is a piece of twenty-first- century mathematics that has, by our good fortune, fallen into our hands in the twentieth century. I tend to think they may be right, and that, whatever happens to string theory as we presently understand it, some of this mathematics will turn out to be essential for the construction of the theory that finally unifies quantum theory with relativity and cosmology.

At the same time, it must be said that, for all its elegance and promise, string theory has not so far led to any new predictions concerning the properties of the elementary particles. The reason is that we pay the same price for unification of the interactions in string theory that we did in the Weinberg-Salam model: an increased arbitrariness, an increased number of free parameters. And we pay it in spades.

The problem arises because while the theory is almost unique in its pure, nine-dimensional form, that uniqueness is lost when one curls up the six extra dimensions to make a theory that describes our three-dimensional world. There are literally tens of thousands of ways to curl up those six dimensions, and each one leads to the prediction of a different set of particles and interactions. Even worse, the number of dimensions that curl up does not need to be six, so that the number of dimensions that remain large can be anything between none and nine. Still worse, additional free parameters appear when the universe is curled up, for one has to specify the radius of the universe in each of the curled up dimensions. As a result, although it may be possible to tune the parameters of string theory to give predictions that agree with the standard model, it is unclear how impressed we ought to be, because one could also tune them differently to realize many different sets of particles and forces, if those were observed.

Each of the different ways of curling up the extra dimensions can be understood as leading to something like a different phase of the theory, analogous to the different phases of water. Just as in the case of the Higgs particles, the elementary particles have different properties in different phases. Also, just as in that case, the fundamental laws do not seem to choose which phase the universe is found in, all are allowed as possibilities and the choice depends on the configuration and history of the universe as a whole.

There has been a long-cherished hope that mathematical consistency, together with the principles of relativity and quantum theory, might constrain the possibilities for the laws of nature so severely that there would be only one possible, consistent, fundamental theory. Unfortunately, at least in its present form, string theory seems to provide evidence for the opposite conclusion. Each of the tens of thousands of different ways to curl up the extra dimensions leads to an apparently consistent set of laws for our world. At the very least, one can say that if consistency constrains the possible forces and interactions, or even the possible dimensions of space, one cannot see it from the way string theory has so far been formulated.

Is there a way out of this bind? String theory comes so close to a unified theory of all the interactions. Is there a way that the terrible price of a ten thousand fold arbitrariness can be avoided? What is clearly needed is a principle that could explain which of the many different phases of string theory the universe chooses to be in.

One possibility is that there is a more fundamental level of string theory, which unifies the descriptions of all the different phases. Indeed, there is good reason to believe that the present formulation of string theory cannot be the final one. This is because as the theory is presently understood it relies on a notion of space and time that has more in common with Newtonian physics than it does with Einstein's theory of relativity. In its present formulation, string theory describes strings moving in a space that has a fixed geometry, much like the space of Newton's theory. This is enough to see that gravity and the other forces can arise from the motion of strings, but it is not enough to realize a complete unification with the theory of relativity, which is based on the idea that space and time are dynamical and not fixed. If gravitation arises from the motion of strings—but if, as Einstein taught us, gravity is an aspect of the geometry of space and time—then in a more satisfactory formulation strings would not move in a fixed space. They would make up space and time.

The main project of string theorists for many years has thus been to reformulate string theory in a way that does not rely on the fiction of a fixed background of space and time. Many string theorists hope that such a theory will bring with it a principle that will tell us which phase the universe chooses. While the problem has not yet been solved, there are hints that such a more fundamental formulation of string theory should exist. Some symmetry principles have been discovered that connect the different phases of string theory. These cannot be understood in the current formulation, so this suggests the existence of a deeper, more symmetrical description. These principles have led also to some understanding of how transitions between the different phases may occur.

The need to invent a new conception of space and time in which they emerge from a more fundamental level is not a problem for string theory alone. It is one of the main questions that must be solved to conclude the revolutions of twentieth-century science. As I will explain in Part Four of this book, Einstein's theory of general relativity achieves such a dynamical conception of space and time. The question which remains open is whether this can be done in a way that is consistent with the quantum theory. To solve this problem, we have to learn to look at space and time differently, as participants in a relational world rather than the stage in an absolute world. Although no such theory has so far completely succeeded, I am optimistic that new ideas are emerging which may lead to its solution. These may lead as well to the deeper formulation of string theory that we need to understand if it is to fulfill its promise. In the last part of the book I will explain these ideas and connect them with the questions we have already discussed.

However, as much as there is good reason to hope for the invention of a more fundamental way to understand string theory, I am not sure that there are equally good reasons to expect that this will provide an understanding of why the universe chooses one way of curling up rather than another. It seems at least equally possible that any more fundamental string theory will also allow these different possibilities. The reason is that we are not dealing with different physical theories, we are dealing with different physical phases, which may be manifestations of the same theory.

No fundamental theory will tell us which phase of water we will find in a particular region of the universe. No theory may, from principles alone, predict which way a pencil balanced on its tip will fall, or who will marry whom. These are all choices that depend on contingent, environmental factors. They do not depend on a deeper understanding of how things are made; they depend instead on the history and configuration of the universe as a whole.

The gauge principle has led to a tremendous deepening in our understanding of the forces that govern the elementary particles. However, there is one thing that should be noticed, which is that while our understanding of the nature of the fundamental forces has continually deepened, there has yet to be proposed even one idea that worked to explain why a single elementary particle—quark, electron, neutrino, or any of the others—has the mass it does. It seems that unification has simply not brought with it any understanding of how the parameters of the laws of physics were chosen.

I first began to worry about this during the summer of 1989, when it began to be clear that string theory would not quickly lead to a unique theory of everything. Henry Tye, a string theorist from Cornell University, had told me of his computer program to produce new string theories. When you run Tye's program, you input a rough description of a universe you would like to describe. You tell it the dimension of spacetime, and something about how the world should look. It outputs all the string theories it can construct that lead to the world you requested, one per page. Sometimes it doesn't succeed, but often one finds a little pile of theories in the output tray, each consistent with the conditions you specified.

I have a small racing dingy which I sail on a beautiful lake near home, and I spent a large part of that summer sailing round and round, wondering why string theory could look so promising as a unification of the fundamental interactions, while failing to lead to any unique predictions for the parameters of the standard model. It took much of the summer for the idea to sink in, even if we succeed in constructing a unified theory, perhaps it still might not determine all of the properties of the elementary particles. I found this a very scary thought. If this is true then how is science possible? How are we to understand where those numbers that characterize the fundamental particles and interactions come from?

We certainly should hesitate before rejecting a methodology that took elementary particle physics from the discovery of the electron to the triumph of the standard model in less than a century. The idea that there is a single principle on the basis of which all phenomena can be someday understood is too close to the general idea that nature is comprehensible to be easily given up. The philosophy of reductionism, which tells us to ask how things work by breaking them up into parts, has also served us well. However, the crisis in elementary particle physics can perhaps be expressed by the simple statement that, at least at the moment, reductionism and the search for unification seem no longer to be working.

There is a story that Einstein was once asked why he did not approach quantum theory with the same positivist philosophy on which he based his early expositions of relativity theory. He replied that a good joke should not be told too often. By this, he did not mean that positivism was a joke, but that any strategy in science that is useful at one stage may be useless, or even lead to misleading results, when applied at another stage.

Suppose we take the current predicament of string theory and grand unified theories as an indication that some wrong assumption about nature is being made. Perhaps one of the methodological assumptions that have led us so successfully to this point may be no longer useful. Which of the ideas on which we have come to rely might it be?

Perhaps there is something wrong, not with string theory itself, but with the expectations that we physicists have brought to the theory. Certainly, if there is to be a theory that unifies all of the interactions, string theory is by far the best candidate that has yet been invented. But particle theorists are really after two separate goals. One is to have a unified theory of all the interactions. The second is to understand how all the parameters of the standard model are chosen. It may be that unification may bring with it an understanding of some of the parameters. But is there any reason to hope that a unified theory must determine the values of all the parameters?

If we look at the history of the development of the gauge theories as we have done here, I think it is possible to discern what the problem is. The problem seems to lie in the intrinsic conflict between the quest for unification and the imperative to understand why the different elementary particles and forces have such diverse and varied properties. The gauge theories manage to resolve that conflict. But they do it by introducing, for the first time, an effect by which some feature of the universe determines some of the properties of the elementary particles. By doing so, the theory takes a step away from the tenets of radical atomism, which hold that the properties of the elementary particles are completely independent of the history and the configuration of the universe.

I think that there is good reason to believe that this is what must happen, because unification requires that the distinction between dissimilar particles and forces cannot be intrinsic to them, while atomism requires that the most fundamental particles cannot be explained in terms of their having further parts. There is, perhaps, no remedy for this conflict except to find that the differences between the fundamental particles and forces arise from their interactions with the environment in which they exist.

In any case, this is how the Weinberg-Salam model resolves the conflict. It is also how the conflict is resolved by string theory, at least at the present time. Further, the direction of the thing seems to be clear: the more dissimilar things that are unified, the more the properties of the elementary particles and forces depend on the effects of the environment. The current conundrum of string theory may perhaps be understood as the logical end of this line of reasoning: everything is unified, but there are tens of thousands of choices for the configuration of the universe as a whole, each of which results in a world of different dimension, with different numbers of fundamental particles, interacting with different fundamental forces.

If this is the case, then perhaps the answer to how the universe chooses its configuration must be sought not in a deeper understanding of the basic laws, but in a better understanding of the history of the universe. Perhaps the thing that must be done is to accept what the theories seem to be telling us, which is that the universe indeed had free choice about a great many of the properties of the elementary particles. Perhaps in this case the right questions to ask are then about the circumstances in which this choice was made.

Finally, when we think of how intricately structured the world is, of how the parameters must be chosen to an accuracy of one part in 10^{229} if the world is to have stars, should we refrain from investigating hypotheses which tie their values to the history and configuration of the whole world? No principle to fix these parameters has emerged from the tremendous advances of elementary particle physics. Perhaps it is not too crazy to search for such a principle in cosmology.

The ideas that I will discuss in Part Two came out of that period of desperate reflection. They are one possible approach to a physics that is at once atomistic and cosmological.

PART TWO

AN ECOLOGY *of*

SPACE *and*

TIME

Is there a unique fundamental theory that determines the properties of elementary particles? Or might the laws of nature themselves have evolved?

SIX

ARE *the* LAWS *of* PHYSICS UNIVERSAL?

The idea that the laws of nature are immutable and absolute goes back to the origins of science itself, in philosophy and religion. Until rather recently most physicists believed that their world was made by a god who existed before and apart from his creation. The laws of nature that they aimed to discover were the fabric out of which god had created the world, which meant that they had to have the same eternal and absolute character as the deity who had devised them.

However, even now that science has been severed from its religious roots, the idea that the laws of nature have an absolute and unchanging character has continued to be a central part of its basic world picture. For this reason it may seem strange to us if someone suggests that the laws of nature might be as much the result of contingent and historical circumstances as they are reflections of some eternal, transcendent logic. But this is exactly the idea I am going to propose here. If we are willing to give up the idea that the laws of physics, or at least the parameters that measure the masses of the particles and the strengths of the forces, are fixed and immutable, we will see that new possibilities open up for understanding the puzzles that I described in Part One.

The main lesson I want to carry over from Part One is that, in spite of tremendous progress in the understanding of the elementary particles and interactions, we lack a workable scientific theory that explains why nature must choose the masses and other properties of the elementary particles to be as we find them, rather than otherwise. Furthermore, we have learned that the actual values of these parameters seem to be very unlikely, and in more than one sense. First, there is the problem of understanding why some of the parameters are incredibly tiny numbers, such as 10^{-60} and 10^{-19}. Second, there is the apparent fact that the world has much more structure than it would were the parameters to take more typical values. These facts are troubling for the radical reductionist dream; even more troubling is that, in spite of all the beautiful results that have come out of string theory and the other attempts at a unified theory, we have so far no evidence to support the conjecture that the requirement that the laws of nature be mathematically consistent, or agree with quantum theory and relativity, constrains significantly the possible masses of the elementary particles or the strengths of the different forces.

Thinking about these questions makes me nervous; it has done so for some time. But, perhaps there is no reason for surprise. As I argued in the last chapters, the programs of atomism and unification have built in conflicts and limitations that sooner or later must lead to an impasse. If so, it may be time to try to find new questions to ask about the elementary particles. If the standard, reductionist agenda makes it hard to understand why the parameters of elementary particle physics take values that fall into the narrow range that allows stars to exist, perhaps there is a scientific explanation for how the parameters are chosen that somehow uses this fact.

But is this possible? How can we have an explanation for how the properties of the elementary particles are chosen that is at once scientific and non-reductionist? How can we explain why the universe has so much improbable structure, without appealing to final causes, such as teleology or the anthropic principle?

The main thing that distinguishes scientific hypotheses from ideas in religion or metaphysics is that hypotheses can, at least in principle, be refuted by observation. The principle that scientific hypotheses must be "falsifiable" was espoused by the Viennese philosopher, Karl Popper, as a criterion to distinguish what he wanted to call "real science" from other activities, such as Marxism, of which he disapproved. I do not mean my use of this principle to be taken as an espousal of

Popper's views in general. But I think it is useful to adopt this principle as a guide for searching for new kinds of ideas to get us out of the impasse in elementary particle theory.

If we restrict ourselves to proposals which are falsifiable, what kind of explanations are available to us? In the history of science there have been two kinds of explanation which generally succeeded: explanations in terms of general principles; and explanations in terms of history. We are used to believing that the former are more fundamental than the latter. If we discover a fact that seems to hold universally, such as that all electrons have the same mass, we believe immediately that the reason for it must rest on principle and not on history. We usually expect a phenomenon to be contingent only if we see that it changes from instance to instance. If asked to justify this, we would say that something that is universally true cannot rest on contingent circumstances, which can vary from case to case. This makes sense, but it is an example of the kind of argument that works well only as long as it is not applied at the scale of the universe as a whole. When we are dealing with properties of the observable universe we no longer have any reason to insist that if something is true in every observable case, it cannot at the same time be contingent. One reason is that we have no justification to assert that the universe we see around us represents a good sample of all that exists, or that has existed, or that might in principle exist. There is in fact no logical reason to exclude the possibility that some of the facts about the elementary particles, which appear to hold throughout our observable universe, might at the same time be contingent.

We have failed so far to find any explanation for the properties of the elementary particles in terms of first principles. Perhaps it is then time to consider the other possibility, which is that the reasons have to do with events in our past. This will be the main task of the chapters that make up this second part of the book.

To search for an historical explanation for the parameters that appear in the standard model, we have to consider two propositions. First, that it is logically possible that the laws of nature could be exactly the same, except that these parameters could take on different values. This means that there can be no condition of principle or consistency that ties down their values. Second, if we are to search for an historical explanation of their present values, we should consider it possible that they might in fact have been different at some time in the past. This makes it possible to imagine that there were processes, satisfying the usual principles of causality, which determined the values of the parameters.

Of course, to suppose this is only to rely on the oldest tradition in science, which is to understand that the state of things now was caused by some events that happened in the past. However, once we consider this possibility, we are faced with a new question: Where do we look to find those regions of space or time in which the parameters were different?

One place they certainly are not different is in any region of the universe that can be seen from Earth. We have quite good evidence that the laws of nature everywhere we can see are the same as on Earth. The main support for this observation lies in the spectra (that is distribution of frequencies or colors) of light coming from distant galaxies. As far away and as far back as we can see, the spectra of atoms were exactly the same as they are now.

It is true that we have not so far seen the whole universe. The universe first became transparent, shortly after the "Big Bang", sometime between eight and twenty billion years ago. Since light has a fixed speed, nothing further than the distance that light could travel in that time-eight to twenty billion light years-can yet be seen. But the universe is bigger than this, so there are regions we cannot yet see that, if we wait long enough, will come into view. It is then always possible that the parameters of the laws of physics are different in some far away region that we will not see until sometime in the future. But this seems unlikely. It would mean that at some time in the future we would be able to see different regions, in which the laws of nature were different. Perhaps it would be best to assume that in the future, as now, the laws of nature will be the same in every region we can see from Earth.

Does this mean that the laws of physics are in fact the same everywhere? One might think that this settles the matter, but it does not, because of a fantastic consequence of general relativity theory. This is that the part of the universe that we will ever be able to see does not include the whole of it. The part of reality we can in principle ever see has boundaries. And there are necessarily regions of space and time beyond those boundaries.

There are two kinds of boundaries to the visible universe: those that lie in the past and those that lie in the future. Let us begin with the past.

By now the basic picture of "Big Bang" cosmology has become so familiar that I think I can assume that most readers know the outlines. Since the 1920s it has been observed that, on the average, the galaxies are moving away from each other. If everything in the universe is, on average, moving away from everything else, there was a time when everything was much closer together. Furthermore, a universal property of expanding systems is that they cool; therefore, on average, as we look deeper into the past, the universe must be getting hotter.

This has a simple consequence, which is that there was a time when the universe was opaque to light. This is because ordinary matter made out of atoms is stable only as long as the temperature is sufficiently low. If we raise the temperature of a gas past a certain point, it will become a plasma. In this state, the electrons are stripped from the nuclei and move around freely. Plasmas are opaque because the free electrons absorb light much more strongly than ordinary matter.

Close to the beginning the universe was originally hot enough to be in such a plasma phase. As it expanded and cooled a transition occurred, not unlike the transition in which ice freezes, at which the electrons were captured by the nuclei and settled down as ordinary gas. This is called "decoupling." It happened about a million years after the "Big Bang", when the visible universe was about a thousandth of its present size.

At the time of decoupling the universe became transparent to light. This freed numerous photons that had, just before, been bouncing around in the plasma. Most of these have traveled freely ever since. These photons from the moment of decoupling are called the cosmic background radiation. When we observe them now we are making a snapshot of our universe at the moment, only a million years or so after its origin, that it first became transparent. An enormous amount of effort is now going into resolving the signal of this radiation because it gives the best clue we may ever have as to the configuration of the universe at such early times. Because just before this the universe was opaque, this is the oldest light we will ever see.

Does this mean we must remain forever ignorant about what happened before the universe became transparent? The answer is no, because the universe was transparent to other forms of radiation, such as neutrinos, for quite a bit further back in time before it became transparent to light. It is then not impossible that sometime in the future, although perhaps not in our lifetimes, great neutrino telescopes will detect a signal from the time before the universe became transparent. But, as we go further back and the universe becomes even hotter and denser, there most likely comes a time when the universe was opaque even to neutrinos.

There is only one form of radiation which must be able, in principle, to travel through any amount of matter. These are waves in the gravitational field. There is good reason to believe that such waves exist, and we may hope that they will soon be directly detected. The reason that matter cannot be opaque to a gravitational wave is that any substance would have to be enormously dense to absorb a substantial amount of the energy it carries. However, before reaching this point, the matter will become so dense that it collapses to a black hole.

As a result of this wonderful property, gravitational radiation does provide one way in which, in principle, we can be assured of seeing all the way back to the origin of our universe. However, it will be some time before we will have gravitational-wave antennas good enough to see radiation from the early history of the universe. Thus, in the absence of good neutrino or gravitational-wave telescopes, to say anything about what the universe was like before the time it became transparent we must rely on theory.

Luckily, this is a question on which theory is unequivocal. If we make only some simple and broad assumptions we can draw precise conclusions about what must have happened before the universe became transparent. We can do this because of certain theorems in general relativity theory that were proved in the 1960s by Roger Penrose and Stephen Hawking. These are called the singularity theorems.

A singularity is a point or region in spacetime at which some physical quantity such as the density of mass or energy, the temperature, or the strength of the gravitational field, becomes infinite. Whenever they happen, they pose serious difficulties for physics because they signal a breakdown in the description of the world in mathematical terms. When a quantity that the equations of physics describes becomes infinite, it means that those equations cease to be useful or meaningful. Let me give a simple example. Suppose I ask you to add up a series of numbers. You are going along, one plus two is three, add seventeen to this you get twenty, add—when all of a sudden I put the number infinity on your list. So you add infinity to what you had before. . . . And then what? You don't know what to do next. Once you have infinity you must stop. Even if the next entry is a nice tame number like 42, you won't know what to do.

This is exactly what happens to theoretical physicists attempting to use the equations of physics to predict how some quantity is changing in time. Once the value of some physical quantity reaches infinity, the equations cease to work; there is no way to use them to find out what will happen afterwards.

General relativity predicts that such singular moments (if we may call them that) occur quite commonly. Indeed, as soon as Einstein had written down the equations that describe general relativity people began to discover solutions to them. Because Einstein's theory describes the geometry of space and time, each such solution is a mathematical description of the space and time relationships inside an entire universe. Actually, Einstein himself had never dreamed it would be possible to really solve his equations and write down a complete description of such a universe; in his own papers he had resorted to various tricks and approximations rather than directly solving them. But Einstein was not the best mathematician around, and others, undeterred by neither the difficulty of the equations nor the war that was ravaging Europe (this was 1916), were able to find solutions. Some of the most important solutions ever found—those that describe the gravitational fields of stars and black holes—were written down by a German officer named Karl Schwarzchild as he lay dying in a field hospital of a skin disease he had picked up in the trenches. Shortly after, a Dutch astronomer named De Sitter and a Russian mathematician named Friedman wrote down solutions that corresponded to expanding universes. What is most important is that all of these solutions contain singularities where the density of matter becomes infinite, making it impossible to tell what happens before or after.

For a long time the existence of these singularities was not taken very seriously. Besides the understandable difficulty in believing that there could be such regions in our universe, the main reason for this disdain was that the solutions which contained singularities were very special in certain ways. To simplify the equations to the point that solutions could be found by hand, it had been necessary to assume that stars were perfectly spherical or that the universe was perfectly homogeneous. While such simplifications make it much easier to solve the equations, it was not clear to what extent the singularities that resulted were general consequences of the theory rather than accidental side effects of the assumption of perfect symmetry. Many physicists believed that general solutions that lacked the symmetries of those that had been studied would be without singularities. As the universe is in fact not symmetrical, this meant that the singularities had no implications for the real world. Unfortunately, for a long time nothing was known about the properties of more realistic solutions, which posited less idealized and symmetrical situations. As a result it was possible for several generations of physicists to avoid thinking about the consequences of the singularities.

This complacency was broken in 1965 when Roger Penrose, then working in London, proved that the singularities are general consequences of the equations of general relativity and are present in most solutions that might describe the real universe. His description of this discovery is, by the way, beautiful. He was walking in conversation with a friend, and while they were crossing a street he had a thought which he immediately forgot on resuming the conversation on the other side. Later, in the evening, he felt happy and elated, but could not understand why. It was only by consciously taking himself back over the day that he could recall the thought he had had in the moment of crossing the street, which was that the singularities would be found in every solution that satisfied a reasonable set of requirements, and there was a way he could prove it.

Very soon after, a young research student in Cambridge named Stephen Hawking heard about Penrose's theorem. He was quickly able to show that it was possible to apply Penrose's method to obtain a proof about what happened before the universe became transparent. What he found was that, under very general conditions, there must be a first moment of time, just before which the density of matter and energy are infinite.

Does this mean that there was actually a first moment of time ten or twenty billion years ago, before which nothing at all existed? There are certainly those who would like to believe this, for example certain theologians. But we must remember that Hawking's result is a mathematical theorem, and as such it will only be true if its assumptions are true. If one or more of the assumptions of the theorem are false, it has no bearing on the world. Hawking's theorem in fact rests on two significant assumptions. The first is that the energy density of matter everywhere is positive. The second is that the laws of general relativity as written down by Einstein hold everywhere, for all space and all time. These certainly seem to be reasonable assumptions, but there is a problem. The second one is simply not true because it makes no mention of quantum mechanics, which, rather than relativity theory, seems to describe the real properties of matter.

There are, of course, many cases in which quantum mechanics will agree with the predictions of general relativity. Were this not the case, Einstein's theory could not have been so successful in the absence of a unification with quantum theory. The question is whether we can trust general relativity to give a correct description all the way back to the moment when things were infinitely dense. The answer is no, for as we turn the clock back we reach a time when the universe was so dense that effects of the quantum theory cannot be ignored. In this case general relativity by itself will not even give an approximate description of what is going on. Only a quantum theory of gravity, that combines general relativity and quantum theory, could tell us whether there really are singularities in the world.

The last part of the book will be devoted to the problem of quantum gravity. However, without going into details, it is easy to divide the possible outcomes into three categories:

A   There is still a first moment of time, even when quantum mechanics is taken into consideration.

B   The singularity is eliminated by some quantum mechanical effect. As a result, when we run the clock back, the universe does not reach a state of infinite density. Something else happens when the universe reaches some very high density that allows time to continue indefinitely into the past.

C   Something new and strange and quantum mechanical happens to time, which is neither possibility A or B. For example, perhaps we reach a state where it is no longer appropriate to think that reality is composed of a series of moments that follow each other in a progression, one after another. In this case there is perhaps no singularity, but it may also not make sense to ask what happened before the universe was extremely dense.

The proposal that James Hartle, Stephen Hawking and others have made—that time somehow becomes "imaginary" when the universe becomes very dense—falls into this last category. There are other approaches to the problem of time in quantum theory that also fall into this category, some of which I will discuss in the last chapter. However, none of these ideas have been developed enough to lead to definite predictions, nor has anyone shown that conventional ideas of time and causality must fail when a singularity is approached. Until the theory forces us to face it, or at least until someone explains clearly what it would mean for time to cease to exist, it would perhaps be best to put this possibility to one side while discussing cosmological problems.

That leaves possibilities A or B. If A is true then we have only the million or so years between the singularity and the moment the universe became transparent for the parameters of particle physics to be set as they are. That may seem like a lot of time, but it is not, because of a certain cosmological puzzle which must now be mentioned.

Imagine for a moment that you could see the cosmic black body radiation. You look up at the sky and see a flash of light from a photon that has traveled around ten billion years, from the time of decoupling to your eye. Now, turn your head a few degrees to the right, and wait again for a photon from the black body radiation to come to your eye. Coming to us from different directions, and traveling for such a long time, these two photons come from regions of the universe that were very far apart when they were created. Even taking into account the fact that the universe has expanded a great deal (about a thousand fold) while they were traveling, it is still true that they were very far apart when they began their journeys.

What is remarkable is that, even if they started out from regions of the universe that were very far apart, all of the photons coming from the time the universe was opaque tell the same story. To an accuracy of about one part in a hundred thousand, the temperature at that time seems to have been the same all over the universe.

This is one of the big mysteries of modern cosmology. How is it possible that regions of the universe that were very far apart at that time had, nevertheless, almost precisely the same temperature? Questions like this are usually not hard to answer. We know that a glass of hot water left in a room will eventually cool to the temperature of the room. As a result of the tendency of things to come to equilibrium, when things are in contact for long enough they tend to come to the same temperature. The simplest possibility is then that all the different regions of the universe had been in contact with each other before the moment of decoupling.

Unfortunately, if we believe in the story of cosmology given by general relativity, this cannot have been the case. As the universe is supposed to have been only about a million years old at the time of decoupling, and as nothing can travel faster than light, only regions that were then less than a million light years apart could have had any contact with each other. The problem is that, according to the theory, the universe at this time was much bigger than a million light years across. This means that when we detect the cosmic background radiation coming from two different points in the sky more than a few degrees apart, we are seeing light that originated from regions that up till that time could not have had any kind of contact with each other.

To emphasize how strange this is, let us suppose that the signal of the cosmic background radiation was modulated like a radio broadcast. And let us suppose that, from every corner of the universe, the tune played by the cosmic background radiation was rock 'n' roll, and not only rock 'n' roll, but the same Cosmic Top Ten: The Beatles, Madonna, Bruce Springstein, Gianna Nannini, etc. How could we account for this? It would be no problem if the different regions had been able to listen to each other, for the appeal of good music (or, if the reader prefers, the economics of cultural penetration) is almost as absolute as the laws of thermodynamics. Indeed, we are not surprised to hear the same music in every restaurant and bar on this planet. But what if the different regions could never have been in contact with each other? It would be as if Hernan Cortes, arriving in the court of Montezuma, heard around him only the songs he had learned in the taverns of Seville. We would then have to believe in a miracle of a thousand simultaneous births of rock 'n' roll, a thousand simultaneous Memphises and Detroits, each totally unaware of the others.

This may seem ridiculous, but it is not much more ridiculous than what is actually seen: many regions which, if we believe the standard theory, could never have been in contact with each other, but in which the temperatures are the same, to fantastic precision.

The reader may be confused by this. Isn't the idea of the "Big Bang" that the whole universe expanded from a point? This is the popular conception, but it is not actually what general relativity says. It is true that if we trace back the history of any particle, we find an initial singularity at which the density of matter becomes infinite. However, what is not true is that all the particles in the universe meet at their first, singular moments. They do not. Instead, they all seem to spring into existence, simultaneously but separately, at the same instant. Just after the first instant of time, the universe already has a finite spatial extent. One million years later, the universe is much larger than one million light years across, leading to the problem we have been discussing.

Of course, it is always possible that all the different regions of the universe were created, separately, with exactly the same conditions. The different regions had the same temperature a million years later, because they were created with the same temperature. This may seem to resolve the question, but it only leaves a different mystery: Why were all the regions created with exactly the same conditions? This does not solve the problem; it only makes it worse by forcing us to imagine that whatever created the universe did it in a way that duplicated the same conditions in an enormous number of separate regions.

Indeed, as long as we believe that the world was born a finite time ago, we have the problem of explaining what the conditions were at the moment of creation. Whether the temperatures were the same everywhere, or whether the pattern of hot spots spelled out "Made in Heaven," we would have the same problem of explaining what the conditions were at the moment of creation.

One escape from this dilemma would be if general relativity were wrong about the early history of the universe. We have already noted that this is quite possible, given that general relativity does not take into account the effects of quantum physics. There are indeed at least two ways that quantum effects might win the universe enough time. The first is called the hypothesis of *cosmological inflation*. The idea is that as the universe expands and cools, it makes a transition between different phases of the sort described in the previous chapter, when I discussed spontaneous symmetry breaking. This transition may have occurred very early in the history of the universe only a fraction of a second after its creation. According to the hypothesis, before the transition the universe was in a phase in which it expanded much more rapidly than it does in its present state. This is called a period of inflation, to contrast it with the present period in which the expansion is much slower. During inflation the universe may double in size every 10^{-35} of a second or so. Because of this, regions of the universe that are now billions of light years apart were initially very, very close to each other. As a result, it becomes possible for all of the regions of the universe we can see to have been in contact with each other in the time since its beginning.

The hypothesis of cosmological inflation turns out to have one basic problem, which is that it requires several careful tunings of the parameters of particle physics. This is necessary not to make inflation happen, but instead to make sure that it stops. It is as if the Federal Reserve Board were trying to tune the interest rates now in order to prevent rapid inflation, not only before the next election, but for the next ten billion years. Perhaps they might then be talking about changes in interest rates of a millionth of a percent, which is at least as finely as the parameters in the theory of inflation must be chosen so that the period of rapid expansion lasts only for a very limited time.

Of course, this is only one more problem in which some parameter must be chosen very delicately if the universe is to be as we find it. As it is far from the only one, this cannot be held against the hypothesis of cosmological inflation. If there were a mechanism to tune the proton mass or the cosmological constant to incredibly tiny numbers, it could possibly do the same for the parameters that determine how long cosmological inflation lasts. What is certain is that if alternative A is correct, so that quantum mechanics does not get rid of the singularity in our past, then inflation seems necessary to explain why the whole universe seems to have been at the same temperature at the moment of first transparency, only a million years after the first moment of time.

But there is a second possibility, which is that quantum effects might completely eradicate the singularity. In this case there would be no moment of creation. Time would instead stretch indefinitely far into the past. Regardless of, inflation, there would have been enough time for all the regions of the universe to come into contact. This would not mean that cosmological inflation is wrong, for there are other reasons one might want to consider it. But in this case we have to ask what happened in the world before the "Big Bang". That term would no longer refer to a moment of creation, but only to some dramatic event that led to the expansion of our region of the universe. In this situation it becomes possible to ask if there were processes which acted before the "Big Bang" to choose the parameters of elementary particle physics.

I will shortly propose an answer to this question, but before doing so I would like to turn to the second kind of boundary to the region of our universe we can see. We have been discussing the boundary that lies to the past, now we must turn to those boundaries that lie in the future.

One of the great mysteries about time is why the past is different from the future. Cosmology only deepens the mystery. We know, for the reasons I've just sketched, that the universe came from a dense, opaque state in the past. It is then natural to ask whether it might return to such a state in the future. Must the universe, which is now expanding, eventually recontract? The answer is that universe will recontract if there is enough matter around for its mutual gravitational attraction to stop the expansion. But if there is not enough matter to stop the expansion, it will never collapse—it will keep expanding forever.

Presently, the evidence is that the universe as a whole will not recontract. But, for our purposes here, it is not important that we settle this issue. This is because regardless of whether the universe recollapses, we know for a fact that there are many small bits of it that have collapsed under the force of their own gravitational attraction. These are the black holes.

It is not hard to understand what a black hole is. They are simple and necessary consequences of two facts. First, the existence of a force like gravitation that is universal and attractive; and second, the fact that nothing can travel faster than the speed of light. To understand black holes we need one simple concept, which is called escape velocity. Imagine that you are stuck on some planet that you are dying to leave. To leave it you must launch yourself up against its gravitational attraction. Because not everything that goes up comes down, if your initial speed is large enough, you will be able to boost yourself completely free of that planet. For each planet there is a certain minimum speed called the escape velocity, required to accomplish this.

The escape velocity depends on both the mass and the size of the planet. For a planet of a given mass, the escape velocity is higher the more compact it is. As a result, if a planet shrinks, the velocity needed to escape from its surface increases. The same is true of a star, or of any other object. A black hole is simply anything that has shrunk to a small enough size that the velocity you need to escape from it is larger than the speed of light. In this situation neither light, nor anything else, can escape.

Each black hole is surrounded by a surface beyond which nothing can escape. Such a surface is called an *event horizon*, or horizon, for short. Each such horizon is a boundary to the region of the universe we can see. Furthermore, one can pass through a horizon, but once inside one can never leave. As long as the black hole is there, one has the possibility of passing through the boundary and becoming trapped behind it. Because of this, we think of the horizons as boundaries that lie in our future.

The solution to Einstein's equations that were discovered by the unfortunate Schwarzchild in 1915 describe black holes. But, in spite of the simple picture I've just described, no one really understood what a black hole was, or even what the solutions they were staring at represented, until the late 1950s. One reason for this was that astronomers and physicists were very resistant to the idea that somewhere in the universe there might be things so contracted as to have become black holes.

Now we know that there are in fact black holes in our universe While they are difficult to see, for the obvious reason that they give off no light, there are some circumstances in which they can be seen by their effects on other stars. Evidence for the existence of a few black holes has been accumulating, and there is good reason to believe that a great many more exist. It is difficult to give a precise estimate, not only because they are hard to see, but because not enough is known about the numbers of stars that become black holes at the end of their lives. A conservative estimate is that there is about one black hole for every ten thousand stars. This means each galaxy contains at least one hundred million black holes.

What lies beyond the horizons of all of these black holes? The story here is very much like the story of the Big Bang, only in reverse. If we assume that Einstein's general theory of relativity gives a correct description of what happens to a collapsing star, then it is quite certain that what lies inside of each black hole is a singularity. This is in fact exactly what Roger Penrose proved, when he found the first of the theorems about singularities.

There is an important difference from the case of the cosmological singularity, which is that in a black hole the singularity lies in the future rather than in the past. According to general relativity every bit of the collapsed star and every particle that falls afterwards into the black hole will end up at a last moment of time, at which the density of matter and strength of the gravitational field become infinite.

However, we do not trust general relativity to give us the whole story about what happens inside a black hole, for the same reason we don't trust it in the cosmological case. As the star is squeezed towards infinite density, it must pass a point at which it has been squeezed so small that effects coming from quantum mechanics are at least as important as the gravitational force squeezing the star. Whether there is a real singularity is then a question that only a theory of quantum gravity can answer.

Many people who work on quantum gravity have faith that the quantum theory will rescue us from the singularities. If so, it may be that time does not come to an end inside of each black hole. At present, despite several very interesting arguments that have recently been invented, the question of what happens inside of a black hole when quantum effects are taken into account remains unresolved.

If time ends, then there is literally nothing more to say. But what if it doesn't? Suppose that the singularity is avoided, and time goes on forever inside of a black hole. What then happens to the star that collapsed to form the black hole? As it is forever beyond the horizon, we can never see what is going on there. But if time does not end, then there is something there, happening. The question is, What?

This is very like the question about what happened "before the Big Bang" in the event that quantum effects allow time to extend indefinitely into the past. There is indeed a very appealing answer to both of these questions, which is that each answers the other. A collapsing star forms a black hole, within which it is compressed to a very dense state. The universe began in a similarly very dense state from which it expands. Is it possible that these are one and the same dense state? That is, is it possible that what is beyond the horizon of a black hole is the beginning of another universe?

This could happen if the collapsing star exploded once it reached a very dense state, but after the black hole horizon had formed around it. If we look from outside of the horizon of the black hole we will never see the explosion, for it lies beyond the range of what we can see. The outside of the black hole is the same, whether or not such an explosion happens inside of it. But suppose we do go inside, and somehow survive the compression down to extremely high density. At a certain point there is an explosion, which has the effect of reversing the collapse of the matter from the star, leading to an expansion. If we survived this also, we would seem to be in a region of the universe in which everything was moving away from each other. It would indeed resemble the early stages of our expanding universe.

This expanding region may then develop much like our own universe. It may first of all go through a period of inflation and become very big. If conditions develop suitably, galaxies and stars may form, so that in time this new "universe" may become a copy of our world. Long after this, intelligent beings may evolve who, looking back, might be tempted to believe that they lived in a universe that was born in an infinitely dense singularity, before which there was no time. But in reality they would be living in a new region of space and time created by an explosion following the collapse of a star to a black hole in our part of the universe.

The idea that a singularity in the future would be avoided by such an explosion is very old, it goes back to the 1930s, long before the idea of a black hole was invented. At this time cosmologists worried about the fate of a universe that neared its final moment of time after expanding and then recontracting. Several cosmologists speculated that we live in what they called a "Phoenix universe," which repeatedly expands and collapses, exploding again each time it becomes sufficiently dense. Such a cosmic explosion was called a "bounce," as the repeated expansions and contractions of the universe are analogous to a bouncing ball.

What we are doing is applying this bounce hypothesis, not to the universe as a whole, but to every black hole in it. If this is true, then we live not in a single universe, which is eternally passing through the same recurring cycle of collapse and rebirth. We live instead in a continually growing community of "universes," each one of which is born from an explosion following the collapse of a star to a black hole.

Recall that we wanted there to be boundaries in the past of our visible universe, when processes might have happened that could somehow choose the laws of physics, or at least select the values of their parameters. What we have learned in this chapter is that almost inevitably, the existence of such boundaries follows given only the simplest ideas about light and gravity and the basic fact that we live in an expanding universe. Furthermore, we have learned that if we accept the hypothesis that quantum effects eliminate the singularity at the beginning of the universe, and eliminate as well the singularities inside of black holes, we have the possibility that what lies beyond the boundaries is much vaster than our own visible universe. So with a few simple and reasonable hypotheses we get inaccessible regions with as much time as we'd like for processes to have occurred to form the laws of physics as we see them around us. The question then is only to see what we can do with them.

To suppose universal laws of nature capable of being apprehended by the mind and yet having no reason for their special forms, but standing inexplicable and irrational, is hardly a justifiable position. Uniformities are precisely the sort of facts that need to be accounted for. Law is par excellence the thing that wants a reason. Now the only possible way of accounting for the laws of nature, and for uniformity in general , is to suppose them results of evolution.

—Charles Sanders Pierce, The Architecture of Theories, 1891

SEVEN

DID *the* UNIVERSE EVOLVE?

The astronomers of ancient Greece and Egypt knew how large their universe was. They knew the distance to the moon and, by estimating from that, they were able to fix the distance to the Sun and to the outermost planet, Saturn. Having no reason to do differently, they enclosed the universe in a sphere just over the epicycle of Saturn, onto which they affixed the stars. From the Sun to this stellar sphere they estimated a span, to them enormous, of 10,000 times the diameter of the earth. This was in fact wrong, but what was important was that their universe had a boundary. Beyond the stellar sphere was nothing. This was very convenient for the Christians who came later, as it provided a place for God and His angels.

For Newton the universe lived in an infinite and featureless space. There was no boundary, and no possibility of conceiving anything outside of it. This was no problem for God, as he was everywhere. For Newton, space was the "sensorium" of God—the medium of His presence in and attachment to the world. The infinity of space was then a necessary reflection of the infinite capacity of God.

I recall being stumped as a child on the question of the extent of the universe. It seemed absurd that the universe be infinite, how could it just go on and on forever? It also seemed absurd that it be finite, for then there would be a wall, and one could wonder about what was beyond it. When one has a question like this, perhaps the only hope is that someone will someday imagine a third option, which gets us out of the paradox. This is exactly what Einstein did, when he turned his attention to what his new general theory of relativity might have to say about cosmology. He found that his theory could describe a universe that was finite, but closed, exactly like the surface of a globe that has finite area but no boundary. In this way, general relativity can resolve, at least for space, the great paradox of whether the universe is finite or infinite. Had Einstein imagined only this he still would be one of the great natural philosophers of our century.

However, we must immediately admit that we do not know if this is how nature resolves the paradox. Because light travels at a finite speed, to look out is to look back. Thus, as long as there is a time before which we cannot see, there is a limit as well to how far we can see in space. For this reason, the question of whether space is finite or infinite cannot be resolved until we confront the same question about time.

But if we ask whether time is finite or infinite, we run up against the same paradox. It seems absurd that time go on and on forever, but it also seems impossible that there be a first or a last moment. It seems so natural to ask what would happen just before or just after. But there is no simple resolution to this dilemma as there is for the case of space. To imagine time as finite but unbounded, with all history cyclical, and the future connected to the past, is not so simple; it raises puzzles that are not present in the case of space.

What we learned in the last chapter puts a new twist on these ancient puzzles. For however the questions of the finiteness of space and time are ultimately resolved for the universe as a whole, we know that the part of the universe that we can actually see is bounded, but it is bounded in time rather than space. The boundaries that we now confront are different from those of the Aristotelians. There are no actual walls; we may travel anywhere, but there are places into which we cannot see. There are also places, bounded by the horizons of the black hole, into which we may travel only at the cost of never being able to return.

The fact of these boundaries makes what may have seemed only a harmless philosophical question into a challenge for the foundations of science. For, if we hope to have a complete theory of the universe we must ask what lies beyond them. Across the horizons of the black holes, and far back in time before the cosmic black body radiation was produced, is there only more of the same, or are there regions of the universe that are different from the one we can see? And, if they are different, is this important for understanding the region in which we find ourselves?

More than one observer has complained that much of current cosmological theory relies on hypotheses about what lies behind boundaries, in other possible regions of space and time. The theory I am about to propose is, in this respect, no different from other ideas such as inflationary cosmologies or the anthropic principle. The complaint is just, one would certainly prefer that a scientific cosmology explain the universe that we can see in terms only of itself. And it is hardly a consolation to imagine that sometime in the future gravitational-wave telescopes might enable us to see as far back in time as there is time. Even then there is nothing that will let us see into a black hole.

Neither is it any help that we may choose never to pass through the horizon of a black hole. A door I never pass through still has something on the other side. So, at least until this moment, the simple childhood puzzle about time has been twisted, but not solved, by Einstein's magic. The universe we can see has boundaries, many of them, beyond which may lie clues we need to grasp the deepest secrets of nature.

I begin this chapter with this meditation because I want to present at the beginning what is, for me, the most difficult thing to accept about the idea I am about to propose. I want the reader to understand that the existence of inaccessible regions of the universe is not my idea, nor is it a circumstance I particularly like. Nevertheless, it seems to be a necessary consequence of the general theory of relativity together with a few simple facts from astronomy. Once we accept them we are facing, not a theoretical or an aesthetic issue, but the reality of perhaps a hundred million of these inaccessible regions swimming around in our galaxy alone.

If one recalls the terror brought on by fantasies of aliens or vampires, indistinguishable from humans and at loose in the population among us, or the fascination of the invisible angels walking among us in Wim Wender's *Wings of Desire*, one may have some idea of the feelings an astronomer must have looking at her star charts and contemplating the mystery of the black holes out there among the stars. The best telescope will not tell her what lies beyond their horizons. Once we accept the reality of black holes, the only question is, Are they merely curiosities, or are they in some way important for the understanding of the constitution of the universe?

In this chapter I will make a proposal about cosmology that makes black holes, and what lies at their hearts, of crucial importance for understanding why the world is as we find it. According to this proposal, the explanation for how the parameters of particle physics are chosen will turn out to have a lot to do with black holes. Furthermore, in spite of the fact that it requires us to posit the existence of regions of the universe that we cannot see, this proposal will lead to predictions about the part of the universe in which we live. This means that it is subject to test, and possible refutation, by a combination of observation and theory.

This theory is based on two postulates. The first of these is the proposal, which I discussed at the end of the last chapter, that quantum effects prevent the formation of singularities, at which time starts or stops. If this is true, then time does not end in the centers of black holes, but continues into some new region of space-time, connected to our universe only in its first moment. Going back towards the alleged first moment of our universe, we find also that our Big Bang could just be the result of such a bounce in a black hole that formed in some other region of space and time. Presumably, whether this postulate corresponds to reality depends on the details of the quantum theory of gravity. Unfortunately, that theory is not yet complete enough to help us decide the issue. In its absence we are also ignorant of the details of such a process. But for the purpose of the theory I am going to propose we will not need many details. We will need to presume only that this explosion, or bounce, is a new effect that happens when matter is squeezed to some enormous density, larger than any we have so far observed. This is no problem; if we let the Planck units set the scale for this density, it is absolutely enormous, about 10^{79} times denser than an atomic nucleus.

If we accept this then we have not only the inevitable inaccessible regions, we have the possibility that these regions could be universes as large and as varied as the universe we can see. Moreover, as our own visible universe contains an enormous number of black holes, there must be enormous numbers of these other universes. There are at least as many as there are black holes in our universe, but surely if we can believe this we must believe there are many more than that, for why should not each of these universes also have stars that collapse to black holes and thus spawn new universes?

This is a mind-numbing possibility. I would not be honest if I did not admit to the reader that this is an idea that I must force myself to confront, even after many years of thinking about it. To the extent that I succeed it is only because I have been unable to defeat the force of the argument that says that if time does not end at black hole singularities it must continue, perhaps forever, in regions inaccessible to us. And then there must be many such regions because there seems to be no way to avoid the estimate that a large number of stars must end their lives as black holes.

Perhaps it makes it a little easier to contemplate this possibility if one recalls that by itself the simple proposal that time never ends forces us already to conceive of an infinitude of events taking place that we, in our finite lifetimes, can never know of. All this picture really does is to rearrange all of these inaccessible moments.

There is no longer a simple linear progression. Instead time branches like a tree, so that each black hole is a bud that leads to a new universe of moments.

In this and the next few chapters I will try to convince the reader that this postulate can be the basis for a real scientific theory, one that makes predictions that are subject to experimental test. In order to succeed, we must have a method that will allow us to deal scientifically with a world that consists of an enormous number of regions like our own universe, only one of which we can directly observe. This may seem impossible, but before dismissing it as absurd we may try to apply the usual tools that we use in science to study large collections, which are reasoning through probability and statistics. Experience in many areas of science has shown that it is possible to draw reliable conclusions about members of a large collection. To do this we must be able to make statistical assumptions that allow us to draw probable inferences about their properties.

To draw any consequences from such a multi-universe scenario that could be applicable for our own universe, we must consider our universe to be a typical member of the collection. We will be able to predict that our region of the universe has a certain property only if we can show that it is true of almost every universe in the collection. Furthermore, in order to claim that our theory is part of science, it must be possible to test whether the property in question is true of our universe. As far as I know, this is the only way to build a scientific theory that is observationally testable, starting from the conjecture that what we observe is only one out of a great many regions of the world.

One might wonder if this is asking too much. Perhaps it would be sufficient if the theory predicted only that somewhere in the collection was a universe that resembled our own. However, theories such as this are not likely to be falsifiable, as they make it possible to explain almost anything. Fortunately, it is possible to construct a testable theory if we add the right sort of assumption about what happens at the "bounces," when the collapsing star inside a black hole explodes to become the seed of a new region of the universe. As what we want to explain are the parameters of the laws of physics, we will have to postulate that these change under each bounce. Just how they change will be the content of the second postulate of the theory.

Presumably, when the quantum theory of gravity is more developed, we will be able to predict exactly what happens at a bounce. For the time being, given our ignorance about physics in the extreme conditions of a bounce, we should make the simplest possible hypotheses, and see if they lead to predictions that can be compared to the real world. The simplest hypothesis I know of is to assume that the basic forms of the laws don't change during the bounce, so that the standard model of particle physics describes the world both before and after the bounce. However, I will assume that the parameters of the standard model do change during the bounce. How do they change? In the absence of any definite information, I will postulate only that these changes are small and random.

Later I will explain what is meant by a small change. For the moment, let me emphasize that the two postulates I have made are the simplest I know of. Once we find that they lead to a scientific theory, it will be possible to consider alternatives in which the postulates are modified. To the extent that these different hypotheses lead to different predictions, they are in principle testable. Furthermore, I must insist that this second hypothesis is in no way inconsistent with the hypothesis that there is a fundamental physical theory. It is inconsistent only with the idea that this theory uniquely determines the observed parameters of the standard model of particle physics.

For example, if we believe that physics ultimately is described by a string theory, we might investigate an alternative form of the second postulate in which the parameters that vary label the possible consistent string theories. I will discuss such alternative possibilities later, in the appendix. For the time being, we will stick to the simplest possibility, which is the one I stated.

The idea that the parameters of physics might change at a bounce is not new. In the context of the Phoenix model—in which the collapse of a whole universe leads to a bounce that gives rise to a single new universe—the idea was championed many years ago by John Archibald Wheeler. He called it the "reprocessing" of the universe. What I am adding is only the hypothesis that the change at each bounce is small.

The two postulates are the basis of the theory I will now describe. I will draw out their consequences in stages, first completely intuitively, and then in a formal setting from which definite conclusions can be drawn. This is because I want the reader to understand why the theory built from it makes definite, unambiguous predictions.

The intuitive argument is easy. If we accept our two postulates, then we know that the parameters of our universe are close to those of the universe it grew out of. And they are also close to the parameters of the one out of which that one, in turn, grew. Indeed because the changes in the parameters at each birth of a new universe are small, we have to go back many generations to find an ancestor universe with parameters very different from ours.

From the discussion of the second chapter we may recall one key fact, which is that, for most values of the parameters, stars could not exist. This means that for most values of the parameters, black holes, if they form at all, do not form by the collapse of stars. From this we can draw the conclusion that the rate at which black holes form is strongly dependent on the parameters. A universe such as ours makes as many as 10^{18} black holes. A universe roughly like ours, but without atomic nuclei or stars, would make many fewer. But, as we discussed in that chapter, the range of parameters for which atomic nuclei, and hence stars, exist is rather small. From this we may conclude that there are small ranges of parameters for which a universe will produce many more black holes than for other values.

Now, I reach into the collection and pick out a universe out at random. It is easy to see that it is much more likely to have come from a universe that itself had many progeny than it is to have come from a universe that had only a few progeny. This is because over many generations the universes that had many progeny contributed many more universes to the collection than those that had few progeny. Because each universe is largely similar to the one it came from, we may conclude there are many more universes in the collection that have many progeny than there are those that have few. But we are assuming that our universe is a typical member of this collection. Therefore, its parameters are most likely to belong to a set for which a universe has many more black holes than would be the case for most other choices of the parameters.

This is the principle we have been looking for. It says that *the parameters of the standard model of elementary particle physics have the values we find them to because these make the production of black holes much more likely than most other choices.*

This was the intuitive argument. It raises, however, several questions that can only be answered by going back through the reasoning more carefully. To do this, let us imagine that the universes in our collection are all progeny of a single, initial universe. This is not, strictly speaking, necessary, but it will help us understand the mechanism by which the parameters are selected. If we assume that there was a first universe, we have the problem of how its parameters were chosen. We have no principle which may help us answer this question, so let us make the best of our ignorance and assume that its parameters are chosen randomly.

Actually, it is not quite this simple; some of the parameters are physical quantities that must be valued in some set of units. So before saying that the parameters are random numbers, we must specify which units we are going to use to measure them. Fortunately, there is only one natural choice of units to use, which is the Planck units. As they are built from the basic constants of relativity theory and quantum theory, they are the only units that make sense in any possible universe. We then will assume that the parameters of the initial universe, when measured in Planck units, are chosen randomly.

As a consequence, it is very unlikely that the parameters of this first universe are finely tuned to values that result in a big universe full of stars, as this requires very unlikely values such as 10^{-19} and 10^{-60}. Instead, it is most likely that the life of this universe will be over in a few Planck times, which is about 10^{-43} of a second. After that, one of two things may happen. Either it inflates very rapidly, so that after a few Planck times it is essentially empty, or it collapses all together.

We would like to avoid the first eventuality, because such an empty universe has no progeny. If we create a universe without progeny the process simply stops. For the sake of this argument, we must restrict the allowed parameters of this initial universe, and all those created from it, so that each one has at least one descendent. This is easy to do; it means we must require that each universe contain enough matter for the gravitational attraction to reverse the expansion, leading to its total collapse and, hence, at least one bounce.

As the initial universe almost certainly lives for only a Planck time, it is unlikely to develop any black holes. It then collapses and bounces, giving rise to a second universe. By our basic hypothesis, the parameters of this second universe differ by only small random increments from those of the initial, randomly chosen one. This universe will then also last only a few Planck times, and then collapse without forming any black holes. It then also gives rise to one progeny, whose parameters differ again by a small random change from the previous one.

For a long time, the world is nothing but a series of tiny universes, each of which grows out of the one before it. Associated with each universe is a set of parameters, each differing from the previous set by small random changes. This must go on for a long time, because for most of the possible parameters, the universe lives on the order of a Planck time and collapses, forming one progeny. What is happening is that, one by one, different possible parameters are being picked randomly and the consequences of each tried out.

It is as if a bacterium had arisen that, rather than dividing, dies and then gives rise, Phoenix-like, to a single progeny. Suppose that each time this happened the DNA of the progeny differed from its parent by a mutation in a single gene. What would happen? The Phoenix-bacteria will follow each other, one after the other, until by mutation a bacterium is born that rediscovers the trick of dividing. After this, the population explodes, and it is not too long before almost all bacteria are descendants of normal rather than Phoenix bacteria.

Similarly, we know that there are ranges of the parameters of the standard model that describe universes that produce black holes, and hence leave more than one progeny.

It is true that these ranges are rather narrow, as the ranges of parameters that give rise to simple universes are much wider than those that give rise to universes large and complex enough to have black holes. But sooner or later, as they try out different values of the parameters, the Phoenix universes will discover the trick of having more than one descendant. It does not matter how much time this takes, as there is no limit to how many times new universes with new values of the parameters may be created, as long as we assume that each universe leads to at least one progeny. Once that happens, the population of universes that know the trick of leaving many copies of themselves explodes. The Phoenix universes don't die out, but they can't keep up. It is only a matter of time before they are completely swamped by the growing population of universes with multiple progeny.

A good way to visualize what is happening is in terms of the space of parameters that I introduced in Chapter 3. As we described it there, this is an abstract space, with one dimension for each parameter. The space of parameters of the standard model has about twenty dimensions. It is hard to visualize spaces of such high dimension, but we can often get an idea of what the important features are by imagining that the space has only two dimensions. This would indeed be the case if it turned out that a theory could be constructed that explained the standard model that had only two parameters. We may then imagine the space of parameters as the plane in Figure 2 Each point in this space corresponds to a possible universe. We are interested in only one aspect of that universe, which is how many black holes—and hence progeny—a universe with those parameter values is likely to produce. Let us suppose that we have a good theory, which allows us to compute how many black holes are produced by a universe with each value of the parameters. We then have a function over the parameter space, which is illustrated in Figure 3. This is like a landscape, the height of which is proportional to the number of black holes a universe with those parameters is likely to produce. In the case that there are only two parameters, we can see this as a real landscape, with plains, mountains, hills, valleys, passes and so on, as we see in Figure 3.

We may imagine that each universe in the collection is a little creature that lives at the point of the landscape labeled by its parameters. Each one reproduces itself, by creating a number of new universes. Each of these has parameters slightly different from its parent, and so comes to live at a nearby point, corresponding to the values of its own parameters. To make the picture correspond to our theory, we must imagine that the number of progeny each creature/universe produces is proportional to the height of the landscape at the place it lives. (For example we might fantasize that the mountain air encourages reproduction.) We then have a picture that corresponds precisely to our theory.

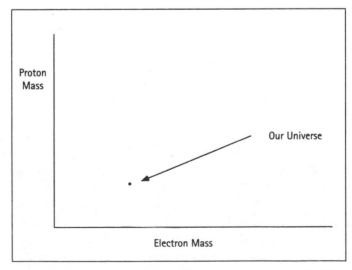

Figure 2 A two dimensional slice of the space of paramenters labeled by the value of the neutron and electron masses.

Most of the landscape is flat, and corresponds to the fact that most values of the parameters lead to universes that have no black holes, and hence have just one progeny. But we know that there are ranges of parameters that lead to universes that produce black holes. Let us begin by making a simplifying assumption, which is that there are only two possibilities for the number of progeny a universe could create, one and ten. The universes that make ten progeny are represented by a mesa of height ten in the landscape. Surrounding this mesa is a vast plain corresponding to the much larger range of parameters that lead to Phoenix universes that only replace themselves.

We began with a single universe whose parameters were chosen randomly. This corresponds to a creature placed randomly somewhere in the landscape. As we said, this will almost certainly be in the vast plain of Phoenix universes. It gives rise to a single progeny, and that to another, and so on. For a long time, creatures create each other, one at a time, and their lives trace a random path across the flat plain that takes up most of the landscape. But sooner or later this random odyssey leads to a point just under the mesa. One more random step may bring its descendant up onto the mesa, into the region of universes that have ten progeny. If this does not happen in the next step, it almost certainly will happen eventu-

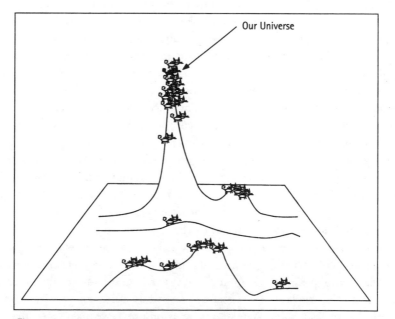

Figure 3 A possible fitness landscape for cosmology on a two dimensional slice of the space of parameters of the standard model. The altitude of the landscape is proportional to the number of black holes a "universe" with each values of the parameters will produce. The cats represent a population of "universes" after a number of "generations."
Drawing courtesy of Saint Clair Cemin.

ally. Then the game changes, because there is now differential reproduction. Since different parameters can give rise to different numbers of descendants, natural selection can act.

That first universe up on the mesa has ten progeny. Some of these will remain on the mesa and have ten progeny themselves. Others may cross back to the region of Phoenix universes and will have only one. But, never mind, it is still overwhelmingly likely that several generations later almost all of the universes that are created have ten progeny rather than one. The number with ten progeny grows exponentially at each step, so it is only a matter of time before they predominate. After a sufficient amount of time the probability that a universe picked randomly from the collection has ten progeny will become very close to one.

Now that the logic is clear, we can generalize and consider the case in which there are more possibilities than either one or ten progeny. To be realistic, we should allow the possibility that there are parameters that give rise to enormous numbers of black holes, as our own universe creates around 10^{18}. We may also assume, as seems to be the case, that the higher the number of progeny the smaller the range in parameter space that creates universes with this number of progeny. The real landscape then becomes very dramatic: there is a vast plain surrounding a mountainous region, where the parameters vary tremendously. Jutting up from the mountain range are narrow peaks which soar to enormous heights.

Despite the complicated landscape, the same logic holds, as in the case of the single mesa. In a population that is growing exponentially, the species with the highest rate of reproduction wins. After a sufficiently large number of generations, a universe picked randomly from the collection will be likely to come from those regions of the parameter space that produce the most black holes.

We began with a single universe with completely random values of the parameters. Given the two postulates—plus some simple, but apparently realistic, assumptions about how black holes are produced—this one universe has given rise to a vast collection, almost all of which have parameters in the narrow ranges that lead to the production of the most black holes. Now, we should notice an important thing, which is that we did not actually use anywhere the assumption that the universe we began with was the first universe. It might have been any universe in the collection; all that we know about it was that its parameters were chosen randomly. What this means is that any universe in the collection, no matter what its own parameters are, is likely to spawn in time a vast family of descendants that after a while are dominated by those whose parameters are the most fit for producing black holes. No matter what assumptions we make about the collection of universes at some earlier time, it will always be the case that after a sufficient time has passed, almost all of them have parameters in the narrow ranges that produce the most black holes.

As a result, the predictions of the theory do not actually depend on what the parameters of the initial universe were. If we like, we may even drop the assumption that there was a first universe. If we do this we also do not need to assume that all universes have at least one progeny. Whatever the details of the ensemble at earlier times, our existence shows that there was at least one line of descent that never died out. Nothing is lost if we admit the possibility of choices of parameters that give rise to universes, which expand forever but never make black holes.

Of course, the simple argument I've given here leaves out many details. To know exactly how the creatures are distributed on the landscape we should know more about the actual topography of the landscape. It is possible to make specific hypotheses about this and study the resulting distributions. But to do this we must know more about the physics at the bounces, and the whole point of what we are doing is to see what conclusions we can draw in our present state of ignorance about physics at such extreme conditions. What is important is that in the absence of such knowledge it is still possible to draw some general conclusions about where our creatures/universes are to be found.

One of these general conclusions is that, while the creatures are found concentrated near the summits, they are not all found exactly on the summit. This is because all the creatures produce progeny that live slightly displaced from them, as their parameters differ slightly from those of their parents. Thus, even a creature that lives exactly at a summit spreads its many progeny around it.

To know how far they are spread, we must take into account both how large the random changes in the parameters are, and how steep the landscape is. This is again a question of detail that cannot be answered presently. All we can say with the information we have is that a typical creature is found near, but not on, a summit. This means that a typical universe is found with parameters that are near, but not at, values that maximize the number of black holes a universe produces.

There are several other issues that we need to discuss. For example, what happens if there is more than one summit in the landscape? Suppose there are a number of peaks, of different altitudes. Does the population of creatures cluster only around the tallest one, or are they found around all the peaks?

The exact answer to this depends on the details of the landscape. But in almost all cases it happens that after a large but finite number of generations, the population is clustered around each of the peaks. More will certainly be found around the highest peak, but there will still be a significant population growing around the lower peaks. Thus, if after a large but finite number of generations, one picks a creature out randomly from the population, one cannot in general be sure that it will live near the largest peak. But it is almost certainly true that this randomly chosen creature was born near one of the peaks. We may then draw our main conclusion: *After a sufficient time, it is probable that a universe chosen at random from the collection has parameters that are near a peak of the production of black holes.*

It is exactly because of this that this theory, based on a collection of unobservable universes, can have explanatory power. We need only make one additional hypothesis, which is that *our universe is a typical member of the collection.* Then we can conclude that the parameters that govern our universe must also have parameters that are close to one of the peaks of the production of black holes.

Given this, we may ask what happens if we take a step away from the parameters of our universe. It may be possible to walk up a little bit to the summit. But in every direction, except that which leads directly to the summit, if we walk a sufficient distance we will go down. Although it is hard to visualize a landscape in a space of high dimensions, the more the number of parameters, the more likely it is that a walk of any finite distance, taken in a random direction from a place near the summit of a mountain, will go down rather than up. Thus, we may conclude that, if the hypotheses made here are true, *most changes in the parameters of the laws of physics will decrease the rate at which black holes are produced in our universe.*

Because of this, the theory I am sketching here is actually subject to observational test. In the next chapters I will explain how this can be done. But before closing this chapter, I would like to make several general comments about the proposal.

First, stars are clearly one way to make black holes. Even if only one out of every ten thousand stars becomes a black hole, then there are lots of black holes in our universe. However, it is far from obvious that the present values of the parameters actually maximize the production of black holes. Why not arrange it, for example, so that all stars become black holes? This is one of the questions we will need to answer in the next chapters.

A second question that could be raised is whether there might not be ways for a universe to make black holes that do not involve stars at all. Is it not possible that such universes would produce more black holes than our own universe? For example, might there be a choice of parameters that leads to a universe so chaotic in its early stages that enormous numbers of black holes are made directly, even before the stars are formed? Might the numbers of these so-called "primordial black holes" outnumber the black holes produced in our universe?

The answer is that even if this is possible, it is not relevant to the question of the testability of the theory. For, as we argued a few paragraphs ago, the theory does not predict that most universes are born around the highest peak in the landscape. It predicts only that most are born near peaks, so long as those peaks tower sufficiently above the terrain around them. If there is more than one such peak in the landscape, the collection may be dominated by several communities, each of which is associated with a single peak.

As a result, we are able to assert only that if our universe is typical, its parameters are close to some peak. But this is enough to deduce the prediction that small changes lead to a decrease in the number of black holes produced by the universe. To check this prediction, we don't have to know anything about the effects of large changes in the parameters; we only need to explore the landscape around us and confirm that we are near a peak. This means that it is irrelevant whether or not there are other very different choices of the parameters that also lead to copious production of black holes. This is good because it is much easier to try to reason about what would happen to our world if we make small changes in the parameters of the laws of physics than it would be to try to imagine what the world would be like for some arbitrarily different values.

Given the picture of the landscape, I can also clarify something that I've so far left vague. I've stated that the changes in the parameters from one universe to another are *small*. We can see from the discussion what is needed: the steps must be small compared to the scale over which one goes from parameters that lead to copious production of black holes to parameters that lead to more limited production. In other words, the steps should be small compared to the size of the mountains in the landscape; one step should not bring us all the way down from a peak to a valley. This is necessary for the progeny of a universe that produces black holes copiously to also produce many black holes.

Natural selection only works in biology because the changes in the organisms that result from mutations and sexual recombination are small. This is necessary not only to preserve the fitness of organisms from one generation to the next, but to make possible the development of greater fitness through the accumulation of incremental changes. If the changes in the parameters of universes are small, then the same will be true in cosmology. It is not enough to assume merely that the parameters change upon creation of the new universe. If the parameters of each universe were chosen randomly, with no relation to the previous universe, then it would not be possible to explain anything. The assumption that the changes in parameters are small is the crucial idea that makes it possible to rest a scientific theory on the ideas that new universes are created from black holes. Without it we have an interesting speculation. With it we have a theory.

The similarity to biological evolution is then not spurious. There is a precise analogy, which depends on the fact that exactly the same formal structure as I've used here can be used to describe the workings of natural selection in biology. To construct this formal analogy, we begin by taking the genes to be analogous to the parameters. The collection of all possible sequences of DNA is then something like the space of parameters.

Certainly, there are many ways in which the genes of real creatures are different from the parameters in the laws of physics. For example, if we want to represent the different possible genes as points in a space, such that those genes that differ by a single mutation are close to each other in the gene space, the gene space must have a very high dimension, since there are many possible mutations of a given DNA sequence. But this is the power of formal analogies: they allow us to abstract from two different systems only the features that actually play a role in the mechanisms in which we are interested.

The key quantity in genetics is the average number of offspring of creatures with a particular set of genes that themselves survive to reproduce. The biologists call this number the fitness. This quantity is precisely analogous to the average number of universes produced by a universe with a particular set of parameters. As this quantity depends on the genes, we may represent it as a landscape over the space of genes, with the altitude proportional to the fitness. This is then called the fitness landscape. Natural selection works in biology because there are genes that lead, on average, to higher numbers of progeny than others that are nearby in the space of genes. Thus, the fitness landscape is characterized by a complicated topography of valleys, hills and ridges, just like the function of the parameters of physics that gives the number of black holes produced.

Further, those combinations of genes that lead to surviving offspring are much less numerous than those that do not, so that the viably reproducing creatures always correspond to small regions of the space of genes. This is completely analogous to the fact that most of the parameter space for physics corresponds to universes that survive only for a few Planck times and do not reproduce more than one copy of themselves. Finally, our rule that in each birth of a new universe the parameters change by a small random step is precisely analogous to the fact that in reproduction the genes of the offspring differ, on average, by a small random change from those of the parent or parents.

We have thus abstracted a formal representation of natural selection that applies equally well to both biology and to our cosmological hypothesis. Any conclusion that can be deduced from it then applies to both domains. Biologists have been using the picture of a fitness landscape since the 1930s. One can find it described in the writings of evolutionary theorists with views as diverse as those of Richard Dawkins and Stuart Kauffman. Biologists have made many studies of the behavior of populations in such landscapes, whose results confirm the conclusions I have drawn here.

It must also be stressed that at this formal level, concepts like "survival of the fittest" or "competition for resources" play no role. What matters is only that rate of reproduction varies strongly as we vary the possible genes. What is responsible for that variation, and what goes into the differential survival rates, is not relevant for how the basic mechanisms of natural selection work. It is, by the way, for this reason that there can be controversy among evolutionary theorists without there being any challenge to the basic theory of natural selection. What is being argued about is what the important determinants of survival rates are, not how or whether natural selection functions in nature.

Actually, the model of the fitness landscape corresponds more precisely to the cosmological theory I've been describing than it does to biology. This is because it assumes that the fitness landscape is fixed for all time. But this is, of course, not really the case. The environment that determines how successful any particular organism will be is largely made of other organisms. It is only a crude approximation to consider that any single species evolves in a fixed environment.

This approximation seems nevertheless to be useful for many purposes. But if we want to understand the evolution of life on this planet as a whole, we must take into account the fact that the different niches are created by the whole community of species as they evolve together. Only by doing so could we understand, for example, why natural selection fosters such a high level of diversity of species. Indeed, while popular accounts of evolution have often stressed competition, looking at the plethora of different ways species have invented to live, it seems that an important theme of evolution might instead be the ability of the process to invent new ways of living, in order to minimize the actual competition among the different species. We will discuss this in more detail in Chapter 11.

Of course, the principle of natural selection will be more difficult to apply in cosmology than it is in biology, as we have access to only one member of the collection. In biology we may study properties of the distribution of genes in a population. This may lead to conclusions about how evolution proceeded, such as the classifications of species by genetic distance or the discovery that we are all descendants of a single Eve. It is hard to see how cosmological analogs of these may be extracted from observations made only on one universe. For example, it is difficult to see how we will be able to tell whether there was a first universe, as I assumed earlier. But this will not prevent us from drawing some testable conclusions from the theory.

Thus, if this theory is right, the universe shares certain features with biological systems. In both cases there is a large collection of distinguishable individuals, the properties of each of which are specified by a set of parameters. In both cases, the configurations that would be realized for most values of these parameters are very uninteresting. In both cases one has the development of structures that are stable over time scales that are very long compared to the fundamental time scales of the elementary dynamical processes. In both cases what needs to be explained is why the parameters that are actually realized fall into these small sets that give us an interesting, highly structured world. And, if the theory I have proposed is right, the explanation in both cases is found in the statistics behind the principles of natural selection.

This is certainly right in biology, how likely is it that it applies as well to cosmology? In the following chapters we will examine this question, but for now I would like to emphasize some rather general reasons for taking it seriously. The key point, it seems to me, is how seriously we should take the observation that our universe is very highly structured, compared to universes with most other values of the parameters. Should we take this as some kind of cosmic coincidence, or should we see it as something that needs to be explained?

I would argue that there are not many options, if we restrict ourselves to an explanation that can be subject to experimental test and that stays within the usual framework of causal explanation. To quote Richard Dawkins, in *The Blind Watchmaker*, "The theory of evolution by cumulative natural selection is the only theory we know of that is, in principle, capable of explaining the existence of organized complexity. Even if the evidence did not favor it, it would still be the best theory available."

The proposal I've made here may not hold up to test. But, once we begin to seek a scientific explanation for the values of the parameters that does not ignore the fact that the actual values produce a universe much more structured and complex than typical values, it may be difficult to ignore the possibility that cosmology must incorporate some mechanism analogous to natural selection.

No matter, never mind
—Gary Snyder, "Turtle Island"

EIGHT

DETECTIVE WORK

No matter how smart she is, no matter how modern her methods and how tricky her reasoning, a detective cannot be a good detective unless in the end the bad guys are found out. It is the same with science. Why science works is perhaps a mystery, but it does work, and often enough, that those of us who do it are content with the notion that, in the end, the only true measure of what we do is the extent to which it stands up against test by observation and experiment. In fact, the experience of most scientists is that most of our ideas turn out, in the end, to be wrong. Many ideas never even get to the point of being testable before being discarded for other reasons. Perhaps one of the reasons that science progresses at all is that there are not a few of us, and we are a stubborn bunch.

In the last chapter I presented a theory, which we may call the theory of cosmological natural selection, intended as a possible answer to the various puzzles of elementary particle physics and cosmology that I discussed in the first part of the book. Perhaps it is sufficient just to know that it is possible to invent a theory of how the parameters of elementary particle physics are determined that does not rely on the standard assumptions of radical atomism. Still, once it is on the table, we may wonder whether there might be some chance that it could be true. I will confess that when the idea first occurred to me I saw it as a kind of prototype. I expected it would be easy to find a simple argument that proved it wrong. But this is not what happened. Instead, as the theory survived a series of tests, I began to be convinced that it is a real scientific theory.

The basic prediction made by the theory is that the parameters in the laws of elementary particle physics are close to a value that maximizes the numbers of black holes made in our universe. It may seem at first sight that it is not possible to test this prediction. We do not have available a laboratory full of universes whose parameters could be set by tuning dials. But this is not the only way the theory might be tested. Even if we do not have access to these other universes, we ought to be able to deduce, from the physics and astronomy that we know, what the effect of a small change in the parameters would be. If there are cases in which we can do this well enough to be able to deduce whether the result would a world with more or less black holes than our own universe, the prediction of the theory is testable.

In fact, there turn out to be many cases in which we can deduce that a change in the parameters would have a strong effect on the processes by which black holes are made. Given that there are at least twenty parameters in elementary particle physics and cosmology, this means that if the theory is wrong, it should be easy to refute it. As each parameter can be either increased of decreased, this gives us at least forty chances to contradict the theory. If the theory is wrong, we have no reason to expect that there is a relationship between the parameters and the numbers of black holes. It is then likely that as many changes should lead to decreases as lead to increases. On the other hand, were all forty changes to lead to decreases in the numbers of black holes, there would be little choice but to take the theory seriously, for the chances of this happening if the values of the parameters and the numbers of black holes have no relation to each other, is about one chance in 2^{40}, which is a huge number.

The present situation lies somewhat in between these two possibilities. As I will describe, there are at least eight cases in which an increase or decrease of a parameter will lead to a dramatic decrease in the number of black holes in the universe. There are a few cases in which there may be a change, but it is not possible with current knowledge to predict the direction of the effect. But I know of no case in which a small change in the parameters would definitely lead to a large increase in the number of black holes.

The basic reason why it is possible to test the theory is that it is not easy to make a black hole. They only form when matter is squeezed to enormous densities, and this is unlikely to occur except under special circumstances. As far as we know, most black holes in our universe form in a particular way when a massive star collapses after it has burned all its nuclear fuel. And if it is not easy to make a black hole, it is also difficult to make a star. They form copiously only in certain circumstances. This means that what the universe is composed of greatly affects how many black holes are made. As I will explain, our universe, with a rich chemistry made possible by the existence of a large number of stable atoms, forms a great many more black holes than would many simpler worlds.

Moreover, only the more massive stars form black holes. Most stars are like our sun: they will burn uneventfully for ten billion or so years and then die quietly, cooling off gradually to become a white dwarf. Those stars massive enough to form a black hole are not very common, apparently because the circumstances in which they form are even rarer than those that lead to the birth of ordinary stars.

A star forms when a cloud of gas contracts under the force of gravity. If the cloud is massive enough its center will become hot and dense enough for nuclear reactions to take place and the new star will begin to shine. However, because gravity is very weak, this does not happen to just any cloud of gas. Like the air in the atmosphere, the gas that fills the universe has a pressure that resists the force of gravity. A cloud has to be dense enough for the force of gravity to be able to overcome its pressure and pull it together to the point that it becomes a star. The cloud also has to be cold, because the hotter it is, the higher is the pressure resisting the force of gravity. In fact, the clouds out of which stars form are very cold, about ten degrees above absolute zero.

This is important because it means that if a universe is going to form a lot of stars, the conditions have to be right for much of the gas that fills it to be able to collapse into the dense clouds that form stars. It is essentially because of this fact that changing the parameters of the laws of physics can have a big effect on how many black holes form.

We can also see right away that there is a built-in obstacle to the formation of lots of stars. What they need to form is a dense, cold environment, but as soon as a star forms, it will warm its surroundings. This heats the gas around it, which makes it more difficult for other stars to form. This is the heart of the problem; the stars themselves disrupt the conditions required for their formation. As a result, the process of star formation is more complex and interesting than might have been expected.

At much earlier times, before the galaxies formed, we suspect that the universe was filled with a more or less homogeneous gas of hydrogen and helium. As the universe expanded that gas became rarer and cooler. It is possible that early in its history—about a hundred million years after its creation—the gas that filled the universe was dense enough for stars to form. Whether or not some stars actually did form at that time is quite an interesting question, on which there has been much speculation. But even if some stars did form then, their numbers could not have been great, as the universe quickly became too dilute.

Since then, the universe has been much too dilute on average for stars to form. But certainly a great many stars have formed since then-they are all around us. How is this possible? The answer is that for a universe to continue to form stars indefinitely, there must be clouds of gas which remain cold and dense, in spite of the expansion of the universe. Moreover, they must maintain this state despite occupying the places where stars are born.

Such places exist—they are in the disks of spiral galaxies. It is an important fact that our world is filled with such galaxies. Because of this, the universe has many more stars—and many more black holes—than it would had the stars been formed only during a brief burst near the beginning. To understand why there are so many stars in our universe, we need to understand the processes by which the galaxies appear and organize themselves to preserve the conditions under which stars form.

The key issue is how these star-forming clouds can cool themselves in spite of the fact that there are stars all around, which tend to heat them. The answer is that hydrogen and helium do not accomplish this alone; the processes by which the clouds are cooled involves other elements, principally carbon. Remarkably, the cooling mechanisms depend on processes in which organic molecules, such as carbon monoxide, radiate excess heat. Indeed, these cold clouds are full of organic molecules. For this reason they are called *giant molecular clouds*. In addition, these clouds tend to be full of dust, which helps further to cool them by shielding them from starlight. This dust is also primarily made of carbon.

Carbon and the other organic elements are of course what living things are made out of. But it seems that they are important for life for a completely different reason: they are necessary for the processes by which the universe makes stars.

Of course, these elements are themselves made in stars. So it must have been possible originally to form stars from the clouds of pure hydrogen and helium that filled the primordial universe. How this happened is a largely unsolved problem, but what is certain is that, however the first stars formed, it was in a very different way than most stars form now. As far as astronomers can tell, most stars in spiral galaxies like our own were born in cold massive clouds cooled and shielded by carbon and other organic elements. Furthermore, it seems likely that if there are stars that form outside of these clouds, they are too small to become black holes. It is thus seems that the vast majority of black holes would not have formed were there not carbon. This means that *any change in the parameters that results in a universe without carbon would result in the formation of many fewer stars, and hence many fewer black holes.*

We may conclude from this that there are many changes in the parameters that lead to a world with fewer black holes. This is because, as we discussed in the third chapter, there are many such changes that have the effect of leading to a world in which there are no stable atomic nuclei at all. Atomic nuclei are only stable because of a balance between the different forces involved. This balance can be upset if some of the masses of the particles or the strengths of the forces involved are changed. Changes that lead to a world without nuclei include either increases or decreases in the masses of the proton, neutron, electron, and neutrino, as well as in the strengths of the electromagnetic and strong interactions. All told, there are five cases in which it is clear that the change leads to a world without atomic nuclei. Given what we have just said, these are all worlds with fewer black holes.

But these are not yet all the changes that may lead to a world without carbon. For it is not enough that carbon can exist; it must be made, and in large enough quantities that it can play a big role in the life of a galaxy. In our world, carbon is made in stars, during one stage of a sequence of reactions in which larger and larger nuclei are fused. As the world is organized, a great deal of carbon is made. But, as the British astronomer Fred Hoyle found many years ago, the processes that form carbon in stars depend on a certain delicate coincidence. He discovered that very little carbon will be made unless its nuclei can vibrate at almost the same frequency as the nuclei of another element involved in the reaction.

To understand this in detail we would need some quantum physics, but it is enough to say that nuclei, like atoms and molecules, can vibrate at certain discrete frequencies, like a guitar string. When two nuclei can vibrate at almost the same frequency, there can be a resonant effect in which one excites the other, just as plucking one string on a sitar can cause others to vibrate sympathetically. This, it turns out, can greatly accelerate certain nuclear reactions.

Hoyle's calculations led him to the conclusion that little carbon would be produced in stars unless there were such a resonant effect between it and the element beryllium. This required that each have a state of vibration at the same frequency. This was, a priori, rather unlikely, but what was worse was that no such coincidence was known. Hoyle thus predicted that carbon should have a state in which it vibrated at a particular frequency. No state of carbon was known with that frequency, but Hoyle was confident enough in his reasoning that he was able to insist it must be there. At his urging a group of experimentalists looked for that state and found it.

A great deal is sometimes made of this story by people who advocate the anthropic principle. Certainly carbon is essential for life, but it also seems essential for other things in the universe, such as continual star formation. In any case what Hoyle did does not require any hypothesis about why carbon is here. To make his prediction, he had only to assume that since there is a lot of carbon in the world it must somehow have been made in stars.

This is still not the end of the story; for it is also not enough that the carbon be made in stars, it must be distributed through the disk of the galaxy. It does no good if it just collects in dead stars. In fact, most of the carbon made in stars is expelled into space as the star is dying. First, powerful winds form at the source of a dying star, which carry away much of its mass. After this, the more massive stars explode as supernovas, which return most of the rest to the interstellar medium. Without supernovas, less carbon would be available to cool the clouds, and fewer stars would be formed.

Supernovas are required for another purpose, as well, which is to provide the energy that the galaxy needs to power the processes of star formation. How this works is itself an interesting story that we will come to shortly. Taking into account both the energy and the mass they return to the interstellar medium, we arrive at an important hypothesis, which is that *a world without supernovas would be one in which fewer stars, and hence fewer black holes, would be made.*

Before addressing the implications of this hypothesis, there is a tricky point that I must mention. A star becomes a black hole if, after it has died, its mass is more than a certain amount. Otherwise it becomes a white dwarf or a neutron star. The mass necessary to become a black hole is called the *upper mass limit,* as it denotes the most massive dead star that is not a black hole. Its exact value is not known precisely; it is something between one and a half and four times the mass of the Sun. However, before a massive star can contract to the point where it becomes a black hole, it first explodes. The force of the explosion throws most of the mass of the star out into the galaxy. The part left over is called the remnant. It is only if this remnant is larger than the upper mass limit that it becomes a black hole.

Thus, even if a star is originally massive enough to become a black hole, it may end otherwise because most of its mass is thrown out in the supernova explosion. This happens in a large number of cases. As a result, if there is a way to change the laws of physics so that there are no supernova explosions, the number of massive stars that become black holes in the end may be increased.

Does this mean that the number of black holes would be increased overall? If so, this would count as evidence against the theory. The answer, however, is no, because, as was just remarked, it is likely that without the effects of the super-novas, many fewer massive stars would be made. Thus, even if a given massive star is more likely to become a black hole without supernovas, so few massive stars might then be made that the result would still be a world with many less black holes.

There is at least one parameter whose strength can be tuned to turn off super-novas. This is the strength of the weak nuclear reaction. The reason is that it is primarily that interaction which triggers the supernova.

The collapse of a massive star is something like the fall of a large autocratic state: things happen from the inside out. The first thing that happens when the star runs out of nuclear fuel is that the inner core contracts under the force of gravity until it becomes as dense as an atomic nucleus. This creates a lot of energy, which must somehow escape. The core is too dense for light to travel far; only neutrinos, which interact very weakly with matter, can get out. Thus, as the star collapses, many neutrinos are created that carry energy out of the core.

On their journey outward, the neutrinos must, however, pass through the remainder of the star, which is itself just starting to collapse. This is enough material that an appreciable fraction of the neutrinos, perhaps five per cent, do interact on the way. When they do, they give up the energy which they are carrying to the atoms of the outer layers of the star. This heats the material tremendously, causing the supernova explosion.

It is rather touching that the little neutrinos, which seem most of the time to have no role at all to play in nature, are the key players in the supernova explosions, which are among the most dramatic things that happen in nature. But what is really interesting about this is that the process by which neutrinos take energy away from the inner core of the star and give it to the outer layers can happen only if the strength of their interactions with matter is tuned to within a small range. If the interactions are too weak, then all the neutrinos pass through the outer layers of the star, giving up no energy. The result is no supernova. But if they are too strong, then they interact too much with the inner core, and they never get out. The result is that there is again no explosion. The core heats up, but, because it is much denser, it does not explode. And then the rest of the star simply falls on top.

Thus, to have a world in which there are nuclei and stars, but no supernovas, we only need to increase or decrease the strength of the weak nuclear interaction, leaving the other parameters alone. Given our working hypothesis, these are good candidates for changes that lead to a world with less star formation, and hence fewer black holes.

However, before we can be sure of this, we must make sure there are no side effects of changing the strength of the weak interaction that might affect the scenario. In fact, there is such an effect. According to the standard cosmological scenario, there was a time when the universe as a whole was as dense as the interior of a star. As long as this lasted, nuclear reactions took place, which fused a certain proportion of the hydrogen into helium. The result is that the world emerged from this era about one quarter helium. Small amounts of other light elements, such as deuterium and lithium, were produced as well.

There is a detailed theory about this era, which gives predictions of how much helium and other nuclei were synthesized, that does impressively well when compared to the evidence. According to this theory, the amount of helium in the end turns out to be rather sensitive to the strength of the weak nuclear interaction. A change in one direction significant enough to prevent supernovas will likely result in a universe that is all hydrogen, while a change in the other direction results in a world that is all helium. Before we draw any conclusions we must then ask what effect these would have on the rate of formation of black holes.

The first possibility—a universe with all hydrogen—would not be very different from our universe, as far as processes involving stars and galaxies go. It is then difficult to see a reason why this circumstance could lead to a new production mechanism for black holes that could make up for what was lost from the supernova. Thus, in this case we do have a change that, given the hypotheses we have made, leads to a world with fewer black holes.

The situation is more difficult for the other possibility, a universe made primarily of helium. The reason is that stars made entirely of helium would burn much faster than stars made mostly of hydrogen, so that the whole history of the galaxy would be changed. There is no reason I know of to expect that such a world would make more black holes than ours, but given how different a helium world would be from ours, I don't think it is possible to be very sure about this. Thus, we are left with one change that does seem to lead to a decrease in the number of black holes, and a second that results in a world too different from ours to draw definite conclusions from present knowledge.

A situation like this is good for the theory because it provides an opportunity to make a prediction. If cosmological natural selection is correct then the universe we are speaking of, with more helium but without supernovas, should not have more black holes than our world. Of course we cannot do an experiment to test this prediction, but it is reasonable to expect that as astrophysics progresses it will reach a point where it will be possible to decide it one way or another.

We have discussed so far two ways in which changing the parameters might affect the numbers of black holes: by eliminating carbon; and by eliminating supernovas. These give us together at least six ways to change the parameters of the standard model that lead to a world with less black holes: one change in the strength of the weak interaction, plus the five ways to make a world without nuclei. I hope that this is sufficient to demonstrate that the theory is testable. We have so far examined only seven out of forty or so possible changes in the parameters. It is quite possible that if we look more closely at the processes that make stars and black holes we will find ways to change the parameters that lead to more black holes. Thus, it is reasonable to expect that if the theory is wrong it should be possible to demonstrate that fact. If, on the other hand, we discover after more investigation that we are unable to find a way to change the parameters that increases the numbers of black holes, I think it should be difficult not to take the theory seriously.

In the last few years I have investigated several other tests of the postulates of cosmological natural selection. Some of these rely on an amount of technical detail that would bore most readers; for this reason they are discussed in the appendix. There the reader will find also discussions of a number of objections, criticisms and suggestions that have been offered concerning the theory.

However, it is clear from the cases we have already discussed that one cannot discuss the testability of the predictions of cosmological natural selection without knowing something about how stars and black holes are formed in galaxies. The next two chapters will thus be devoted to describing what a galaxy is and how it works.

NINE

THE ECOLOGY
of the GALAXY

Did you ever look at the stars and wonder at how separate, how alone, each of them is? Is there not something invisible, some secret weave, that ties the jewels of the sky together? Did you ever feel like a star, on a solitary journey amidst the other silent stars?

The picture of an endless space filled with stars has been with us since the time of Newton, when the stellar sphere was smashed and Bruno's mystical dream of an infinite universe filled with suns became the setting for the scientific picture of the world. At the same time, atomism was revived and replaced the Aristotelian idea that the world was built from five essences. This simultaneous rise of atomism and the infinite universe could not have been coincidence. What is the main image of atomism if not an endless space filled with atoms each separate and discrete? Thinking about it, I find that this mingling of the atomistic picture of matter with the Newtonian theory of the cosmos has been with me as long as I have been thinking about these things. Still in the back of my mind is the picture, which must have been given to me by a grade school teacher, of the atom as a tiny solar system, with the nucleus a tiny sun and the electrons planets.

It is perhaps just a further coincidence that at the same time as the atomistic Newtonian view of the cosmos became popular, political and economic theorists began to write about society as a collection of individuals, each of which exists and acts more or less independently of the others. The idea of a society of individuals, each of which enters into association with the others by way of a contract in which he or she agrees to the preexisting laws of the society is very like the idea of a universe of atoms, each of which enters the stage of the Newtonian cosmos as an individual, independent of the existence of any others, and then moves and interacts according to universal and pre-existing laws of nature.

But there is something very misleading about the image of the stars as a collection of separate entities, each lost to the others in the depths of infinite space. In spite of the great gulfs of distance between them, an individual star is as unusual a thing in our universe as an individual tree.

`It is only since the 1920s that people have known that the stars are not distributed uniformly through space, but are collected into the great stellar communities we call the galaxies. And it is only much more recently that we have begun to develop a view of a galaxy not as a collection of individual stars, but as a system, or even as an ecology.

We are used to thinking of the stars as isolated because we have been taught that the huge reaches of space between them are empty. But this is not true. There is instead an incredibly diffuse medium that fills most of the disk of the galaxy. Made up of clouds of gas and dust, this interstellar medium is essential to the life of the galaxy. Stars do not live forever; they are born and they die, and what it is that they are born out of, and what it is that they return the bulk of themselves to when they die, is the interstellar medium. Furthermore, as we shall see presently, this medium is more than the soil out of which the stars are born, for, as I am about to describe, it is processes in the medium that dictates the rate at which the stars are born, and so ensures that the beauty of the galaxy, while composed of the more effervescent fires of the stars, is seemingly forever.

In this chapter we are going to take a detour from the main line of the argument in order to describe what a galaxy is and how it works. This is necessary to complete the discussion about testing the theory of cosmological natural selection. But, beyond this, the story of what a galaxy is embodies the idea that we may gain a great deal of understanding about the universe if we can learn to see it more as a self-organized system and less as a machine. More than anything else I have learned since I began to study astronomy, the story of what a galaxy is has changed my expectations of what the universe might be like on astronomical and cosmological scales.

One word of warning before we proceed. The subjects of star formation, supernovas, the interstellar medium, and galactic evolution are presently developing very rapidly. A great deal of new observational evidence relevant to these subjects has been gathered in the last twenty years, and a great deal more is expected in the near future. A consensus about some aspects of how a galaxy works is emerging, there is controversy at some critical points. To do full justice to this subject would require a book as long as this one. Since this is not my main aim, what I have done here is to give a necessarily selective reading of the current astrophysical literature on these topics.

Let us now turn to a description of what may be the largest coherent system, short of the universe as a whole, of which we are a part: the galaxy.

Looking from the largest known scales, galaxies are the primary form of the organization of matter in our universe. There are about 10^{11} of them in our observable universe, and they each contain about as many stars. They are typically about one hundred thousand light years across and, unlike stars, the distances between them are not very much larger than the distances across them. Thus, the closest sizable galaxy, Andromeda, makes an image on our sky as large as the sun or moon and can sometimes be just barely made out with the naked eye.

Galaxies come in several types. Our galaxy is of the type we will be mainly concerned with here, the spiral galaxies. These are the ones that one usually sees pictured, and it is the familiar spiraling patterns that of course give them their name.

To begin with, a spiral galaxy is a system that contains both stars and an interstellar medium consisting of gas and dust. There are processes by which the matter of the interstellar medium is converted into stars and there are processes by which matter is returned from the stars to the interstellar medium. To understand what a galaxy is, and especially to understand it as a system, is then primarily to understand the processes that govern the flow of matter and energy between the stars and the interstellar medium.

To describe how a galaxy works is then a bit of a chicken-and-egg problem, for we miss something if we look at it either from the point of view of the stars or from the point of view of the medium. (Of course, there really is no chicken and egg problem; certainly there were eggs long before there were chickens.) The picture becomes clearest if we look at the galaxy as a process which plays out over time, and not as a static thing. The time it takes for things to happen in the galaxy may seem long on a human scale as the important processes take place over periods of tens of thousands to tens of millions of years. However these times are short compared to the lifetime of the galaxy as a whole, which is on the order of ten billion years. The time it takes a star to form might, indeed, be called a day in a galactic life. In a human life, there must be some stabilizing factors which guide the vicissitudes of our days into a form that can be recognized as the life of an individual. In the same way, the apparent stability of the appearance of the spiral galaxies is the result of stabilizing influences that form slowly changing patterns out of ephemeral processes.

The problem with describing how a complex system like a galaxy works is that it is, well, complicated. However, there are often facts which, if you know them, make the picture simpler. In the case of the galaxy it will help to know a few elementary things about stars. The first is that stars come in a variety of masses, from a minimum of about one tenth of the mass of the Sun to a maximum of at least one hundred times the mass of the Sun. The way in which a star participates in the system of a galaxy depends more on its mass than anything else, for a few simple reasons. First, the brightness of a star increases strongly with its mass. The rate of increase is approximately as the cube of the mass, so that a star twice as massive is eight times brighter. More dramatically, a star thirty times as massive as the Sun shines almost as brightly as thirty thousand suns.

The second fact is that the more massive a star is, the less time it will live. This is because the more massive stars are throwing out energy at such a rate that even though they have larger reservoirs of fuel, they burn out very rapidly. The rate that the lifetimes decrease with increasing mass is also quite dramatic—it varies inversely to the square. This means that a star twice as massive lives one fourth the time, while a star thirty times more massive lives one thousandth the time. Put in terms of numbers, the sun will live about ten billion years. A star thirty times as massive as the Sun will live only about ten million years.

It is these very massive stars that explode as supernovas at the end of their short lives. Thus, for the short period of its existence, a massive star has a tremendous effect on its surroundings. While it is burning it pours forth an incredible amount of energy into the interstellar medium. Then it explodes, injecting even more energy. With this explosion the star also returns most of its mass to the interstellar medium. The material that the supernova disperses contains a higher proportion of the more massive elements, such as carbon, oxygen, and iron, than the material out of which the star was originally formed. Finally, it is these very massive stars that will become black holes.

Although they have such dramatic effects, the giant stars are much less common than stars with masses near that of our sun. This is not just because the massive ones burn out quickly. Many fewer of them are made, to begin with.

From these facts we can draw several conclusions that are fundamental for understanding how galaxies work. First, the massive stars are seen only for relativity short times after they are made. If one looks at a region of a galaxy in which no stars have been made for more than about a hundred million years (which is about the time it takes a star to revolve once around the disk of the galaxy) one will see only the smaller, long-lived stars.

It follows from this that we see massive stars in our galaxy because star formation has taken place very recently. This is a fact of fundamental importance; it means that in our galaxy, and in galaxies like it, new stars are continually being created. One might have imagined that the stars that make up a galaxy were all made when the galaxy was formed, but this is not the case. This raises some basic questions, such as what determines the rate at which a galaxy is making new stars?

This is indeed a big mystery. Under the right conditions, it takes only about ten thousand years for a cloud of gas to collapse under its self-gravity and form a star. Why is it then that five or ten billion years after the galaxy was formed, there remains plenty of gas to form new stars?

There are in fact many galaxies in which there is little or no formation of new stars currently going on. These galaxies are made up entirely of low mass, long-lived, stars; any massive stars that may have been originally present have long since burned out. There is not much in these galaxies out of which new stars could be made for they seem to contain much less dust than our galaxy or other spiral galaxies. What gas they contain has been heated to the point that makes its collapse into new stars very unlikely. Such galaxies lack the disks and spiral arms when new stars are formed. They are just more or less spherical collections of old stars moving together under the influence of their mutual gravitational attraction. For this reason they are called elliptical galaxies.

This heightens the mystery of star formation. If there are galaxies in which the formation of stars ceased long ago, what is special about the spiral galaxies that has allowed them to maintain an apparently constant rate of star formation until the present time?

On the other side, there are some galaxies in which the rate of star formation seems to be, at least for short times, much higher than in our galaxy. These are called star-burst galaxies and they are producing new stars at a rate that cannot possibly be sustained over a very long period of time. The existence of these heightens still further the mystery of the spiral galaxies. Why is it that the spiral galaxies are able to sustain a steady and moderate rate of forming new stars, while other galaxies are swept with waves of star formation that seem to consume all the available dust and gas the way a forest fire sweeps through a California hillside?

To begin to answer these questions, we will must first draw some simple consequences from the basic facts about stars we have just discussed. In the regions where they exist, and only for the brief time between their formation and explosion, the massive stars dominate what is going on around them. Even though there are many less of them, they outshine their less massive and longer lived cousins. Thus, the regions of a galaxy in which massive stars have recently formed shine much more brightly than those quiet places populated only by older, smaller, stars.

If one looks at a galaxy in which star formation continues to take place, those regions where massive stars have recently been born should stand out from the other regions like the lights on a Christmas tree on a dark night. This is exactly what we see. The bright spiral arms that one sees in most pictures of galaxies are exactly those regions in which star formation is now taking place. These regions, lit up by the bright young stars, form the characteristic swirling patterns that we associate with the spiral galaxies.

This is one of those amazing and simple facts that all astronomers know which seems to have failed to filter out to the general public. Those beautiful spiral patterns that one sees in pictures of galaxies are not, in most cases, the patterns of where the stars are. In many cases, if one looked at a picture of where the stars are actually distributed, one would not see a spiral pattern. The spirals are only the regions in which new stars are currently being formed. As a result, while it is true that spiral galaxies rotate, it is not true that the spiral structure, which are only the traces of the process of star formation, rotates with the stars of the galaxy. Instead, observations suggest that they move through the galaxy, dissolving and reforming on scales somewhat slower than the rotation of the galaxy.

Still more questions are raised by this. Why do the regions in which stars are formed make spiral patterns? This question is the key to unraveling the mystery of the spiral galaxies.

One of the wonderful things about the spiral galaxies is the variety of different ways in which the regions of star formation express the basic pattern of a spiral. Like clouds, there is a morphology of different types. In some cases there are two strongly symmetrical spiral arms. In others, one may speak not so much of spiral arms but of a kind of a spiral fluffiness, with many bright regions streaming out in a kind of random spiral pattern. Also, sometimes, but not always, there is a single, rectangular bar-like region, out of which the spiral arms emanate.

In spite of all of this variety, there is a basic structure, common to all spiral galaxies, that is not difficult to describe. Looking at a galaxy from the outside, one sees first a large, spherical halo of stars that surrounds the flat disk which one usually sees in pictures. This halo is made up of smaller, long-lived stars, which were made many billions of years ago. Their motion seems to be random, so that there is no apparent overall rotation. And although it is the dimmest component of the galaxy, it is believed that the halo contains most of the matter of the galaxy.

Embedded in the halo one finds a disk of stars, gas, and dust rotating slowly around an axis through the halo's center. This rotational motion is definitely not random; in any region of the disk the velocities of nearby stars differ by not more than ten percent from the overall rotational speed. One of the interesting and unexpected facts about the disk is that it does not rotate rigidly, like a merry-go-round or a top. Instead, the stars and the clouds of gas rotate with roughly the same velocity no matter how far they are from the center, so that those farther out take more time to complete a rotation.

The constancy of the rotation speed of the stars in a spiral disk is one of the spectacular scientific discoveries of the second half of the twentieth century. This is because it is possible to use Newton's laws to deduce the distribution of matter in a galaxy, given only knowledge of the velocities of the stars. In most galaxies, the result of this is a very different distribution of matter than is seen in the stars and gas. Typically, between 80 and 90 percent of the matter of a galaxy is found to be spread out beyond the disk and is not in the form of visible stars and gas.

All we know about this matter is that it is does not give off or reflect a great deal of light. The dark matter, as it is called, may be old burned-out stars, or black holes, or very cold dust, or some combination of all of these. It could also be something more exotic, such as neutrinos, or some kind of so far undiscovered particle. At some point, if we are to understand how the galaxies formed we will have to know what it is and why it is there. At the same time, besides the work it does holding the galaxy together, the dark matter is likely not relevant to the problem of how spiral galaxies work. The 90 to 95 percent of humanity no longer living have made us who we are, but we may, and in some ways we must, ignore them if we are to find a way to make peace among all of us who share the planet now. Likewise, the dark matter is essential to understand the history of the galaxy but, except for its gravitational effect, it can be ignored if what we want to understand is how a balance among the different components of the galaxy is maintained.

The galaxy's disk is where we find the continual star formation that is the distinguishing characteristic of the spiral galaxies. Indeed, several of the main features that distinguish the disk from other parts of the galaxy are directly tied to the presence of star formation. First, there is a variety of different kinds of stars, of all different ages and masses. Because it is possible to measure the age of a star by analyzing its spectra, we know from the proportions of stars of different ages that the rate of star formation in the disk of a typical spiral galaxy has been more or less constant.

Most significantly, it is also only in the disk that we find a lot of dust. This dust consists primarily of carbon, silicon, iron, and other elements that have been made in stars and either blown off their surfaces by stellar winds or thrown out into the interstellar medium by supernovas.

Given that star formation has been going on continually for many billions of years, there is a surprising lot of dust and gas in the disk. In a typical spiral galaxy, including our own Milky Way, at least ten percent of the matter in the disk is gas and dust. The gas is not spread uniformly; rather, it collects in a thin disk that sits inside the disk of stars. In that thin disk there is as much matter in gas and dust as there is in stars. It is in this thin disk that one finds continually going on the processes that make a galaxy much more than a collection of stars.

To see the interstellar medium for what it is—as something like the atmosphere of the galaxy—one must drastically adjust one's notion of scale. The interstellar medium is incredibly dilute; its most vacuous regions contain around one atom for every thousand cubic centimeters, while its denser clouds range from several hundreds up to a million atoms per cubic centimeter. Thus, from the most to the least dilute regions, the density of the interstellar medium can vary by a factor of as much as a billion, which is greater than the difference in density between air and rock. Still, even at its densest, the interstellar medium is a much better vacuum than can be created on Earth in a laboratory.

The key point about the interstellar medium is that it is truly a medium, in which chemical processes are taking place. It is just that these processes are, because of dilution, drastically slowed down. To appreciate the chemical processes of the interstellar medium as analogous to the processes in our atmosphere, one must think of ten thousand years as if they were a second. In other words, to see the interstellar medium in its functional role we must think of it in terms of the time scale of the life of a massive star, which is ten million years from formation to supernova. Then, to understand the whole system, we will have to see the life of that star as a day in the life of a galaxy that lives at least ten billion years and rotates once every few hundred million years.

More precisely, it is best to see the interstellar medium not just as the atmosphere of the spiral disk, but as its atmosphere, ocean, and ice cap, all mixed up. For, like the water on our planet, the interstellar medium consists of clouds of material in different phases, analogous to ice, water, and gas. As they have been mapped by astronomers, the interstellar mediums of spiral galaxies are quite complex. The different phases of the medium, which differ dramatically one from the other in density, temperature and composition, coexist side by side. One of these phases consists of the very cold and dense giant molecular clouds in which stars are formed. Very different from this is an extremely hot plasma phase, in which the electrons and nuclei have become disassociated. Still another phase consists of normal atomic gas, with rather moderate temperatures extending up to room temperature.

Let me dwell for a moment on the significance of the fact that the interstellar medium consists of components of quite different properties. How, we may ask, is it possible for such different phases to coexist for long periods of time? Why do they not mix? Why does heat not flow from the hot regions to the cold regions? Why does matter not flow out from the dense regions to fill the space around them?

Our expectation that the density and temperature of a gas must be homogeneous is based on the law of increasing entropy, which tells us that a gas isolated from outside influences must come to a state of equilibrium. The fact that the interstellar medium consists of different components of widely different temperatures and densities means that it is not in equilibrium. Since the law of increasing entropy applies to any isolated system, this means that the interstellar medium cannot be isolated; there must be significant exchange of energy and material with something outside of it.

That something else is the stars. Even before we address details, we can deduce, from the basic fact that the interstellar medium consists of many different phases, that there must be flows of energy and material that keep the medium out of equilibrium. These flows must be driven by a source of energy. Their source is primarily the intense starlight and supernovas of the massive stars. Thus, the stars and the interstellar medium are truly bound together into a system; the stars are made from the medium by means of processes which are powered by the energy the stars themselves produce.

The interstellar medium is not only far from thermodynamic equilibrium, it maintains that state indefinitely. It is true that the giant molecular clouds are continually forming and dissolving, so on small scales there is a lot of interchange of materials among these different phases. But if we average their proportions over the whole disk we find that, as far as can be determined, they change only very slowly in time.

How is it possible for a system to maintain itself far from equilibrium while keeping stable proportions of its different components? This is a question that has been studied from a general point of view by chemists and physicists over the last several decades. While it is still a developing subject, some general wisdom about the question has emerged. In particular, two features seem to characterize systems that maintain themselves in stable configurations far from equilibrium. The first is that such a system must be characterized by processes that cycle the materials among its different components. The second is that the rates of these processes must be determined by feedback mechanisms. These are necessary to keep the different processes in balance with each other so that the overall amounts of material in each component do not change in time. Both of these features characterize the processes that go on in spiral galaxies.

Astronomers have found that the material in the interstellar medium is in as many as seven distinct phases. This may seem complicated, but we should remember that to describe the role of water on our planet, we need to consider the three different phases of water: gas, liquid, and solid. Just as a particular water molecule will spend part of its time on Earth in any of these phases, a particular atom in the galaxy spends its time cycling between stars and the different components of the interstellar medium.

The hot plasma phase has a temperature of several million degrees, but it is quite sparse, containing only about one atom per thousand cubic centimeters. What keeps the hot plasma hot? Perhaps the reader can guess. The answer is that the energy to heat the medium to these temperatures comes from the supernovas. Each supernova explosion pours so much energy into the medium that a cloud of hot gas forms and quickly expands, heating the gas it passes through and stripping off its electrons. The result is a bubble that expands rapidly, sweeping the material it encounters in front of it and leaving behind it a very dilute and hot plasma.

One reason that it may have taken so long to discover the interstellar medium is that we are sitting inside of such a hot bubble, which is about three hundred light years across. This bubble has been mapped, and found to have a rather irregular shape. Very recently a neutron star has been identified that may be the remnant of the supernova whose explosion created the bubble.

It is actually not so unlikely that we find ourselves inside one of these hot bubbles, because they take up most of the volume of the disk of the galaxy, perhaps around 70 percent. Supernovas occur in our galaxy at a rate of one every thirty or forty years and, by continually sweeping out these regions, they provide the energy that keeps the whole interstellar medium under a constant pressure.

The image of the warm living Earth in the depth of cold empty space is thus not only misleading, as I suggested in the first chapter; it is simply wrong. Most of the volume of the galactic disk is taken up by a medium which, although it is incredibly dilute, is hot, much hotter than it is here on Earth! To appreciate this, and the processes that heat the medium and govern its properties, is to begin to glimpse the huge system of the galaxy that we live inside of.

However, if most of the space is in these hot regions, this is not where most of the matter is. Most of the matter is in the giant molecular clouds I discussed in the last chapter. With temperatures as low as ten or twenty degrees above absolute zero, these are the coldest clouds in the galaxies. They are also quite dense, with as many as a million atoms occupying each cubic centimeter. Thus, while they contain most of the disk's material, they take up only about one percent of its volume.

The gas in the giant molecular clouds is mainly in the form of molecules. The hydrogen, which is of course the dominant component, is mostly bound into molecules each containing two hydrogen atoms. But there are also large amounts of carbon, oxygen, nitrogen, and other organic elements in the clouds. The carbon is present partly in the form of dust grains, and partly in various organic molecules.

The giant molecular clouds are among the strangest and most singular structures in the galaxy. Each typically contains a million times the mass of the Sun, spread over a volume tens of light years across. (Recall, by comparison, that the nearest star to our Sun is four light years away.) Within the cloud the matter is irregularly distributed in dense clumps and filaments, each containing perhaps a thousand times the mass of the Sun. These in turn contain even denser cores, each containing perhaps ten times the mass of the Sun.

The most common organic molecule found in the interstellar medium is carbon monoxide. But, in what is certainly one of the most surprising developments of the last twenty years, a large variety of organic molecules have been found including some containing as many as twenty atoms. More than sixty different organic molecules have been discovered in the interstellar medium, including many common organic materials, such as ammonia and various alcohols. There are also controversial claims that much larger organic molecules, containing as many as one hundred atoms each, are present in large amounts.

It is no accident that most of the organic molecules that have been found in the interstellar medium were seen in the giant molecular clouds; their delicate molecular bonds survive best in the very cold temperatures that are found there. Furthermore, as I discussed earlier, these molecules play a large role in cooling these clouds. However, even given these cooling mechanisms, it is rather surprising that the giant molecular clouds are so cold, surrounded as they are by stars whose light can easily heat clouds of gas to far higher temperatures. Also mysterious is how so many molecules form, as starlight can break apart the delicate bonds that hold them together. For significant numbers of molecules to form, and certainly for such complicated molecules as have been discovered, something has to shield them from starlight.

That shield is the dust found in these clouds. Dust absorbs light, so that the molecules inside a dust cloud are shielded from most of the star-light that would break them apart. The dust also seems to catalyze the chemical reactions that form the molecules, as the surface of the dust provides a place for the atoms to concentrate.

In the last several years we have begun to develop a coherent picture of the processes by which stars form in the giant molecular clouds. This is primarily because it is now possible to make detailed observations of star-forming regions. From this wealth of evidence theorists have begun to piece together a rough picture of the process of star formation. While there is still healthy disagreement over some of the details, the outlines of the picture are clear.

The process that leads to the formation of a star begins when a small, especially dense portion of a giant molecular cloud begins to contract under the force of its gravitational self-attraction. More material is then drawn by its gravitational field, which falls on top of the original core. As the core grows, its center becomes hot due to the high pressure caused by the weight of the matter that has fallen on top of it. At some point the temperatures and pressures become hot enough to cause nuclear reactions, fusing hydrogen into helium to ignite in the center. The result is the birth of a new star.

When it ignites, our new star cannot be seen directly because it is still surrounded by a cocoon of dust which is being drawn to it by its gravitational field. The first thing that the new star must do is throw this matter off; otherwise its mass will continue to grow to a point where it becomes unstable and collapses right away to a black hole. The star accomplishes this by heating the cloud around it, which generates a wind blowing outwards from its surface, driving off the cloud of gas that has gathered around the star.

Actually, recent evidence suggests that the wind that blows the cloud off comes not from the new star itself, but from a whirling disk of matter that forms around it. These disks are important for another reason: the planets are believed to form from them, as they cool. Thus, planets such as our own form from the same soup of gas, dust, and organic molecules that are the wombs and cradles of the stars.

The outgoing wind stops the accretion of matter, so that, as the cloud disperses, the new star may be glimpsed peeking out from behind the dense clouds in the region it formed. It is, by the way, a wonderful thing that it is exactly the ignition of the star's nuclear burning that drives off the remainder of the collapsing cloud of gas, because the result is that the typical mass of a star ends up being around the mass which is optimal for the nuclear burning of hydrogen into helium. This is right around the mass of the sun. Thus, we may say that the process by which stars are formed regulates itself so that most stars come out to be the optimal size to burn hydrogen into helium.

One important fact we would like to understand is why it is that there are many more small stars than massive stars. Because only massive stars become black holes, it is important to understand why there are not more of them. Although the answer is not completely known, it is certainly important that massive stars have a considerable effect on their surroundings. When a small star—say about the mass of our Sun, or less—is formed then the cloud just around the newly formed star is blown away. But this does not affect other regions of the cloud, in which new stars continue to form. The result is that new stars are typically formed in associations, with dozens to hundreds of stars all forming from the same dense cold region of the interstellar medium. But when a massive star forms, it spews out so much energy that it heats up the whole cloud in which it formed, bringing the process of star formation to a halt.

One result of this is that the efficiency of the star formation process is quite low. Only about one percent of the mass of a giant molecular cloud will be converted to stars before the cloud is heated and dispersed by the energy pouring forth from the most massive of its progeny. This, more than anything else, accounts for why the star formation process did not long ago devour all the gas in the galaxy, turning it into stars.

However, this is only one of the important effects that the massive stars have on the process of star formation. While they halt the star formation process in their parent cloud, *they catalyze the process in nearby giant molecular clouds.* This is the key to the whole picture.

The contraction of dense clumps in the giant molecular cloud—which accounts for the formation of stars—does not happen very easily. The cloud is more or less stable as it is, because various effects, such as magnetic fields and turbulent motion, stir up and support the cloud, making it more difficult for the force of gravity to cause it to contract. So, independently of any outside force, the spontaneous contraction of a part of the cloud leading to the formation of a star is not common. Instead, what seems to occur most often is that the beginning of the collapse of the cloud is induced by an outside force.

One possible cause of such an event is a supernova explosion, from a massive star nearby the cloud. As I described before, this causes a bubble of hot gas to expand rapidly in the interstellar medium. In front of this bubble a wave forms, called a *shock wave*, which pushes the matter it encounters in front of it, similar to the bow wave of a boat. As the wave moves out, it may encounter another giant molecular cloud. The result is quite dramatic. The wave compresses the cloud, and by doing so it can catalyze the collapse of its densest regions, thus beginning the star-forming process in several parts of the cloud simultaneously.

This process is called self-propagating star formation. It is easy to see that this is an effect that will propagate through the galaxy as long as there are cold dense clouds of gas. At each step, a supernova from a star born around ten million years earlier disperses the cloud out of which it formed, but then catalyzes the formation of new stars in one or more nearby clouds. Ten or so million years later, one of those stars will supernova, and a new shock wave then moves out and catalyzes the formation of more stars in a giant molecular cloud neighboring to it. And so on.

This is the motivation for the working hypothesis I made in the last chapter. Without supernovas, the phenomenon of self-propagating star formation would be less likely to occur. It should be said that there seem to be cases in which shock waves form without supernovas; they are driven instead by the energy radiated by very massive stars. But the evidence presently is that supernovas are necessary, without them there would not be enough energy to continually regenerate the process of star formation.

In a spiral galaxy such as our own, this process apparently recurs perpetually, causing waves of star formation that sweep continually through the medium of a spiral galaxy. The spiral patterns that flow through the disk of a galaxy are then only the most visible manifestations of the workings of a continually renewing process.

Thus, a galaxy is a system, in which the processes of star formation occurs continually, as part of an apparently stable cycle of energy and material. If one knows nothing about the interstellar medium, a galaxy may seem just a static collection of stars, but in reality it is much closer to an ecosystem. And, although the spiral galaxies are certainly much simpler than the Earth's biosphere, they may very well be the most complex naturally occurring systems which are not living. What is more, they are ubiquitous, for the universe is filled with them. We cannot think of the universe as a simple homogeneous gas in dead equilibrium if its most common features are enormous self-organized systems of great complexity and beauty.

In scientific investigations, it is permitted to invent any hypothesis and, if it explains various large and independent classes of facts, it rises to the ranks of a well-grounded theory.

—Charles Darwin,
The Variations of Animals and Plants Under Domestication

TEN

GAMES *and* GALAXIES

In spite of the fact that the disk of a galaxy is made only of stars, gas and dust, we have just seen that it may be interesting to think of it as analogous to an ecological system. But we may wonder, is this just a romantic idea, or might it really be helpful to study the physics of the galaxies with the same concepts and tools we use to understand living things? In this chapter I would like to explore this question, and see the extent to which conceiving a galaxy as something like an ecological system may help us understand it more deeply. This approach will be useful for testing the theory of cosmological natural selection, as we need a good theory of how a galaxy works to tell us what happens to the production of black holes if we change the parameters of the standard model. But even beyond this, if a galaxy can really be seen as an ecological system that will be very interesting from the point of view of the larger themes of this book, concerning the relationship between physics, cosmology and biology.

Let me begin by summarizing the evidence that the disk of a spiral galaxy is like an ecological system. First of all, it exists continuously in a state far from thermal equilibrium, as is shown by the existence of many distinct phases in the interstellar medium. As in the case of the biosphere, a stable nonequilibrium state is maintained by large flows of energy. The flows cycle material from the stars through the various phases of the interstellar medium, and then back again. The energy to drive the flows comes from the massive stars, whose creation and destruction are themselves stages in the cycles. The catalysts for the production of the massive stars, which are dust and organic molecules, are created by the massive stars and dispersed by the same supernovas that power the system. Thus, there can be no doubt that the system of a spiral galaxy is characterized by autocatalytic cycles of energy and material, of the same general kind that underlie the ecology of the biosphere.

We may think of star formation as analogous to the chain reaction that powers a nuclear reactor. The neutrons released by the fission of a nucleus have the possibility of catalyzing other nuclear decays, just as a supernova may catalyze the formation of more massive stars. But to run at a steady rate, a reactor must be carefully controlled. If there are too many new neutrons the reactor explodes, not enough and the reaction dies out. We need to understand what controls the rate at which giant molecular clouds form, or supernovas explode, so that the star formation process in a spiral galaxy runs to neither extreme, but continues to renew itself at a steady rate.

This is, indeed, the key question for understanding what a spiral galaxy is. There are hundreds of billions of spiral galaxies in our visible universe. That each manages to find a stable configuration in which star formation continues at a steady rate cannot be an accident. For star formation to continue in each for billions of years there must be some process that balances the rate at which gas in the medium is being turned into stars with the rate at which the medium is being replenished by material from the stars. This is a fantastic thing, it suggests that there must be feedback processes that operate on the scale of a whole galaxy to regulate the rates of the key processes and keep them in balance.

If the notion that galaxies are governed by feedback processes seems outlandish, I should remark that the energy production of each star is regulated by a thermostat. The mechanism is simplicity itself. The rate that a star burns nuclear fuel is proportional to the pressure of the gas at the center. But the energy produced in the star causes an outward pressure on the gas, as the photons produced in the nuclear reactions seek to escape to the surface. This acts to decrease the pressure at the center. The result is that each star discovers a stable equilibrium in which these two competing processes are in balance. The star maintains its energy production at a steady rate around this equilibrium by a process that works just like a thermostat. If the energy production happens to increase, the result is a decrease of the pressure, which slows the burning down. Conversely if the energy production slacks, the pressure will increase, which brings the rate of nuclear burning back up.

The most direct evidence that feedback mechanisms govern the processes of the galaxy is simply that the rates of the flows of matter in different directions are approximately in balance. It has been estimated that in a typical spiral galaxy an amount of material equal to about three to five times the mass of the Sun is each year converted from gas to stars. On the other side, the estimates are that each year the stars return, on the average, at least half of this same amount of material to the interstellar medium, through stellar winds and supernova explosions. Given the kind of errors inherent in astrophysical measurements, these measurements can be considered to be in approximate agreement.

A number of different proposals have been made concerning feedback mechanisms that could govern the rate of star formation in our galaxy. Given the present state of the art, these are most likely oversimplifications that leave out some of the complexity of the actual processes. While it is unlikely that the processes of the galaxy have anything near the intricacy of living systems, it seems certain that several different feedback processes act over different scales of space and time to control the processes of the galaxy.

One of the most remarkable proposals for such a mechanism was made by a Venezuelan astronomer named Parravano. His idea is based on the fact that the process of the formation of dense cold clouds out of the interstellar medium is an example of a phase transition, like the transition that causes water to condense to ice. A key point about such a transition is that, like the freezing of ice, it can only happen at certain critical temperatures.

What Parravano proposed is that the interstellar mediums of spiral galaxies are always in the critical state at which the giant molecular clouds condense out of the medium. He and his colleagues have found evidence that this is the case in a large number of spiral galaxies. This is an amazing fact that, if true, requires an explanation. It is as unlikely as it would be if the temperature of the Earth's atmosphere were always near zero degrees Centigrade, where water freezes. The only plausible explanation is that built into the mechanisms of the galaxy there is a thermostat, a feedback process that governs the conditions of the medium of a galaxy to keep it always in this critical state.

A thermostat keeps a house at a fixed temperature because there is a process that turns the heat on if it gets too cold and turns the heat off if it gets too hot. According to Parravano, the galactic thermostat works the same way. The role of the heater is played by the hot young stars, because they give off large amounts of ultraviolet radiation that heats the interstellar medium. The gas of the interstellar medium acts as a thermostat because, as it is heated, it makes it harder for cold dense clouds to form. This diminishes the rate of star formation. As fewer new stars are made, and as the older ones die out, the medium soon begins to cool. This is because the hot stars that heat the medium live for very short periods. As the medium cools, it passes the point at which the giant molecular clouds start to form. Then many more stars are made. Among them are hot, massive stars that heat the medium. We are back at the beginning. The result is that the medium as a whole is kept just at the critical point of the transition for the formation of the giant molecular clouds.

If this theory is right it explains why there are always giant molecular clouds around so that new stars can continually form, in spite of the fact that each individual cloud is dissolved in the process of star formation. But there is more. We have still not explained why the clouds in which stars are formed tend to make spiral patterns. This also, as we shall see, is due to a feedback process.

In Japan I am told there are indoor pools for surfers. A machine at one end generates a wave that runs the length of the pool before crashing into the other end. This must be great fun, but like many of the thrills of life it is imperfect, because too soon one comes to the end of the pool and the ride is over. A better idea would be to build the pool in the shape of a ring, like a running track for surfers. The wave could go round and round and we could surf forever. Unfortunately, there is a problem, which is that any wave loses energy as it travels, so after a few times around it will die out. To be able to surf forever the machine must continually regenerate the wave. And it must do it right. Each time around it must add exactly the right amount of energy to the wave; no more and no less than it lost on the trip around. If it adds too little the wave will die out; if it adds too much it will grow uncontrollably in time, until something breaks.

The problem of spiral structure has a lot in common with this, as star formation can be thought of as a wave that is sweeping continually through the disk of the galaxy. The wave doesn't penetrate the center, so the disk is like a running track around which circle the stars and clouds. There is, of course, a big difference which is that there is no machine outside of a galaxy to generate the wave. Somehow, the galaxy must generate the wave itself. And it must continually feed energy to the wave at the right rate so that it neither dies off nor grows uncontrollably.

In many spiral galaxies, the spiral structure is only a trace of the pattern made by the regions where new stars are forming. If one looks at how the old stars are distributed, one doesn't see spiral arms. But there are other galaxies in which it seems that there really is a wave, very much like a sound wave, sweeping through the galaxy. In these galaxies the spiral arms are not only places where stars are being made; they are also places where there are more stars. These include the most dramatic types of spirals, in which two clearly delineated arms grow symmetrically out of the center.

The first theories of spiral structure that were invented were meant to apply to

these kinds of galaxies, in which something like a sound wave appears to be traveling through the disk of the galaxy. Star formation was thought to be a by-prod-
uct of the wave, which can squeeze a cloud as it passes, with the result that the
cloud collapses and forms stars.

The problem with this density wave theory, as it is called, is the same as besets
our circular surfing pool. First, something external has to get the wave going.
And then, after only a few times around the wave loses energy and dies out. There
are some cases in which the passing of a nearby galaxy might have given the
bump that is needed to get such a wave going. But this cannot be the whole story
as there are many spiral galaxies that are too far from their neighbors for interactions to have been the cause of their spiral waves.

In more recent theories of spiral structure, the energy for the wave comes from
the galaxy itself, from the very intense radiation of the massive stars and from their
supernova explosions. In the 1970s two astronomers named Mueller and Arnett
proposed that rather than sound waves, what moves through the galaxies are simply waves of star formation, which move from cloud to cloud by the processes of
self-propagating star formation. But can such a process really lead to a permanent
spiral structure? How is it that the bright stars and supernovas add exactly the right
amount of energy to the wave as it travels through the galaxy, so that over long
time periods the wave neither dies out nor grows uncontrollably?

One very interesting, although controversial approach to the problem of spiral structure was made by Humerto Gerola, Larry Schulman and Philip Seiden,
three physicists who worked at IBM in the 1970s and 80s. They likened the spread
of star formation in a galaxy to a very different phenomenon, the spread of a virus
through the population it infects. A virus has a problem very similar to that of a
wave of star formation. If it infects and kills too many people, it dies out. But if it
infects too few it also dies out. To live continually in a population, a virus must
not kill too many of its hosts, and it must infect new victims at a steady rate, not
too high and not too low.

A great part of the art of physics is the talent to ignore details and proceed by
analogy. A good physicist can sometimes perceive relationships between two phenomena that specialists in each would never have seen. It is probably because neither Gerola, Schulman nor Seiden were biologists that they were able to imagine
the gas and dust in the disk of a galaxy as analogous to the population of the earth.

Just as the Earth's population is concentrated in cities and towns, the gas and
dust is collected into clouds of various sizes. The point of the analogy is to see star
formation as a process that infects a cloud, just as a virus infects a town. The virus
travels from town to town, carried by travelers or people fleeing the epidemic.
Similarly, the infection of star formation also travels from cloud to cloud, carried
by a shock wave which is the result of the energy produced by the stars made in an
infected cloud.

In both cases we need to understand a similar question. How is the rate at which the infection spreads between clouds controlled so that over a long time it neither peters away nor grows too much and kills its hosts?

The solution to the problem, at least from the point of view of the virus, involves several aspects. First, not everyone who is infected dies, or even gets sick. Second, whether someone gets sick or not, once they are infected they develop an immunity. This may seem to be bad for the virus, but it is actually what makes it possible to limit the infection in the short term so that the hosts, and hence the virus, can survive in the long term.

Because of the immunity, the virus must move on and infect another region if it is to survive. The immunity would eventually be bad for the virus, were it permanent. But by continually mutating, a virus is generally able in time to overcome the immunity . The immunity you gain to this winter's flu virus may not protect you from next year's. As a result, if a virus can keep moving through a population, and mutate slowly as it goes, it can live forever.

The key to the survival of the virus is then the possibility that the host develops a temporary immunity. By analogy, if star formation is to go on indefinitely in a galaxy, there must be something like temporary immunity in this case as well.

The point of the theory of Gerola, Schulman, and Seiden is that the immunity comes from the fact that a new massive star heats and dissolves the cloud in which it formed, making that material unable to form stars. This has the effect of limiting the efficiency of the star-forming process, so that only a small proportion of each cloud is turned into stars. Only after a long time has passed may the gas cool and return to a state in which it can form stars.

These simple ideas may be attractive, but do they really work? To show that they really might capture what happens in a galaxy, Gerola, Schulman, and Seiden invented a delightfully simple game that models the star-forming process with a few simple rules. They did this by making a few changes in the rules of a game that already existed, which is called the game of life. Invented by John Conway, a mathematician, this game is played on a simple chessboard, each square of which can be thought of as living or dead. The game proceeds through a series of steps. At each step a decision must be made for each square of the board as to whether it is to be alive or not. There are some simple rules by which this is determined. A square will be alive on the next step if it has next to it some, but not too many, squares that are now alive. This is like how an infection spreads. One has to think that the squares are the hosts and what is alive are the viruses that can live in them. To be infected, you must have at least one neighbor who is already infected, but too many infected neighbors can convey an immunity, and prevent an infection. What is good about this game is that ,with a few simple rules, a large variety of beautiful structures are produced, which continually appear and dissolve in a kind of dance.

To make a simple model of how a galaxy works it was necessary only to make a few simple changes in the rules of the game. The board is now a ring of disks, which rotate like the disk of a galaxy. A square, once alive, will have an immunity to being infected that lasts for a certain number of moves. Finally, (and this is probably not significant) they also made the rules probabilistic, so that an infected neighbor increases a square's probability to be also infected. With these changes they had in the end a simple game in which spiral patterns of infections move continually around a rotating disk. Like the flu virus, the star-forming infection moves continually through the disk of the galaxy, always present somewhere. By adjusting the rate of rotation of the rings to match that seen in real galaxies they were able to reproduce rather well the spiral patterns that one sees in many of the galaxies.

This simple model seems to capture the essence of the ecology of the spiral galaxies. Most importantly, it explains to us how feedback processes, such as those that are ubiquitous in the biological world, act to control the star formation process so that the waves of star formation neither die out, nor grow uncontrollably, but propagate at exactly the right rate to persist in the galaxy indefinitely.

Of course, the theory is too simple. It leaves out the real motions of the stars and clouds, as well as the gravitational forces between them. As such there are many things it cannot describe. Among them are the galaxies in which there seem to be real waves in the density of stars and gas sweeping through the disk. The theory fails to explain the large, definitive, spiral arms that seem to be waves moving through the medium of the galaxy.

The whole truth, then, must involve both the effects of the gravitational forces between the stars and clouds and the dynamical effects of the star forming process. There really are waves that move around, at least in the medium of some galaxies. To capture both these and the feedback processes that control the processes of star formation, it is necessary to construct more elaborate models that capture both kinds of phenomena. Such models have been constructed. Perhaps the best of these, so far, was developed by Bruce Elmegreen, an American astronomer who also works for IBM, together with a Danish astronomer named Magnus Thomasson. By adjusting the parameters of their model they are able to mimic the feedback processes that are captured in the simpler model of Gerola, Schulman, and Seiden, and account for the presence of spiral arms in all kinds of galaxies.

This story of the attempts to understand the reasons for the spiral structures of galaxies tells a lot about the changing styles of explanation among physicists. It is very interesting to compare the older density wave theories with the newer theory of Gerola, Schulman, and Seiden. Both of them are tremendously oversimplified; each throws away most of the complexity of the real situation. What is interesting is what is kept and what is thrown away in the two cases. The density wave theory ignores all of the processes, such as star formation, taking place in the galactic disks, and models the galaxy as a simple medium or fluid through which waves may travel. The model of the physicists from IBM is of a quite different sort; it throws away all the material phenomena, such as densities, pressures, and temperature, and instead models the process of star formation as a discrete on-or-off process.

The kind of mathematics used in the two cases is correspondingly different. In the first case the matter is modeled by continuous quantities such as densities and pressures, which satisfy complicated differential equations. The other model is much simpler, star formation is represented as something that is either on or off, like a light bulb or a bit in a computer's memory. It spreads through the galaxy in a process generated by the repeated application of some simple rules, like the flow of information through a computer.

It is not an exaggeration to say that these two theories illustrate how tastes about what constitutes good science have been changed by the advent of the computer. A generation ago, good theories were thought to be those like the density wave theory, in which continuous quantities evolve according to nonlinear differential equations. These days, the theories of interest to physicists are often like the game of Gerola, Schulman, and Seiden: defined by simple rules that many fifteen-year-olds could model on a personal computer. Rather than having to solve equations, and then study the meaning of the solutions, one may simply write a program and see the consequences presented directly as patterns on a screen. In an interesting way, the computer is not only serving as a tool which allows people to play easily with ideas—it is itself serving as a metaphor.

Further, the fact that the new model is based on an analogy to a biological phenomenon is not fortuitous. What is so attractive, but perhaps also frightening, about this new kind of theory is how it suggests analogies between biological systems and other complex physical systems of a sort that could not easily have been made using the old kind of mathematics. For it is not an accident that simple computer games can model the processes in biological populations. The key to both biology and computer games is that the right set of simple rules, repeated over and over again, can lead to the formation of enormously complex patterns and structures that reproduce themselves continually over time. These are phenomena that are difficult, if not impossible, to capture with the kind of mathematics that is usually used to model the flow of waves through fluids. But they are well described by the newer mathematical games that are described in terms of algorithmic systems such as the cellular automata that define the game of life and the Gerola-Seiden-Schulman model of galactic spiral structure.

This is not to say that the old mathematics is not necessary for a complete understanding. The logic of the growth of science is much more a logic of *AND* than it is of *OR* . As the story I have been telling illustrates, the simple game captures a lot, but it also leaves out important phenomena that are captured more simply in the older kind of theory. But, even so, beyond the simple question of the evolution of scientific tastes, I believe that there is something significant about this difference between the new and the old kind of mathematics. The old style of mathematics, represented by differential equations, keeps track of where the matter is. It speaks a language of continuous functions, such as densities and flows. The new style of mathematics is discrete, and speaks a language of on or off, alive or dead, infected or uninfected. At the level of its basic vocabulary, it may seem relatively impoverished. But it easily captures structure and organization that is difficult or impossible to encode in the old, continuous mathematics. It can do this because it is a language, not of substance, but of information.

The fact that a simple model built in this new language captures the basic phenomenology of the spread of star formation through a galaxy suggests that the analogy between biological and astrophysical phenomena may not be spurious. The success of this model means that it really is the case that the same logic is governing both the spread of star formation through the disk of a galaxy and the spread of a virus through a population. Thus, the answer to the question I posed at the beginning of the chapter is: Yes—ideas from biology and ecology do help us to understand what is happening in the disk of a galaxy.

Of course, even if this is granted for galaxies, it is a much larger step to imagine that processes analogous to natural selection or the self-organization of non-equilibrium systems might apply on the scale of the whole universe. However, notice that the picture of how spiral galaxies work developed in the last two chapters greatly strengthens the case for the testability of cosmological natural selection. For, given this picture, it is difficult to escape the conclusion that a small change in the parameters that led to a universe without either copious production of carbon in stars or supernova would have far fewer black holes than our universe. Taking into account side effects, that gives us six cases in which a change in a parameter lowers the number of black holes. At the same time, I am aware of no change that leads to the production of more black holes. There are also several cases in which it is not possible with current astrophysical knowledge to tell what the outcome would be. This is, indeed, not bad, as each of these cases leads to a prediction: if cosmological natural selection is right they must in the end not lead to more black holes.

A complete discussion of all the possible tests of the theory I am aware of is contained in the appendix. The result, for the reader who does not care to know the details, is two more cases that clearly lead to fewer black holes, and several more where the result is inconclusive, leading to more predictions. I also discuss in the appendix several ways to modify the postulates of the theory in order to lead to different theories, which may be distinguished from each other, as well as from the original, by their predictions. Indeed, there are even versions of cosmological natural selection that can be ruled out, given what we know now. The result of all of this is that the theory of cosmological natural selection is clearly a scientific theory. It is testable, and stands ready to be falsified on several points. Given that few other hypotheses about the origin of the parameters of the standard model lead to testable predictions, it is then a theory that should be taken seriously.

But, beyond this particular theory, the fact that one can construct a general scientific theory about cosmology and particle physics using the logic of natural selection suggests that the physics of self-organization may be as much needed to understand what is happening in the sky as it is to understand the tremendous intricacy of the ecology of which our life is a part.

PART THREE

THE ORGANIZATION

of the

COSMOS

---

*Is it accidental or necessary that the universe have
such a large variety of structure? Why is the universe so interesting?*

Life sends up in blades of grass
its silent hymn of praise
to the unnamed light
　　　—Tagore, Fireflies

ELEVEN

WHAT IS LIFE?

The classical physicists of the 18th and 19th century likened the universe to a great clock, by which they meant to suggest that everything in it obeys simple and deterministic laws of motion that are as inescapable as the ticking of a pendulum. Indeed, the science of motion, which they thought of as the deepest part of their understanding of nature, was called by them *mechanics*, which is, of course, the word associated with knowledge of machinery. But, a clock requires a clockmaker, and most physicists and philosophers of the last three centuries have had no trouble imagining that the universe was the creation of an intelligent and eternal God. Perhaps it would not then be too much to suggest that the image of

the universe as a mechanical clock was, for many of those who came after Newton, a religious idea.

But behind the metaphor of the universe as a clock there lay, almost hidden,
three images which contain the seeds of the obsolescence of the Newtonian world
view. First, the image points back to the roots of Newtonian physics in Greek science, for in a clock complicated motions are produced from combinations of simple circular motions, just as in the Platonic and Ptolemaic models of the universe.
Then there is the idea of a universal and absolute time, which was both the idea
that held Newtonian physics together and the key to its unraveling.

Finally, there is a way in which the image of the clock represents both this
imagined absolute time and its opposite, for any real clock, made up of gears, or of
silicon, eventually runs down. Indeed, Newton worried whether the planets in
their orbits might lose momentum and spiral into the sun, and to preserve eternally the workings of the comic clock he was willing to contemplate the necessity
of the clockmaker stepping in from time to time to give the planets little nudges
to keep them in their orbits.

Only much later, in the 19th century, was it proved that the solar system is in
fact stable. However, at around the same time the science of thermodynamics
was born, which, with the law of increasing entropy, or disorder, suggested that
there was, indeed, a great danger of the universe running down. This law says
that any system must eventually come to thermal equilibrium, in which case all
structure and all regular motion must dissipate. Thus was born the idea of the
necessity of the death of the whole cosmos—the "heat death" of the universe, as
it was called. This idea represented the logical termination of the path scouted
out by Newton and his contemporaries; it implies that the final state of any universe described by Newton's laws must be a featureless equilibrium, and that all
change, all structure—indeed all life—must represent only improbable and
transient fluctuations.

In fact, the idea that the natural state of the universe is a featureless chaos, to
which it would decay were it not for the imposition of a god's intelligence, is
much older than thermodynamics or mechanics. In his worry about the solar
system running down, Newton was expressing an ancient anxiety. The roots of
this idea, and of much that has led to the modern crisis in cosmology, can be
found in the very moment that people first began to conceive of the universe as
arising from primordial chaos. Let us listen as the Stranger tells Young Socrates
how the world was made, in Plato's dialogue, *The Statesman*:

> And now the pilot of the universe let go the handle of its rudder, as it were, and
> retired to his observation tower. . . . With a jerk the universe changed its rotation,
> driven by an impulse in which beginning and end reversed their positions. . . . Then,
> after the interval needed for its recovery, it gained relief from its clamors and con-
> fusions and, attaining quiet after the tremors, it returned to its ordered course and

continued in it, having control and government of itself and of all within it and remembering, so far as it was able, the instructions it had received from its maker and father. At first it fulfilled his instructions more clearly, but as time went on, more carelessly.

And what made the universe careless? The Stranger goes on to tell us:

> The bodily element in it was responsible for its failure. This bodily factor belonged to it in its most primitive condition; for before it came into its present order as a universe it was an utter chaos of disorder. It was from him that composed it that it has received all the virtue it possesses, while from its primal chaotic condition all the wrongs and troubles that are in heaven arise in it, and these it engenders in turn in the living creatures. When it is guided by the pilot, it produces much good and but little evil in the creatures it raises and sustains. When it must travel on without him, things go well enough in the years immediately after he abandons control, but as time goes on and forgetfulness settles in, the ancient condition of discord also begins to assert its sway. At last, as this cosmic era draws to a close, this disorder comes to a head.

Does it not seem that with this *myth of the reversing cosmos* we may be near the original scene of the crime, for it is all here; in this dialogue we can see how Western cosmology and political theory arose together from the opposition of the spirit and the body, the eternal and the decaying, the externally imposed order and the internally generated chaos. But what is perhaps most interesting is how the central metaphor of this story ties together the endless cycles of the circular motions of the things in the sky with the endless and eternal cycle of birth and death. What then surely is most new about our modern understanding of life is the idea of evolution, for it enables us to see life not as an eternally repeating cycle, but as a process that continually generates and discovers novelty. And, by the same token, what is most new about modern cosmology is the discovery that the universe is also evolving. Whatever questions remain open, observations show us that the universe arose out of a state that it may never return to, and that each era in its evolution since then has been unique.

It is then most significant that one of the things that the cosmologies of Plato and Newton have in common is that they lack the notion of evolution, in either the biological or the astronomical sense. Stuck with a universe in which past and future cannot fundamentally differ from each other, we see how those things that we now understand as born and bound in time are instead set as timeless oppositions. Thus, both Plato's myth and Newton's universe are framed in terms of a duality in which the intelligence of a god who exists outside the universe is forever opposed to the imagined tendency of material things to disintegrate to chaos.

In a universe without evolution, this duality permits an almost paradoxical conception in which the universe persists eternally in spite of the danger of it running to chaos. Indeed, this is a problem that must be solved, for if the universe is forever running towards death and chaos, why do we see around us a world full of life and variety? Without the idea of evolution, the only solution is to find a way to embed the decaying universe in a cyclical cosmology in which the cosmos is forever created anew. Thus, in Plato's myth, each new cycle is imposed by "the pilot," when *"Beholding it in its troubles, and anxious for it lest, racked by storms and confusions, it sink and be dissolved again in the bottomless abyss of unlikeness, he takes control of the helm once more."*

Given this background, it is fascinating to note that Newtonian physics also requires that the world be cyclical. This is because in a Newtonian universe that lasts eternally, any possible event, no matter how improbable, must reoccur an infinite number of times. This is best understood if we imagine that there are only a finite number of configurations in which the world might be found. There is a small probability that the random motions of the atoms generated by heat put the world in any of these configurations. But once the universe returns to some configuration, determinism requires that it proceed as before.

Thus, in a world of atoms governed by deterministic laws, chance alone plays the role of the pilot who returns the universe from time to time to an ordered state that makes life possible. All that is required for the universe to return to something like the present state is enough time, for although such a configuration is very improbable, this only means it will be formed less often by the random motions of the atoms in the chaos of equilibrium. But in eternity there is enough time for anything.

Ludwig Boltzmann, a late 19th century physicist who more than anyone else was responsible for the triumph of atomism, understood very well that in a universe of atoms governed by Newton's laws, it is extraordinarily improbable that a phenomenon as complex as life would arise. His solution, and indeed the only solution possible without recourse to a god, was that such an improbable configuration must arise from time to time in eternity as a kind of interruption of equilibrium, as a result of the random motion of the atoms. Thus, in the Newtonian picture of the cosmos, the life we see around us is not only transient, but suffers also the banality of eternal repetition. In this picture our life lacks significance, both for its being a temporary excursion from the normality of dead equilibrium and for the fact that even as such it lacks the dignity of uniqueness.

Friedrich Nietzsche drew his notion of the eternal return from the writings of the eighteenth century physicist Ruggiero Giuseppi Boscovich who, even before Boltzmann, had appreciated that one implication of the determinism of Newtonian physics is that any configuration of the world must reoccur at some moment of infinite time. Of course, the nihilism and alienation of Nietzsche and other late 19th- and 20th-century philosophers cannot be blamed entirely on the Newtonian picture of the physical universe; its roots lie in the dualistic opposition of intelligent spirit and degenerate matter that seems to have captured the Western mind by Plato's time. But I suspect that the image of our Earth as lost in a dead and hostile universe, while certainly not its only cause, has to some small extent fueled the pessimism in the art and literature of this troubled century.

Certainly the validity of a physical theory cannot be judged from how it makes us feel. At the same time, there is no denying that, apart from its use to predict the results of experiments, a good scientific theory may function as a metaphor that captures and expresses what we think is essential in the world. We must be able to separate the question of the empirical validity of a theory from the ethical and spiritual implications of its central metaphor; they are not the same thing, even if the metaphor may gain authority from the success of the theory while it, in turn, shapes our understanding of the theory's meaning. While beautiful to the educated eye, there is no escaping the fact that the central metaphor of Newtonian physics is a pessimistic one, as it provides an image of a dead world into which we do not fit.

But the failure of Newtonian physics to describe a world in which living things have a natural place is more than a philosophical issue. It means the theory fails scientifically, just as much as it does for its inability to provide an explanation for the existence of stars. For this reason it is important to appreciate that the science that gave rise to the dark metaphors of the clock universe, the universal heat death, and the eternal return is now itself quite dead. Instead, the science that is slowly growing up to replace the old Newtonian physics may offer an image of a universe that is hospitable, rather than hostile, towards life.

At the same time, there is no reason to believe that either a galaxy or the universe as a whole remotely approaches the complexity and intricacy of the organization of a single living cell. Thus, I want to warn against the suggestion, attractive as it may be, that we simply pronounce the galaxy or the universe to be "alive." It is not that I want to separate science from poetry; indeed, it may be that the tendency towards such separation is a consequence of Plato and Descartes' separation of spirit and mind from the material world, an idea which may properly die with the science it spawned. But poetry and science alike require for their healthy practice both precision and an appreciation for subtlety and complexity. Thus, rather than simply making the ultimately boring statement that the universe is alive, I think what would be really interesting is to state just exactly what characteristics living things and the larger world in which we find ourselves might share.

To do this we need a definition of life.

In biology textbooks one reads that a living thing is something that shares the characteristics of metabolism, reproduction, and growth. There are, however, two problems with such a definition. The first is that it is not very insightful; it tells us nothing, for example, about why those characteristics are often found together, or about why things with these characteristics exist in the universe. The second problem is that any definition of life that may be applied to a single organism gives the false impression that a solitary living thing could exist in our universe.

In the first chapter, we examined the image of the warm, living Earth in the midst of a cold and dead cosmos, and we have since seen the extent to which this is an absurd idea. The same problems hold, even more strongly and clearly for the notion of a living thing in isolation. Certainly on Earth we never find a tree or an animal living alone on an otherwise dead island. Instead, we know of no place on, or even near, the surface of the Earth that does not contain life of some kind. Thus, the one planet we know which is not dead is not just a rock decorated with life in a few corners. It is a planet teeming with life.

Of course, we don't have access to any other life except on our own planet. But it is impossible that a single individual of any of the species with which we are familiar could live alone on any planet. It is almost equally difficult to imagine a planet populated by only one species. The reason is that each species plays a role in the great cycles that circulate material around the biosphere. We breath in oxygen and exhale carbon monoxide. Plants do the opposite, freeing the oxygen in carbon dioxide for our later use. We could not survive very long without plants for the elementary reason that all of the free oxygen now in the biosphere was rather recently produced by them.

This holds, not only for the oxygen we breath, but for the nutrients we eat, and for the other gases in the atmosphere: the nitrogen, carbon, and so forth. The life of any plant or animal cannot then be usefully conceived, except as embedded in the great system of the biosphere. This is particularly true if what we are interested in is a conception of life that could be useful for our project of understanding why life exists from the framework of physics and cosmology. A definition of life that focuses on individual organisms might be useful if we want to debate whether a virus or a coral shell is alive, but will not be of much help for this project.

The basic understanding that life on this planet constitutes an interconnected system must be considered to be one of the great discoveries of science, perhaps as profound as the discovery of natural selection. In a sense, natural selection tells us how a given species is related to the whole system over time, while ecology tells us how each species is connected to the others in the present. While there are aspects of our understanding of the system of the biosphere that are still unsettled , such as the Gaia hypothesis of James Lovelock and Lynn Margulis, there can be little doubt that it is necessary to understand life on this planet as an interconnected system to have any sense of what life is and why it is here. Thus, what is needed is a definition of life that focuses, not on individual organisms, but on the whole system of life as we know it on this planet. I will propose one such definition, which will allow us to discuss what the biosphere, the galaxy, and the universe have in common.

To understand the definition of life I will propose here, it will be necessary first to understand some of the basic ideas of the science of thermodynamics. This is the set of ideas that lead to the law which says that the entropy, or the disorder of a system, must always increase. Thermodynamics is something like a scientific counterpart of Kali, the great Hindu Goddess of death, for it is that part of science on which is grounded our understanding of the necessity of death and the temporariness of anything we construct. At the same time, thermodynamics is also the science of life, for it may tell us what are the necessary conditions for overcoming the tendency of things to run to disorder. Seen most generally, thermodynamics is the science of both the organization and the disorganization of things in the universe.

A simple example may illustrate the usefulness of thermodynamical ideas for understanding the place of life in the universe. Thermodynamics provides the simplest way to distinguish clearly between a planet with life, such as Earth, and dead planets such as Mars and Venus. The reason is that the atmosphere of the Earth is permanently in an enormously improbable state, very far from thermodynamic equilibrium.

What would the atmosphere be like were it in equilibrium? To find out one might do the following: Take the various elements that make up the atmosphere; seal them in a container at the temperature of the atmosphere and wait until chemical reactions bring the system to equilibrium. If one did this experiment with the Earth's atmosphere, one would find that the equilibrium state is very different from what we breathe; for example, there would be no free oxygen. This is because the mixture of gases in the Earth's atmosphere are very reactive, and could not last long in isolation. Oxygen, especially, is chemically very volatile, left alone it would burn with the carbon and nitrogen to make carbon dioxide, nitrous oxide, and water.

On the other hand, the atmospheres of Venus and Mars are very close to their equilibrium configurations. In spite of being at very different temperatures and pressures, their atmospheres contain very similar mixes of gases. Moreover, both their atmospheres are rather like what the Earth's atmosphere would be were it in equilibrium. It is then very striking that the earth's atmosphere has not come to equilibrium. We may conclude from this that there must be some outside agents that are maintaining the earth's atmosphere permanently in such an unstable state, far from equilibrium. There are such agents, they are the living things of the biosphere. For the great cycles that continually replenish the oxygen, carbon, and other elements of the biosphere are driven by the metabolic processes of living things. These cycles are the largest and most visible manifestations of life's domination of our planet.

This simple but powerful argument was, to my knowledge, first raised by James Lovelock, the British chemist who is the inventor of the Gaia hypothesis. As one may read in his books, he invented this reasoning when serving as a consultant to the NASA mission that was to search for life on the Martian surface. A consequence of his point of view is that it is not necessary to go to the surface of a planet to search for life. If a planet has life, one can see it easily by determining whether its atmosphere is in thermal equilibrium.

This means that we can search for the presence of life outside the solar system without leaving Earth. If we can get a spectrum of the light from a planet, we can see immediately if it has life or not by analyzing it to find the composition of its atmosphere. This may not be a pipe dream. Given that several planets have recently been detected around nearby stars, and given the power of telescopes that are now planned, it is not impossible that life will soon be detected in this way.

One might ask whether it is possible for a planet to have a little bit of life, perhaps a few plants and a few animals, or a few bacteria or algae, without so dominating a planet as to remake its atmosphere. It is possible that a planet might be found to be in such a condition for short periods of time. But evolution guarantees that life will eventually spread into every available niche. All that is required is that each living cell leave, on average, more than one progeny. Given only this, a population grows exponentially, so that in only a few generations it fills whatever niches are available to it. The basic mechanisms of natural selection thus imply that any planet with only a little life must be in a transient stage: any stable occupation of a planet by life must involve the whole planet. This is the essence of Lovelock's observation, which is that having life or not having life is a property of the whole planet and can be read easily from the mix of chemicals in its atmosphere.

As it is only one step from what we have said so far, let us detour for a moment to consider the full Gaia hypothesis of Lovelock and Margulis. Put most simply, this is that natural selection among early species of bacteria led to the generation of organisms that could, by the roles they play in the chemical cycles of the biosphere, regulate the contents of the atmosphere and the oceans. Not only could they develop mechanisms that regulate their internal chemistries, which every living thing does, they could develop mechanisms to regulate how much oxygen is in the atmosphere, what the average temperature on Earth is, and how salty the oceans are (to name just three examples). From this general principle, it is possible to make detailed hypotheses about specific mechanisms by which organisms could be acting to regulate the biosphere. Each of these is a good scientific hypothesis, subject to test and refutation, by observation. Because of this, the Gaia hypothesis is a proper scientific idea: the imputations of mysticism to it are certainly specious.

My understanding is that at the present time some of these hypotheses have stood up to observational test, although perhaps not yet to the point that the general idea can be considered to have been completely confirmed. Even so, I must confess I find it difficult to understand why this idea is so controversial. Let me mention several points that lead to its plausibility.

First, from a physical point of view, the stability of the conditions of the biosphere is impressive. The fact that the average temperature, level of oxygen, average salinity of the ocean and so forth, have been stable for hundreds of millions of years, while the cycling times involved are drastically shorter, is something that needs explanation. If one adds to this the fact that the energy received from the sun has changed by a factor of about 30 percent over the same period, there is certainly something to be explained.

As in similar cases such as the disks of spiral galaxies, the only possible explanation for the stability of such a non-equilibrium system is the existence of feedback mechanisms that control the rates of the various cycles involved. The only question is then the nature of these feedback processes. According to the Gaia hypothesis, they involve biological organisms; their critics must then maintain that they do not. It is possible that the temperature of the atmosphere and the composition of the atmosphere and ocean are maintained by feedback mechanisms that involve only dead things such as water and rock. But if this were the case what would need to be explained is why it happened that these processes keep the conditions of the biosphere in a state optimal for life. This is not impossible; after all there are undoubtedly many planets, and it might be that there were a few that accidentally were maintained by such processes in states hospitable for living things. But it seems, a priori, at least equally plausible that the feedback effects that maintain the environment in a state hospitable for life involve the living things themselves. At the very least this would be the more parsimonious and elegant explanation.

Critics of the Gaia hypothesis sometimes argue that it is impossible that microorganisms could evolve that would play a role in regulating the environment, for they would be at a selective disadvantage compared to others that did not take on this burden. While this criticism has been made by evolutionary theorists I have the highest respect for, it seems in the end to be unconvincing, and to rest on a failure to appreciate the extent that collective effects may play a role in natural selection, even within the standard neo-Darwinist paradigm.

This criticism relies on the assumption that the "fitness landscape," which describes the relative selective advantages and disadvantages of different genomes, is fixed for all time, independently for each species. But this picture, in which each individual species evolves independently to fit a preexisting niche, while useful for certain purposes, is too naive to address issues such as the Gaia hypothesis. Instead the niches—and the environment in general—are created by the species as they evolve. In such a situation, one species cannot mutate without there being an effect on the fitness of a number of other species. As a result, there may be collective effects in which two, several or many species evolve together. It is then natural to try to understand the evolution of the biosphere as the self-organization of a complex system with many interacting components. Very interesting studies of this kind have been made by Per Bak , Stuart Kauffman, and their collaborators. Some of the things they have found are relevant for the question of the Gaia hypothesis.

Most species are around for only a finite time before becoming extinct. In some cases they are replaced by a new species filling the same niche, in other cases the niche itself changes. What is important is that, in either case, there will be an effect on other species. For example, when a species becomes extinct, those that eat it are in big trouble, as are those that live in it, while those it eats are suddenly in a different situation. In many cases, this is all; only a few other species are affected by the extinction. But in some cases many species will be affected; for example, if that species produced a waste product, like oxygen, that is necessary for the life of many other species.

By modeling the effects of mutations and extinctions in such a complex network of relationships, Bak, Kauffman, and others have found that collective effects dominate the pattern of extinctions and successful mutations, so that the evolution of the biosphere can only be understood as a single, coupled system. The behavior of these models is fascinating. One sees that these systems spend most of their time in stable configurations, in which the different species are in balance. From time to time the balance is upset when a species becomes extinct. The result is that a wave of mutations and extinctions can sweep through the ecosystem, until some new stable configuration is found.

Such a wave of evolution can involve any number of species, for it will go on until a new stable balance has been discovered. Most of the time it will involve only a few species, as in the case of a competition between a predator and its prey. But in rare cases an avalanche of births and deaths of species may engulf the whole system, so that a significant portion of the species are changed by the time a new balance is achieved. These may be events in which the whole balance of the environment is renegotiated.

The Gaia hypothesis fits naturally into this view of evolution. During a very large avalanche of mutations and extinctions, there will be selective advantage for species that can regulate, and hence stabilize, the biosphere through the byproducts of its metabolic processes. When such a species arises it can contribute to the stability of the whole system. By doing so it may bring to an end an unstable period, in which the rate of extinctions has been very high; and by doing that it contributes to its own survival.

It must be emphasized that Per Bak, Stuart Kauffman, and their collaborators are not denying that individual species evolve according to the classical neo-Darwinist mechanisms. What is new is the idea that when one has many species that evolve together in an ecosystem, new collective effects emerge which determine things like the rates at which old species become extinct and new ones appear.

There is some evidence for this view of evolution. It implies that if we look at the evolutionary history of any particular species, there should be long periods when nothing happens, punctuated by moments of a great deal of change. This evolution by punctuated equilibria is what is actually observed, as pointed out some time ago by Niles Eldridge and Stephen Jay Gould. This new view explains why this is the case. For most of the time the system as a whole is stable, and each species is well matched to its niche. During such stable periods, mutations are less likely to succeed, and extinctions are less likely to occur. It is only at those moments when an instability has been introduced due to some rare event that several or many species evolve quickly until a new stable equilibrium is established. This theory also predicts that from time to time there will be large avalanches of changes which sweep through the system. This also is observed, for there are rare events in which many species become extinct and many new species arise in a very short time.

One of these events has been tied to the collisions of a large meteor with the Earth. This is not inconsistent with this new view; what is new is the prediction that, even without such an outside stimulus, large waves of extinctions may from time to time sweep through the system.

What Bak, Kauffman, and others have accomplished by taking a global view of evolution is the beginnings of a theory in which the Gaia hypothesis is only the rare and extreme case of a completely general phenomenon, which is that the evolution of the different species are coupled to each other. At one extreme are the rare events in which the whole system, having been upset, searches for a stable balance, and many species appear and disappear. During these times species whose activities affect large numbers of other species may arise. At the other extreme, and certainly much more common, are the classic cases studied by evolutionary theorists in which only two species evolve together, such as in predator-prey and host-parasite relationships.

Between these two extreme cases there will also be events in which a small number of species co-evolve together. It can happen, for example, that several species can evolve together to reach stable symbiotic relationships. Lynn Margulis has proposed that such events may account for the origins of both eucaryotic cells and multi-celled creatures. These theories fit naturally into the new picture, as intermediate cases between the very rare cataclysmic events and the more common cases of coevolution.

This example teaches us that when we are dealing with a self-organized system, we cannot afford to look only on one scale. If we look only at single species, we capture a necessary part of the truth, but we miss the collective effects that can determine the rate at which evolution is occurring. If we look only on the largest scale, we also capture part of the story, but we miss the basic mechanisms of evolution, which certainly happen at the level of the genomes of each species. It is only by finding a viewpoint which allows us to see that something interesting is happening on all the possible scales, from the smallest to the largest, that we are able to really comprehend the whole of any really interesting system.

This same lesson applies also to the question of how it is that living things come to exist. While the problem of the origin of life is still unsolved, it is clear that the first step must be to understand generally how it is that the chemistry of a planet can organize itself into units complex enough that natural selection can take place at all. For this we need a general theory of how self-organization can proceed from some random starting point to the origin of life.

The basis of such a theory must be in thermodynamics itself. To see why, we may begin by going back to the strange notion of the heat death of the universe. In fact, the nineteenth-century physicists who speculated that the law of increasing entropy requires the death of everything in the universe were wrong, and a good starting point for understanding why there is life in the universe is to understand why they were mistaken.

The main point is that the law of increasing entropy only applies to systems that are isolated from the rest of the universe, so that neither matter nor energy can enter or leave. In these cases, and in these cases alone, the law of increasing entropy holds. The surface of the earth, however, is not a closed system. Energy enters the biosphere all the time, primarily in the form of light from the sun. It leaves primarily in the form of heat, which is radiated into space. It is this constant flow of energy that makes life possible.

A flow of energy is essential for any process of self-organization, because on any scale there is a tendency for things to become disordered due to the effects of the random motion of the atoms. This is because of a simple fact, which is that there are many more configurations of atoms that are disordered than there are configurations that are organized in any interesting way. A collection of atoms, each moving about randomly, is much more likely to come to a disordered state than to an organized configuration, simply because there are so many more of the former. This is why the disordered state is the state of equilibrium, for once such a state is reached, it is very unlikely that the system by itself will revert to a more ordered configuration. This is the essence of the law of increasing entropy.

Living things depend delicately on the information coded into the sequences of their DNA and proteins. These configurations of atoms are very improbable; if you put a collection of atoms together in equilibrium, you are very unlikely to find a DNA molecule. Because of this, the random motions of the atoms constantly break the DNA and amino acid sequences, threatening the functioning of the living cell. The whole biological world would run into chaos in a short time had evolution not developed means to check and repair the information in the DNA of each living cell. Thus, a living thing cannot be static—it must constantly reconstruct and repair itself. This requires a constant source of energy.

It is generally the case that any process that resists the tendency of random motion to disorganize things requires energy. This is the reason that no isolated system can be self-organized; they instead run inexorably to randomness and equilibrium. Only in an open system, through which energy flows at a steady rate, is there the possibility that processes of self-organization may naturally arise that keep it permanently ordered.

It is also important that not just any input of energy will lead a system to organize itself. The energy must arrive in a form that is useful, and at a rate that is neither too fast nor too slow. Adding random energy in the form of heat only serves to increase the disorder of a system. If I put my computer in the oven, I give it a lot of energy, but the result is only to disorganize it. To organize itself, my computer requires energy in a particular form: a steady electric current in a particular range of voltages, entering through a particular wire. Similarly, if we did nothing but heat the biosphere we would kill it. Light from the Sun can be utilized by processes of self-organization because it arrives predominantly at a frequency that is just right for stimulating organic chemical reactions.

But energy must not only enter a system in the right form to organize it. Having done its job, most of it must be able to quickly leave. Some of the energy that drives the formation of complex molecules is locked up in the molecular bonds, but a great deal is transformed into heat. This heat must be able to leave the biosphere; otherwise the temperature would go up, disordering and killing everything. Thus, the biosphere requires not only a neighboring star to provide a flow of photons of the right frequency; it requires some very cold place where it can send the heat it generates. Luckily, such a cold place is nearby: space.

As we saw in the first chapter, life is possible on the Earth because the universe itself is not in thermal equilibrium. There are hot stars radiating into cold space. The biosphere can organize itself because it finds itself in the middle, and is able to harness the flow of energy to drive the processes that continually form the complex molecules that are necessary for life.

From the point of view of a physicist, this is the right way to understand what the biosphere is, and what it is doing in the universe. Life must be a special, perhaps extreme, example of a processes of self-organization that can spontaneously appear when there is a steady flow of energy through a system. This line of thought suggests that there should be a general theory of self-organization, which is based on the thermodynamics of systems that are far from equilibrium because they are infused by a steady flow of energy. Such a theory might tell us that in these situations the level of organization, rather than of entropy, increases steadily in time.

Of course, such a theory could not be the whole story. A system must have the potential for organization. Life is possible in our universe because carbon, oxygen, nitrogen, hydrogen, and the other organic elements may be arranged in extremely large and complex molecules. In a state of equilibrium these configurations are extremely unlikely, but they are still possible. What a flow of energy does is to change the game, so that improbably structured configurations become probable.

A few people have been looking for such a general theory of self-organizing systems, the best known of which are Per Bak, John Holland, Stuart Kauffman, Harold Morowitz, and Ilya Prigogine. Several interesting ideas, which may play a role in such a theory, have been put on the table, although my guess is that this is a subject in which the great discoveries are yet to be made. It is not an easy subject, as a few moment's reflection will show.

There is, for example, the problem that any general theory that explained on simple physical grounds why there is life on Earth might easily also explain why there is life on Venus and Mars. The energy flux reaching them is not all that different from that which reaches us. It must either be that Venus and Mars do not quite have the right conditions for life to begin, or that history has played an important role in determining where life appears and flourishes. We have no idea which is the case, and this ignorance tells us that we have as yet only an incomplete understanding of the conditions required for life to begin.

Even so, some interesting ideas have emerged, which seem to be steps in the right direction. Morowitz, Prigogine, and others have found that systems with a steady flow of energy through them do generally reach steady states that are very far from equilibrium, in which the distributions of elements, both spatially and with respect to chemical compositions, are very far from random.

To my mind, the work of Harold Morowitz is especially interesting. He finds that organization arises by the formation of cycles of chemical reactions. These cycles carry energy through the system from the mode in which it enters to the mode in which it exits. For example, energy may be absorbed by the biosphere when a photon catalyzes a chemical reaction which bonds two molecules together. The energy is stored in the bond until a further reaction releases it in the form of heat, allowing it to leave the system. The molecules must then cycle through a series of such reactions, back to their original form, otherwise they would be all used up, and the process would come to a stop.

Such cycles underlie the basic processes of the biosphere, and they involve all of the basic elements of life. Morowitz proposes that these cycles are more fundamental than life; and will arise in any chemical system that has a steady flow of energy through it. According to him, the formation of these cycles may have been the first step in the self-organization of the biosphere, occurring perhaps even before the evolution of the proteins and nucleic acids.

While necessary, the existence of such cycles cannot be enough to explain or to characterize life. For example, it is crucially important to know how fast the material is cycling through the system. Living things require not only that there be cycles of energy and the basic elements, such as carbon and oxygen, but that those cycles proceed quickly enough to maintain their delicate internal organizations.

In some sense, living things are a particular type of process which has emerged on top of the flows of energy and cycles of materials that characterize such open systems. Life perhaps might be seen to have evolved a way to ride these flows and cycles the way a surfer rides the flow of energy in water waves. But life has also taken over control of the flows of energy and materials that may have previously existed on the Earth.

To get to life, then, we need to add several more elements to the basic picture of how open systems may organize themselves. The first is that living organisms always have clear boundaries between themselves and the outside world. At the smallest level, each cell is surrounded by a membrane which allows it to control the exchange of energy and materials between it and its environment. This membrane is also necessary if the interior of the cell is to be maintained for a long period in a state far from thermal equilibrium, in a medium which is at, or at least much closer to, equilibrium. Were there no such barrier, diffusion and heat flows would quickly result in a mixing of the matter and energy between the inside and the outside of the cell, killing it. Instead, the cell is able to control exchanges between its interior and exterior to its own advantage, in order to maintain a high level of internal organization.

At the level of multicellular creatures, there is always a skin which serves the same purpose for the whole organism. And, at the level of the biosphere as a whole, the material that makes up the biosphere is kept isolated from the rest of the universe by the action of the Earth's gravitational field, while the atmosphere and ozone layer serve partly to control its exchange of radiation with the outside universe.

A second point is that if a system is to reach a steady state then there must be some stability, so that small bumps in the flows of energy or materials don't result in strong changes in the way material is cycled through the system. Generally, a process will be stable if there are feedback effects which tend to reverse the effects of small bumps and return the system to the original process. So we expect feedback to be a ubiquitous element of far-from-equilibrium, open systems, in which a stable configuration has been reached. This is certainly the case for living organism, as we know that the regulation of the internal environment of all cells and organisms is accomplished by feedback loops.

It will be useful if we pause here and give a name to systems that have all the characteristics of living systems we have so far discussed. I would like to call a *self-organized, non-equilibrium system* one which is:

> a distinguishable collection of matter, with recognizable boundaries, which has a flow of energy, and possibly matter, passing through it, while maintaining, for time scales long compared to the dynamical time scales of its internal processes, a stable-configuration far from thermodynamic equilibrium. This configuration is maintained by the action of cycles involving the transport of matter and energy within the system and between the system and its exterior. Further, the system is stabilized against small perturbations by the existence of feedback loops which regulate the rates of flow of the cycles.

It is clear that this definition may be applied to a variety of systems, including a living cell, a plant or animal, and the biosphere as a whole. It is also interesting to note, following our discussion of Chapters 9 and 10, that the disks of spiral galaxies seem to be systems of this kind. It is interesting to have a definition of a category of systems that can include both living cells and galaxies. I think it is a nontrivial fact about the world that such two widely disparate systems share so many characteristics. However, we should not make the mistake of believing that they therefore share all characteristics—a galaxy is a self-organized structure, but it is not alive. For something to be alive it is certainly necessary that it be part of a self-organized, non-equilibrium system. But no definition of life could suffice that ignored the role that information and control play in the workings of a living cell.

To completely characterize life as we find it here on Earth, we must add three more points to the definition, which are each connected with the role played by DNA and RNA in living cells. The first is that in a living cell the rates at which its chemical processes take place are controlled by enzymes, which are proteins. The second is that the synthesis of the enzymes is made and controlled by information that is coded symbolically in the structures of certain nucleic acids. The third is that the cell can reproduce itself, and that, when it does so, the nucleic acids coding it also reproduce themselves.

We may then make the following definition: *A living system is*

A   a self-organized non-equilibrium system
    such that
B   its processes are governed by a program which is stored symbolically
    and
C   it can reproduce itself, including the program.

What I think most recommends a definition such as this one is that it grounds the existence of life in physics—and in the right physics—far from equilibrium thermodynamics. At the same time, it makes clear to what extent living things have properties which, at least at the present time, cannot be understood purely in terms of the general theory of non-equilibrium systems. The definition implies that living things are among a category of systems all of which have a capacity for self-organization, but most of which do not achieve the fantastic levels of organization and structure that characterize living things. So all living cells, as well as all multicellular plants and animals satisfy this definition. At the same time, while spiral galaxies may satisfy part A, they do not satisfy the rest of the definition.

It is interesting to ask whether the biosphere as a whole satisfies the whole definition. We have seen that part A is satisfied. Part B is really equivalent to the basic statement of the Gaia hypothesis, which is that there are feedback effects that control the climate and the constituents of the atmosphere and ocean which involve microorganisms and are thus under the control of the genetic programs of those organisms. Part C is interesting. At first, we might be inclined to say that the biosphere as a whole does not reproduce itself. On the other hand, there are proposals to "bring Mars to life" by introducing microorganisms and plants that would have the effect of constructing a biosphere on that planet. This has, of course, been a staple of science fiction for years, but given that it might be contemplated seriously, perhaps we should say that there is at least a theoretical possibility that the biosphere could reproduce itself—with us acting as the agents of its reproduction. If this happened I think we would have to say that the biosphere satisfied the conditions of the definition of a living system.

What about the universe as a whole? While it may be tempting, I don't believe we should try to stretch analogy to say that the universe is alive. As to the first part of the definition, it may very well be reasonable to regard the whole universe as a self-organized system. But to push the analogy to the point that the universe fits the definition of a living system, we would have to regard the laws of nature themselves as a program. I don't think that this could be sustained. For one thing there is, as far as we know, no sense in which the laws of nature could be represented symbolically, as something analogous to a computer program.

One characteristic of a computer program is that it must be possible to specify it precisely by a finite amount of information. Because of this, such systems are portable, in the sense that they can be realized arbitrarily well on any computer, given only that the computer has enough memory. Because of the discrete nature of the genetic code, it is the case that the genetic information in a living organism could be represented as a program on a computer. It is not at all clear that the same is true of the laws of nature. It is true that small systems can, to very approximate degrees, be modeled on a computer. While this necessarily involves some approximate representation of physical laws in terms of a computer program, the fact that this is possible does not stand as an argument that the laws of nature themselves may be represented to arbitrary accuracy as a computer program. It may turn out in the end that the laws of nature are representable by an algorithmic system, but I do not know any reason why this must be so. Furthermore, there are very interesting arguments, such as those raised by Roger Penrose in his recent books, that this should be impossible. For this reason I think it would be, at best, premature to try to force the definition of a living system onto the universe as a whole.

With a definition of life available to us, we can also return to the question I raised in the first chapter: Why the conditions of the universe are such as to allow for hospitable living systems in it? We may divide this question into parts, corresponding to the different parts of the definition. Starting from the last two points, the existence of all of the intricate machinery which makes the coding and expression of discrete genetic information possible is due to the fact that carbon chemistry allows an enormous range of complex molecules. But we have already seen how carbon and its complex chemistry plays a necessary role in the star formation process. It is thus fascinating to note that a universe that has the capacity for efficient star formation is already going to have the basic ingredients necessary to turn self-organized non-equilibrium systems into living systems.

We may then turn to the first part of the definition and ask what conditions are necessary for the universe to contain self-organized, non-equilibrium systems. The answer is that either their existence is transitory, so that sooner or later the whole universe will come to equilibrium, or the universe as a whole must itself be a self-organized, non-equilibrium system. The reason for this is that it is impossible to have a self-organized, non-equilibrium system which exists permanently inside of a larger system which is itself in thermal equilibrium. It is not hard to see why. Part of the definition of a self-organized, non-equilibrium system is that it has a flow of energy through it. The energy enters the system at one point from the outside, which we may call the source, and leaves at another, which we may call the sink. Now, it follows from elementary ideas about heat that the source and the sink must be at different temperatures; in particular the source must be hotter than the sink. This is because of the simple fact that heat flows from hot regions to cold regions.

This means that the source and the sink cannot themselves be parts of a single system in thermal equilibrium because, if they were, they would be at the same temperature and no heat would flow. As the source and the sink are parts of the environment surrounding our self-organized, non-equilibrium system, this means that the environment cannot itself be in equilibrium.

This is the case with every living organism on Earth. We live because we can take in energy that is at a higher temperature than the heat that we relinquish to our environments. For plants, this energy comes from the Sun; for animals, from the molecular bonds of living tissue. It is also the case with the biosphere as a whole, which can exist because the Sun is much hotter than space into which it radiates heat.

But there is more to it than this. We have not yet taken into account one element of the self-organized, non-equilibrium systems, which is time. Part of the definition of a non-equilibrium living system is that it must maintain its state of self-organization for very long times. This means that not only must the source and the sink be at different temperatures; these temperatures must not change too suddenly. Otherwise the organization of the system would likely be destroyed. This is again something that is true both on the level of individual organisms and on the level of the biosphere as a whole. It is necessary that the non-equilibrium environment in which they reside maintain conditions that are stable—or only slowly changing—for arbitrarily long periods of time.

This means that it is not sufficient that the environment be an arbitrary non-equilibrium system, because in such systems the fluctuations in temperatures, densities, and so forth are typically very large. There must then be some mechanisms that stabilize the conditions of the environment. This will be accomplished if the environment itself is a self-organized, non-equilibrium system, for these are systems that maintain constant, stable conditions over long time scales.

This is certainly the situation of living organisms in the Earth. But what about the biosphere itself? Is there a sense in which we can say that the biosphere is situated inside a larger self-organized system? In Chapters 9 and 10 we found that there are good reasons for why the galaxy in which we find ourselves might be considered to be a self-organized system. We are now asking whether this might be to some extent essential for the existence of life on Earth. Strange as it may seem, the answer to this question may be yes. In a spiral galaxy there are a lot of organic elements around; carbon and oxygen are common elements in the interstellar dust. That the Earth has generous quantities of these elements reflects the fact that our Sun and its planets were formed from a medium that contained them. To the extent to which the process of star formation is part of the process of self-organization of the disk of a spiral galaxy, the Earth inherited the organic elements necessary for its own self-organization from that larger self-organized system that formed it.

It then seems that our life is situated inside a nested hierarchy of self-organized systems that begin with our local ecologies and extend upwards at least to the galaxy. Each of these levels are non-equilibrium systems that owe their existence to processes of self-organization, that are in turn driven by cycles of energy and materials in the level above them. It is then tempting to ask if this extends further up than the galaxy. Must there be a non-equilibrium system inside of which sits our galaxy? Is there a sense in which the universe as a whole could be a non-equilibrium, self-organized system?

We see that there are, essentially, two choices. Either the universe evolves after some long time to a uniform equilibrium distribution, or it does not. If it does, we must at some point in the hierarchy encounter a system at equilibrium, in which case the time scale over which life can exist in the universe is limited to the period over which the largest system out of equilibrium survives. Unless this time scale is equal to the lifetime of the universe itself, the heat death must eventually come, and life must be viewed as a transient phenomenon in the history of the cosmos.

On the other hand, if there is no time at which the universe will come to equilibrium, then it might be useful to view it permanently as a self-organized non-equilibrium system. We might then want to say that life may exist in the universe, over its whole lifetime, because the universe, by being itself a non-equilibrium system, creates through its own processes of self-organization conditions that are hospitable to the evolution of life.

The question why there is life in the universe takes on a very different light in such a postulated non-equilibrium universe than it did in the old picture of an equilibrium universe. In the old picture, the existence of life is an anomaly, or at least an enormous improbability, which thus can only be the result of a statistical fluke. In the postulated non-equilibrium picture, the universe remains permanently in a non-equilibrium state. Such a state is a necessary condition for life to exist indefinitely in the universe. We see that in this picture living things share in some ways, and extend in other ways, the basic properties of non-equilibrium self-organized systems that seem to characterize the universe on every scale, from the cosmos as a whole to the surface of planets.

We will see in the next chapter the extent to which what we know about cosmology allows us to consider the universe as a whole to be a self-organized system. The evidence, as we will see, is presently inconclusive. But still, it is important to begin to investigate the question, as a great deal is at stake. To put it most simply, if it is possible to construct a new picture of cosmology based on non-equilibrium rather than equilibrium thermodynamics, it will give us a picture of a universe in which the existence of life might be comprehensible and natural. But, even more than this, the possibility of conceiving the universe, as a whole, as a self-organized system, in which a variety of improbable structures—and indeed life itself—exist permanently, without need of pilot or other external agent, offers us the possibility of constructing a scientific cosmology that is finally liberated from the crippling duality that lies behind Plato's myth. It is clear that if the natural state of matter is chaos, an external intelligence is needed to explain the order and beauty of the world. But if life, order, and structure are the natural state of the cosmos itself, then our existence, indeed our spirit, might finally be comprehended as created naturally, by the world, rather than unnaturally and in opposition to it.

TWELVE

THE COSMOLOGY *of* *an* INTERESTING UNIVERSE

Is it not one of the great blessings of life to live in a beautiful place? And does it not seem that beauty is so strong a need that many people will travel whatever distance their circumstances allow to stand and gaze on a beautiful scene, whether it is a vista of mountains or seashore or the panorama of a beautiful city? But, what makes what is beautiful, beautiful? What is it about these scenes that draws us to them?

I am not an aesthetitician, and I would not try to give a full answer to this question. But certainly part of what makes a scene beautiful is that there is so much to look at. Apart from the austere beauty of a desert or the cynical appeal of certain contemporary art, a beautiful scene will hold our gaze for so long, and so many times bear revisiting, because one sees at every scale, from the overall composition to the smallest details, so much that is both novel and harmonious with the whole.

If I think of the most beautiful places I know, I am constantly reminded of the ways in which pattern and harmony repeat over a whole ascending scale of relations. From a beach in Big Sur in California one sees first of all the great curve of the coast, the mountains ascending in the distance, then the waves of the ocean, rising and falling slowly and majestically as if the bay is breathing. But, approaching the beach these fragment into the smaller, more impatient and more insistent breakers, which after their climaxes become so many piles and streams of water, running to and fro. And one can stand for many minutes and watch the little patterns etched and then erased in the sand by the fingers of water that just reach the furthest part of the beach. The integration of pattern on so many scales continues in time as well as in space, as the incursions of the water on the sand move up and down the beach following the rising and falling of the tides. And, over much longer times there is change as well, so that returning from season to season one finds the color of the landscape altered by the effects of the many individual births and deaths of the leaves and the grasses that live there.

So part of the beauty of the scene is that over every scale, from the majestic to the tiny, and over any interval of time, from the second to the year, there is something happening, some harmony to notice or some structure being formed or erased.

It is the same also with a beautiful and ancient city like Verona, which was already a thriving center of culture and commerce when the Romans first came upon it. From a tower or a garden overlooking the city one sees the great curves of the river Adige, in the elbows of which are nestled the different parts of the city, built in different eras, by what were almost different civilizations. And what is most striking is the way the same red tile roofs cover such a variety of shapes and sizes of buildings, from the medieval churches and palazzi to the modern stores and office buildings. Descending, one comes to streets that curve gracefully, with the rhythmic patterns made by the balconies and windows blending harmoniously the styles of houses built over ten different centuries, until one stands before a door or gazes on a medieval wall, and sees there the carvings and the frescoes made by artisans long passed. And then one goes through the door into a gallery or a boutique to see what strange tastes the modern inhabitants of this ancient place now fancy.

Imagine, by contrast, those man-made landscapes that we find most ugly, the suburban deserts with every house simple and similar, the American shopping center, the great monoliths of soviet architecture, or the unfortunate office towers and hotels based on economized postmodern styles. Certainly, what most of these lack is a variety of interest and harmony occurring over a large range of scales, so that so many of them look like models or computer images of themselves. In most of these the planners have taken care only of how things look on one scale, so that with one glance one takes in all that is to be seen there.

The universe we live in is beautiful. And it is so at least partly for the same reason a beautiful landscape or a beautiful city is, because a multitude of phenomena are taking place on a vast array of different scales. So the question of why the universe is beautiful is, indeed, closely related to the question of why it is interesting. For as we have asked several times before, why is the universe not more uniform, not more like a gas evenly filling space without structure or organization or beauty?

Indeed, in our universe we not only find structure on a variety of scales, *we find structure on every scale we have so far explored.* To emphasize the point, let us imagine that we scan the universe, looking from the smallest to the largest scales we have been able to observe. What are the characteristic structures we see, out of which our universe is built?

At the upper end, the largest scales we have been able to probe are about half a billion light years, which is roughly 10^{59} times the fundamental Planck length. The smallest scale we have so far been able to probe is about one hundredth the diameter of the proton, which is 10^{18} Planck lengths. Thus, from the largest to the smallest phenomena we have yet studied, the known world spans forty-one orders of magnitude.

The smallest things we know of are quarks, electrons, and neutrinos. The quarks sit inside the protons and neutrons, which then combine into about a hundred different kinds of nuclei. One hundred thousand times larger than these are the atoms, which then combine into molecules that contain any number from two to millions of them. The hierarchies of structures in the living world then proceed from the level of the molecules to the levels of organelles, cells, creatures and ecosystems up to the level of the whole planet. From quark to biosphere we have passed through a continuous hierarchy of structures stretching twenty-four orders of magnitude. But we still have a long way to go.

The smallest stars are about twenty times the size of the Earth, while the largest are a thousand times bigger. Stars are organized into galaxies, whose characteristic size is 10^{38} times the scale on which we began. But between stars and galaxies are many intermediate levels of organization, such as the globular clusters; and we must not forget the structures in the interstellar medium that span scales from the dense cores that condense to single stars to the spiral waves of the whole galaxy.

Is this the end, or is there still more structure above the scales of the galaxies? For most of the seventy or so years since the existence of galaxies was established, cosmologists assumed that the they are randomly distributed in space. Only recently astronomers have been able to map the three dimensional structure of the galaxies. The surprising result is that there is a great deal of structure above the scale of individual galaxies. Great voids have been found where almost no galaxies are seen. These voids are as large as a hundred million light years, roughly a thousand times the size of a galaxy. Then, separating the voids there appear to be great sheets in which the galaxies are concentrated.

So far, astronomers have observed structure in the distribution of galaxies that are about as large as the scales they have been able to map. It is not known how far up in scale the structures extend. There is also no consensus about how this structure was formed, although this is a question under intensive study at the present time.

So, over all of the forty-one orders of magnitude that can now be probed, we see structure. Of course, there is some distance to go, at both the upper and lower end. At the upper end, the largest possible visible structure is the universe itself, whose scale, for the purposes of viewing, must be taken to be the distance a photon of light could have traveled in the time since the big bang. Amazingly, this is only a factor of a hundred or so larger than the scales we are currently probing. So, at the very least, the universe is structured up to one percent of its visible diameter.

There is some evidence about the structure of the universe on the largest scales that comes from the cosmic background radiation. This tells us that at the time that the radiation was produced, which was the moment the universe first cooled enough to be transparent to light, the universe was at the same temperature everywhere, up to an accuracy of a few parts in a hundred thousand. The present time is a very exciting one for cosmology, as it seems we are caught between the almost conflicting messages of these two great discoveries. There is structure in the distribution of the galaxies at scales which reach up to at least one percent of the radius of the observable universe. But when the universe first became transparent, it was at a uniform temperature to an accuracy of a few parts in a hundred thousand. The present task is to invent a history for the evolution of the universe that is consistent with all of this evidence.

The fact that we live in a universe with structures at so many scales is so commonplace, so manifest, that it is easy to miss its significance. After all, scientists, as well as everyone else, are not usually inspired to a sense of awe by our everyday surroundings. But awe is certainly what we should feel when we confront the manifold structures of our cosmos because a universe as structured as ours is extraordinarily unlikely to have come into being, had the parameters of physics and cosmology been chosen randomly.

Leibniz, in a passage that has often been misunderstood, speculated that our universe might be the most interesting, the most varied one, possible. We certainly do not know enough to know whether a universe could be imagined that had much more structure or variety than our own. But it is incredibly easy to imagine universes that are less interesting and less structured than ours.

For most of this century cosmologists have taken the view that the universe is inherently simple when viewed with a perspective so large that the individual galaxies are like the atoms of a gas. That view seems more and more imperiled, as structure on larger and larger scales is discovered. In recent years, the key problem in cosmology has been to understand how the galaxies, as well as the patterns that they make in space, formed out of a universe that seems, initially, to have been so uniform. At stake in this problem is the extent to which the structures of the cosmos can be understood as small bumps or perturbations in an essentially symmetric and uniform universe. At stake here also, as we saw in the last chapter, is the ultimate answer to the question of whether life is an intrinsic and necessary manifestation of the cosmos, or only a temporary brightening of an otherwise dead world.

The idea that the universe is simple has roots that go deeper than twentieth century cosmology, to the ethos and practice of physics itself. For the great pride of science—that nature is comprehensible—has often been taken by physicists to mean that nature is simple, when looked at in small enough pieces. More than this, the training of a physicist consists to a large extent of learning how to perceive in complex phenomena a few simple and dominant forces which may then be modeled and comprehended easily.

This is illustrated by a joke that I heard more than once as a physics student: A theoretical physicist moonlights by taking a job on a dairy farm, where his job is to study the processes of milk production and suggest ways for improvement. After a year's study and many calculations, he produces a report, which begins: "Consider a spherical cow of radius R and mass M, which ingests a steady stream of grass at a constant rate G, and produces. . . ."

The joke was funny, at least to physicists, because most of the tools that we learned as students to attack problems in the real world involved the construction of models which drastically simplified real phenomena to the point that they could be described by simple mathematical equations.

Perhaps the idealization of a cow to a sphere of a mass M seems laughable, but this is not very different from the way that theoretical cosmologists have pictured the universe since Einstein in 1916 first applied general relativity to it. The result has been a succession of what are called "cosmological models" in which the universe is idealized as a completely uniform gas of matter and radiation in thermal equilibrium. As such, it is completely described by a few parameters that describe the average distribution of matter and energy. The advantage of such a simple model is that there are only two things that a gas can do without violating uniformity: it can expand or contract ; and it can heat up or cool down. This results in a mathematical model with two variables corresponding to the volume and the temperature of the universe. Such models are easy to study, and they have provided the framework on which almost all of theoretical cosmology has been based, from Einstein's first paper on cosmology to the recent inflationary models.

As we look around us, we can see that the universe is not a uniform gas in equilibrium. But most cosmologists have believed that while the universe may have structure such as trees, stars, and galaxies, there is some scale above which the distribution of galaxies becomes uniform. From this traditional point of view, any structure that we observe in the universe must be considered a secondary effect, a kind of perturbation or ripple on what is essentially a perfectly homogeneous and uniform universe. To understand why modern cosmology developed as it did, emphasizing uniformity rather than structure, it is important to appreciate that the main outlines of the modern cosmological models were drawn during the first quarter of our century, during the period of transition between Newtonian and modern physics. As such, modern cosmology owes more to nineteenth century science than might at first be thought. The first cosmological theory based on general relativity was proposed by Einstein in 1916, which was not only before astronomers understood that stars were organized into galaxies; it was before it was generally appreciated that quantum physics would lead to the complete overthrow of the Newtonian world picture. The model, which is still the basis of modern cosmology was published by the Russian meteorologist Friedman in 1922. It was an improvement on Einstein's first cosmological theory, but it did not challenge the notion that the universe could be modeled as a homogeneous gas in equilibrium.

Two ideas from nineteenth-century science may have particularly shaped the expectations of the early twentieth-century cosmologists. The first is the mistaken idea that the university must necessarily evolve to a "heat death", after which it would be devoid of life or structure. A second reason for the prejudice that the universe is unstructured on large scales might be that this idea is a vestige of the old Newtonian cosmology based on absolute space. Absolute space is supposed to be absolutely uniform and, while it is invisible, in any cosmology based on it there is a tendency to assume that the matter is more or less distributed uniformly, mirroring the uniformity of absolute space.

When one reads the cosmological literature from the 1910s and 1920s, one gets the impression that the main question was understood to be which absolute space should replace that of Newton—should it be the one proposed by Einstein, the one of De Sitter, or perhaps the one of Friedman? It does not seem to have been appreciated then that the essence of general relativity is that space itself has no fixed structure, but is dynamical, so that its geometry is always changing in time.

There seems to be nothing in general relativity itself to prevent the universe from having structure on every scale. The traditional expectation that there will be a scale above which the universe appears uniform may thus be criticized as arising from an idealized view of cosmology that has its roots in Newtonian physics. We may then consider an alternative point of view, which is that the universe consists of a hierarchy of structures on larger and larger scales, extending up to the scale of the whole universe. This suggestion was made in the early years of the century by Charlier and was championed as well by the French astronomer de

Vaucouleurs, who was one of the first to propose that the galaxies are organized into clusters, which are in turn organized into superclusters. Recently, the idea that structure extends upwards to all observable scales has been argued by several physicists such as Coleman and Pietroneiro.

Given that both alternatives are conceivable, whether the universe has structure on all scales, or only up to a certain scale above which it appears uniform, can only be settled by observation. At present, the evidence favors the traditional point of view. The most recent maps of the distribution of the galaxies in space do seem to suggest that there is a scale above which the universe is uniform. However, new surveys are also underway that will greatly extend the scale out to which we have detailed maps of the distribution of matter in the universe. Given this, we may expect that this question will be definitively settled in a few years.

If the traditional view prevails, we face two very interesting questions: Why is there a scale above which the universe appears to be uniform? And, how did the structures that we see form? A simple answer to these questions is that the universe was initially almost completely symmetrical. It could not have been precisely symmetrical, for then structure would never have formed. But if the initial symmetry was marred so that there was some distribution of regions of slightly greater or lessor density, then these might have grown as the universe expanded, resulting in the structures that we see today.

This is more or less the standard view in present day cosmological theory. If it works it will explain why the universe seems symmetrical on very large scales, because the structures grow bigger in time. By any given time there will be a scale above which structures have yet to form. But it leaves open three further questions: Why was the universe initially so close to perfectly symmetrical? What caused the initial deviations from symmetry? And, how have these grown to form the structures we see?

The first two questions are very deep, and we are not very sure of the answers. Certainly, the initial state of the universe was very special, and this specialness is akin to the specialness of the parameters of elementary particle physics. The solution may lie in some cosmological mechanism, like natural selection. Whether this can work is a problem for the future.

The problem of how the structure formed—given small inhomogeneities in a symmetrical universe—is one we should be able to answer, given present knowledge. A lot of work has been devoted to it by theoretical astronomers in the last twenty years. Detailed models have been invented and studied on supercomputers, and the results are quite impressive. They do succeed at describing a universe in which galaxies form amid structures that extend up to much larger scales, and when the models are run, they produce pictures of a universe that is remarkably similar to ours.

However, there are still open issues, and the problem of how structure formed in our universe is not yet settled. There is a lot of freedom in the choice of these models—the properties of the dark matter that comprises at least 90 percent of the matter may be specified freely—as most of it has not been observed. Also, the models do not work quite as well as we would like in reproducing all of the observations, which include not only the distribution of galaxies, but the measurements of the distribution of the black body radiation from decoupling. It has also become clear that structure formation is not a simple process and that strongly non-linear effects play a major role. While these effects may be modeled in a supercomputer, we are lacking a good conceptual framework to understand what is happening in general terms. It is also possible that highly energetic processes involving supernovas and star formation may play an important role in the formation of the galaxies.

The picture that structure forms from small perturbations in an otherwise symmetric universe may be true in outline, but even if it is true it is incomplete. There must be some good reason why a universe that starts off as symmetrical and simple ends up developing such complexity, and if we are honest, we must admit we do not yet know what it is. In the remainder of this chapter, I will describe a set of ideas that might contribute to the development of a fuller picture of how structure formed in our universe. They are based on developments that have been as important for the twentieth century understanding of matter as the study of equilibrium thermodynamics was for the nineteenth century. The systems nineteenth- century thermodynamics studied show no structure whatsoever above the levels of the individual atoms. The systems that have more recently been the focus of study in thermodynamics are quite different; they are systems *whose defining characteristic is that they show structure on every scale.* These are called *critical systems.*

Many things in the universe come in particular sizes. All house cats are more or less the same size, as are all guitars, stars, galaxies, elephants, planets, trees, and automobiles. For each of these there is a reason they do not come very much larger or smaller than the standard size. These reasons make, in some cases, quite fascinating stories, but these are not the stories I want to tell here. Instead, I want to speak about those things in the world that do not come in any particular size. Such things are perhaps not as common as the things that come in definite sizes, but there is something important that we may learn from them.

For example, the features of a coastline have no definite size. If one looks at a jagged coast from an airplane, one sees a pattern in which the coastline moves in and out in an apparently irregular way, with peninsulas and bays interspersed with smaller features. There is no characteristic size for a bay or a peninsula; they come in all sizes from the continental to the intimate. Looking from an airplane window, one can see features that extend from the horizon to the smallest discernible detail.

As your airplane descends, the larger features pass out of view over the horizon, but it becomes possible to see still smaller ones. The result is that the characteristic in-and-out pattern of a coastline looks more or less the same, no matter from which altitude it is viewed. Indeed, one cannot see the largest features until one is looking from space, and these are limited only by the size of the Earth itself. And the smallest features are those fingers of water moving up and down the sand that can be seen only when one stands on the beach.

Another example of something which has no scale is the pattern of peaks and valleys in a mountain range. It is true that there is a characteristic scale for the largest mountains. This is set by the competition between the strength of the gravitational field of the Earth and the strength and weight of the rocks that form a mountain. But at every smaller scale, down to the size of a few meters, one can see features in a mountain vista.

One can also find examples of things that have no characteristic size in our social and economic organizations. Human society is itself such a thing, for one finds social organizations on every scale from a marriage up to the billions involved in the United Nations, the telephone system, or the audiences of Hollywood movies.

Coastlines and mountains are examples of a particular type of patterns which have no characteristic scale, called fractals. These are defined to be patterns that repeat their general features over a wide range of scales so that, looking at them with any magnification, one sees essentially the same pattern. Fractals are often produced by critical systems, because a system that has no particular scale will look the same when examined at any magnification. This means that any patterns they produce will also look the same when viewed at any magnification.

What kinds of systems produce such patterns? One class, which is very well studied by physicists, consists of systems that are undergoing changes of phase, such as the change from ice to water, from a liquid to a gas, or from a magnetized to an unmagnetized piece of metal.

It is not hard to understand why such a transition might involve phenomena that have no particular scale. The reason is that nothing at all happens to the individual molecules when such a transition takes place. What instead happens is a rearrangement of their positions and motions. This rearrangement must take place over the whole material, which is many orders of magnitude larger than the individual molecules. This is possible because at the precise temperature and pressure that the transition occurs both phases may exist simultaneously, just as ice floats in water at the freezing point. In such conditions, clusters of atoms of both phases will form which are of any size, from a few atoms to a region that can be seen with the eye. If one looks with a microscope at the patterns made by the two phases as a transition is taking place, one finds a distribution spread out over a large range of scales, as in the case of fractal patterns such as coastlines and mountain ranges. This is the reason that ice crystals and snowflakes have such beautiful patterns; they have features on all scales.

Our universe, as I've emphasized, is a critical system with structure spread across many scales. More than one physicist has been led to ask whether this might be the result of a phase transition. Might the beautiful patterns of the galaxies, for example, have been formed in some process analogous to that which forms snowflakes?

To answer this question we must think about an obvious but important fact about phase transitions, which is that they require certain precise conditions. Water only boils at a particular temperature, and only freezes at another. The universe is believed to have cooled from extraordinarily high temperatures as it expanded. It then seems possible that, as it cooled, it may have passed through one or more changes of phase. A number of physicists and astronomers have speculated that this may have led to the formation of structures distributed through the universe in something like a fractal pattern, with no definite scale. These might then have been the seeds from which the galaxies formed. Such a theory would explain both the existence of galaxies and the fact that their distribution in space seems to have structure over a wide range of scales If it worked, it would be a concrete realization of the idea of explaining the structure of the universe in terms of critical systems.

There are, however, many things in nature that have no definite size, but whose formation does not seem to be associated with phase transitions. Actually all of the examples we have so far given are of this class. These are all different from phase transitions in that they seem to form spontaneously, without the need for the temperature to be tuned to a precise value.

We may ask how such critical systems can be formed. One thing is clear: this cannot happen in any system in thermodynamic equilibrium, for nothing very interesting happens in such systems apart from phase transitions. The formation of structures over a wide range of scales, apart from a change of phase, is then something that can happen only in systems which are far from thermodynamic equilibrium. Non-equilibrium systems are, as I described in the last chapter, central to our understanding of self-organization, and life itself, and they have been studied for a long time. But it is only recently that someone noticed the significance of the fact that the structures they generate are often of no particular size. This someone was Per Bak, and with this realization was born the notion of a *self-organized critical system*. It is one of those ideas that, when you first come upon it, you can almost hear the sound of the key turning in the lock.

In contrast to the transitions between phases, which cannot happen unless certain conditions are precisely met, a *self-organized critical system* can occur spontaneously. All that is required is a system that is not in equilibrium because there is a flow of energy or materials through it. In such a case, as Per Bak and his collaborators discovered, there are general mechanisms that form structure over large arrays of scale.

One reason why self-organized systems are often critical systems is that the process of self-organization is hierarchical. This is because the process by which the components of a system become interrelated through the formation of cycles can, once it is begun, repeat itself on a larger scale. Thus, the system formed by the original components become the components in a still larger system. In a sufficiently complex system one finds many layers of organization, each of which is tied together by the cycles and interrelationships that characterize stable self-organized systems. In the most complex system we know—the biosphere—there are at least eight such levels of organization: the organelles of cells; the cells; the organs of a body; a plant or animal; a community of like organisms; a local ecosystem; a larger system such as a continent or ocean; and the biosphere as a whole. There are similarly many such levels in human society. Thus, a city has many interlocking levels of organization, which are reflected in the many scales over which its life may be viewed.

Given its universal applicability, it is very tempting to ask if self-organized criticality might be the key to the formation of structure in the universe. This is a question that has only been asked recently, and very little detailed work has so far been done on it. But it is very tempting to believe that the answer is yes. The reason is that systems that are held together by gravity do in fact have a natural tendency to become organized over time, in a way that is reminiscent of self-organized criticality.

A fact of the first importance for the question of structure in the universe is that, quite generally, systems held together by gravity do not share the general tendency, dictated by the law of increasing entropy, to evolve over time to uniform and unorganized configurations. This is a consequence of the fact that the gravitational force is universally attractive and has infinite range. When one has a system of many stars it is always possible to let some of them be drawn more closely together by their mutual gravitational attraction. This releases some energy which can be taken up by the other stars, which makes them move faster. Thus, as time goes on the system separates into different components. One group of stars will fall towards the center while others gain the energy they lost and move further out. So, unlike systems that come to equilibrium, a system held together by gravity tends to become more and more heterogeneous as time goes on. To put it in a provocative way, such systems develop variety; they become more interesting, rather than more homogeneous, as time goes on.

It is true that systems held together by gravity sometimes reach what are called quasi-equilibrium states, which are stable for relatively long times. Examples of such states are the spherical configurations characteristic of elliptical galaxies and globular clusters of stars. But such states are never the states of maximal entropy or probability, and in the long run they are all unstable.

The tendency for systems held together by gravity to organize themselves may be understood directly in the context of the self-organized non-equilibrium systems we discussed in the last chapter. We saw that such systems require that there be a steady flow of energy through them, which over time can drive processes of self-organization. But all systems held together by gravity are, to some extent, systems of this kind. This is because there is an enormous supply of energy available to them, which is their gravitational potential energy. This is how galaxies work, and it is happening, to some degree or other, in all systems held together by gravity. Thus, every such system is, to some extent, a self-organizing non-equilibrium system.

The key question for cosmology is whether this applies to scales larger than galaxies. Might clusters of galaxies be considered to be self-organized systems? Might the whole of the universe be such a system? At the present time we do not know enough to decide if such a process of self-organization, or the occurrence of a phase transition as the universe cooled early in its history—or something else altogether—is the right explanation for the structures we observe in the large-scale distributions of the galaxies. But regardless of the final outcome, it is clear that the effort to explain why we live in a universe of galaxies is driving cosmology from a nineteenth-century view of a static and dead world to a modern view of a dynamic, non-equilibrium, structured universe.

However, while we are speaking of the organization of the universe on large scales, we are forgetting an important lesson from the earlier parts of the book. This is that the structures at the largest scales owe their existence to the improbability of the parameters in the laws of physics. In particular, it is the large ratios in the parameters—such as the fact that the mass of the proton in Planck units is 10^{19}—that are responsible for the occurrence of enormous structures, such as stars and galaxies.

Thus, any attempt to use ideas from the study of critical systems to account for the structure in the universe must explain where these enormous ratios come from. To complete the story of how we might understand the universe as a whole as a self-organized critical system, we must return from cosmology to the problems of elementary particle physics we discussed in Part One.

In most universities this is a walk between nearby floors, or at worst neighboring buildings, from the astronomers to the elementary particle theorists. If we make this trip we are rewarded with a pleasant surprise, for the concept of a critical system is very familiar to the elementary particle theorists. Since work more than twenty years ago by a group of wonderful theorists led by Kenneth Wilson, the concept of a critical system has come to play an essential role in the physics of the elementary particles. It provides an enlightening perspective for talking about the puzzles of elementary particle physics.

Critical systems are relevant for the physics of the very small because of the basic fact that there is an enormous gap between the scale of nuclear and atomic physics and the fundamental Planck scale. As a result, it is possible to see the system of elementary particles as a single critical system that is in some ways analogous to a material undergoing a change of phase. In each case there is an enormous separation between a tiny fundamental scale and a much larger scale, at which interesting phenomena are observed. In the case of a material undergoing a phase transition, the fundamental scale is the scale of atomic physics; the larger scale is the everyday macroscopic world in which we can see, with our own eyes, the material melting or boiling. In the case of the laws of elementary particle physics the fundamental scale is the Planck scale, while the large scale is the nuclear and atomic domain. Just as boiling water produces bubbles, which are much larger than the atoms, the hypothesized fundamental laws of physics acting at the Planck scale produce protons, electrons, and nuclei that are twenty orders of magnitude larger.

The concept of a critical system has played an essential role in the modern understanding of elementary particle physics over the last twenty years. However, what has not usually been stressed is that with this there arises a possibility of a form of explanation for elementary particle physics that is mechanistic without being naively reductionist. The reason is a remarkable fact about critical systems that has been central for our growing understanding of them.

What has been discovered is that many physical systems behave in certain simple and universal ways when they are in the process of a phase transition. This means that it is much easier to predict how a system will behave when it is undergoing a change of phase than it is normally, for one need know almost nothing about it. The reason for this is closely connected to the fact that such systems exhibit phenomena that have no characteristic scale. This means that there is nothing special about the atomic scale; seen with any magnification, the system looks the same.

Kenneth Wilson, one of the most influential living theoretical physicists, developed this insight into a powerful tool for the study of critical systems known, for technical reasons, as the renormalization group. Using it, one can make detailed predictions about how any substance will behave during a change of phases, in many cases without knowing anything about the atoms or molecules out of which the thing is made.

Of course, in the case of a real material, the atoms obey some fundamental laws, even if this information is not needed to understand what happens in a phase transition. But, in recent years the suggestion has arisen that the same picture can be applied to the supposedly fundamental laws themselves. Several physicists have been asking whether it might be possible that some properties of our world, conventionally thought to require explanation by recourse to a fundamental theory, can actually be explained by the hypothesis that at very small scales the world is a critical system. One of these, whose ideas I think are particularly interesting, is a Danish physicist called Holger Nielsen.

What Nielsen imagines is that the whole cosmos is just at the point of a transition between two phases. He and his colleagues, such as Don Bennett, try to demonstrate that many of the observed properties of the elementary particles arise simply from this fact, independently of whatever the fundamental laws of physics are. They want to say that, just as bubbles are universally found in liquids that are boiling, the fundamental particles we observe may be simply universal consequences of the universe being balanced at the point of a transition between phases. If so, their properties may to a large extent be independent of whatever fundamental law governs the world.

When one first hears this idea one wants to ask, "O.K., but still, what is the substance out of which the universe is made? Even if it doesn't matter, there still must be something substantial from which the protons, neutrons and electrons arise." Nielsen's answer is to insist that we shouldn't try to ask irrelevant questions. If all theories will manifest the same universal behavior when their parameters are tuned to describe a phase transition, we have no means to answer this question. The only way we could tell what the fundamental theory actually was would be to probe the Planck scale. But that is impossible, at least for the foreseeable future. To make the point, Nielsen proposes that if one fundamental theory is as good as another; what we should do is to choose the fundamental theory completely randomly. He then proposes a physical theory called *random dynamics*, in which the laws of nature are to be picked entirely by chance.

Although it may sound frivolous, this proposal is entirely serious. In fact, Nielson and his colleagues do claim some successes for the hypothesis of random fundamental dynamics. Among them is the fact that all the fundamental interactions must be gauge interactions, of the type described by Yang-Mills theory and general relativity. This means that the world would appear at large scales to be governed by these interactions, whether or not they are part of the fundamental description of the world at the Planck scale.

This last claim is, in fact, rather well accepted among particle theorists. It has been independently confirmed by Steven Shenker and others. But there are further claims made by Nielson and his colleagues on which no consensus has so far been reached. They claim to be able to demonstrate that the hypothesis of random dynamics is sufficient to show that matter is composed of particles much like electrons and quarks, that the number of dimensions of space is three, and that the laws of quantum mechanics must hold. They also claim success in understanding the structure of the standard model, as well as the value of three of its parameters, those associated with the strengths of the strong, weak, and electromagnetic interactions.

It is certainly much too early to judge the eventual success of Nielsen's program of random dynamics. However, like the hypothesis of cosmological natural selection, its very possibility demonstrates that the naive reductionist program is not the only way that the parameters of the standard model of elementary particle physics might be explained.

But there is one question that we must ask right away, if we are going to take such an idea seriously. If the properties of the universe are, at least to some extent, determined by the fact that the universe is in a state of transition between two phases, what keeps the universe in that state? We don't normally find systems balanced at the point of a phase transition. For two phases like water and ice to coexist some parameter, like the temperature, has to be tuned just right. So this line or reasoning, rather than allowing us an escape from the problem presented by finely tuned parameters, returns us to them.

It must be said right away that this is not only a problem for Nielson. It is a problem that is shared by virtually all approaches to elementary particle physics. The reason is that it is very unnatural that any quantum theory of fields predicts the existence of stable particles whose masses are spread out over a wide range of scales. Virtually any theory of the elementary particles which is based on quantum mechanics has trouble explaining why there are protons, neutrons, and electrons in a world in which the fundamental scale is twenty orders of magnitude different from theirs. It must also be said that this is a problem as much for string theory as it is for the grand unified theories. The question of how it comes about that our universe has phenomena spread out over such a variety of scales is a problem for elementary particle theory in general.

In Part Two, I proposed one possible answer to this question, which is that the parameters are chosen by a process analogous to natural selection. Whatever the ultimate fate of this theory, it must be stressed that it does give a natural answer to the question we have been discussing. It then stands as a model for the kind of explanation that we may hope eventually works to explain the basic fact that our universe is structured hierarchically. In general, mechanisms of self-organization manifest themselves in the development of critical systems with structure spread out over a great many orders of magnitude of scale. Perhaps it is only accidental that our universe has this characteristic. But for the present it may be said that one thing that favors the general hypothesis that the parameters of the laws of physics are set by a process of self-organization over the hypothesis that they are set uniquely by some fundamental theory is that the first class of theories is naturally able to explain the existence of large hierarchies in the scales of a system, while the second is generally not.

Let us imagine that such an explanation will in the end turn out to be right. We will then be able to say that the beauty of the universe will have, in the end, a similar origin to the beauty of coasts, mountains, and cities. Would this not be a pleasing answer to the question of why it happens that we find ourselves in a universe that is both interesting and beautiful? Indeed, perhaps the problem of scales in fundamental theory is not too different from the problem of scales in art and architecture. For in the end, what are the differences between those human constructs that we find beautiful and those we find ugly? What accounts for the difference between the beauty of an ancient city and the ugliness of a modern shopping center? Certainly, these esthetic differences reflect the different ways in which these places were made. For what the beautiful city and the beautiful seascape have in common is that they were built up over long periods of time, by many different forces and, in the case of the city, by many different peoples. In each case, one cannot say that a beautiful city was planned to be beautiful, but at the same time the resulting effect is nothing if not the intention of the inhabitants of the city.

In contrast, it is exactly the ugly places made by our civilization that were planned. This is not to say that design is not necessary. But it is to say that a process of evolution in which different designs and different intentions interact, leading to the discovery of harmonious compromises, can produce something more beautiful than the design of any single planner.

We look around and see that our universe is beautiful and that, with its enormous variety of phenomena spread out over every scale from the nuclear to the cosmological, it resembles more the ancient city than the modern shopping center. Could this beautiful universe be the result of the construction of a single planner? Certainly, it is difficult to imagine any human planner choosing the laws of nature carefully enough to result in a universe with such a variety of phenomena. Indeed, as we saw in earlier chapters, to choose the laws of physics so that such a variety of phenomena results, let alone so that the universe is not simply a gas in equilibrium, requires that many parameters be finely tuned, some to as many as sixty decimal places. Of course, God is imagined to have infinite power, and we cannot limit what might be possible for him. But exactly for this reason, if we believe in the picture of a universe made by the providential choice of an eternal and fundamental theory, must we not also believe in God?

On the other hand, perhaps for the first time in human history, we know enough to imagine how a universe like ours might have come to be without the infinite intelligence and foresight of a god. For is it not conceivable that the universe is as we find it to be because it made itself; because the order, structure and beauty we see reflected at every scale are the manifestations of a continual process of self-organization, of self-tuning, that has acted over very long periods of time? If such a picture can be constructed, it may be possible to understand the fact that the universe has structure and phenomena at every scale, not as some enormous accident or coincidence requiring the fundamental theory to be so finely tuned, but merely as evidence that the maker of the universe is nothing more or less than the random and statistical process of its own self-organization.

If you are asked what you mean by the necessity of the laws of nature (that is to say, by the necessity of the most necessary relations), you can legitimately respond only by laying out the substance of your cosmological and other scientific ideas. People who appeal to fixed conceptions of necessity, contingency, and possibility are simply confused.

— Roberto Mangabeira Unger, *Social Theory*

THIRTEEN

THE FLOWER *and the* DODECAHEDRON

From Pythagoras to string theory, the desire to comprehend nature has been framed by the Platonic ideal that the world is a reflection of some perfect mathematical form. The power of this dream is undeniable, as we can see from the achievements it inspired, from Kepler's laws to Einstein's equations. Their example suggests that the goal of theoretical physics and cosmology should be the discovery of some beautiful mathematical structure that will underlie reality.

The proposals I have been discussing here, such as cosmological natural selection or the idea that processes of self-organization may account for the organization of the universe, go against this expectation. To explore these ideas means to give up, to some extent, the Platonic model of physical theory in favor of a conception in which the explanation for the world lies in the same kind of historical and statistical explanation that underlies our understanding of biology. For this reason, if we are to take these kinds of ideas seriously we must examine the role that mathematics has come to play in our expectations of what a physical theory should be.

It is mathematics, more than anything else, that is responsible for the obscurity that surrounds the creative processes of theoretical physics. Perhaps the strangest moment in the life of a theoretical physicist is that in which one realizes, all of a sudden, that one's life is being spent in pursuit of a kind of mystical experience that few of one's fellow humans share. I'm sure that most scientists of all kinds are inspired by a kind of worship of nature, but what makes theoretical physicists peculiar is that our sense of connection with nature has nothing to do with any direct encounter with it. Unlike biologists or experimental physicists, what we confront in our daily work is not usually any concrete phenomena. Most of the time we wrestle not with reality but with mathematical representations of it.

Artists are aware that the highest beauty they can achieve comes not from reproducing nature, but from representing it. Theoretical physicists and mathematicians, more than other kinds of scientist, share this essentially aesthetic mode of working, for like artists we fashion constructions that, when they succeed, capture something about the real world, while at the same time remaining completely products of human imagination. But perhaps even artists do not get to share with us the expectation that our greatest creations may capture the deep and permanent reality behind mere transient experience.

This mysticism of the mathematical, the belief that at its deepest level reality may be captured by an equation or a geometrical construction, is the private religion of the theoretical physicist. Like other true mysticisms, it is something that cannot be communicated in words, but must be experienced. One must feel wordlessly the possibility that a piece of mathematics that one comprehends could also be the world.

I strongly suspect that this joy of seeing in one's mind a correspondence between a mathematical construction and something in nature has been experienced by most working physicists and mathematicians. The mathematics involved does not even have to be very complex; one can have this experience by comprehending a proof of the Pythagorean theorem and realizing at the same time that it must be true of every one of the right triangles that exist in the world. Or there can be a moment of clarity in which one really comprehends Newton's laws, and realizes simultaneously that what one has just grasped mentally is a logic that is realized in each of the countless things that move in the world. One feels at these moments a sense of joy and also—it must be said—of power, to have comprehended simultaneously a logical structure, constructed by the imagination, and an aspect of reality.

Because of this an education in physics or mathematics is a little like an induction into a mystical order. One may be fooled because there is no ceremony or liturgy, but this is just a sign that what we have here is a true mysticism. The wonder of the connection between mathematics and the world has sometimes been spoken about. For example, Eugene Wigner, who pioneered the use of the concept of symmetry in quantum theory, wrote about the "unreasonable effectiveness of mathematics in physics." But no one ever speaks of the experience of the realization of this connection. I strongly suspect, though, that it is an experience that everyone who becomes a theoretical physicist is struck by, early and often in their studies.

Of course, as one continues in one's studies, one shortly learns that neither Newton's laws nor Euclidean geometry actually do capture the world. But by that time one is hooked, captured by the possibility that a true image of the world could be held in the imagination. Even more, the ambition then rises in our young scientist that he or she may be the one who invents the formula that is the true mirror of the world. After all, given that there is a mathematical construction that is the complete description of reality, sooner or later someone is going to discover it. Why not you or me? And it is the ambition for this, the ultimate moment of comprehension and creativity, even more than the need for the admiration of one's peers, that keeps us fixed on what we write in our notebooks and draw on our blackboards.

Of course, what is both wonderful and terrifying is that there is absolutely no reason that nature at its deepest level must have anything to do with mathematics. Like mathematics itself, the faith in this shared mysticism of the mathematical scientist is an invention of human beings. No matter that one may make all sorts of arguments for it. We especially like to tell each other stories of the times when a beautiful piece of mathematics was first explored simply because it was beautiful, but later was found to represent a real phenomena. This is certainly the story of non-Euclidean geometry, and it is the story of the triumph of the gauge principle, from its discovery in Maxwell's theory of electrodynamics to its fruition in general relativity and the standard model. But in spite of the obvious effectiveness of mathematics in physics, I have never heard a good a priori argument that the world must be organized according to mathematical principles.

Certainly, if one needs to believe that beyond the appearances of the world there lies a permanent and transcendent reality, there is no better choice than mathematics. No other conception of reality has led to so much success, in practical mastery of the world. And it is the only religion, so far as I know, that no one has ever killed for.

But if we are honest mathematicians, we must also admit that in many cases there is a simple, non-mathematical reason that an aspect of the world follows a mathematical law. Typically, this happens when a system is composed of an enormous number of independent parts, like a rubber band, the air in a room, or an electorate. The force on a rubber band increases proportionately to the distance stretched. But this reflects nothing deep, only that the force we feel is the sum of an enormous number of small forces between the atoms, each of which may react in a complicated, even unpredictable way, to the stretching. Similarly, there is no mystery or symmetry needed to explain why the air is spread uniformly in a room. Each atom moves randomly, it is just the statistics of enormous numbers. Perhaps the greatest nightmare of the Platonist is that, in the end, all of our laws will be like this, so that the root of all the beautiful regularities we have discovered will turn out to be more statistics, beyond which is only randomness or irrationality.

This is perhaps one reason why biology seems puzzling to some physicists. The possibility that the tremendous beauty of the living world might be, in the end, just a matter of randomness, statistics, and frozen accident stands as a genuine threat to the mystical conceit that reality can be captured in a single, beautiful equation. This is why it took me years to become comfortable with the possibility that the explanation for at least part of the laws of physics might be found in this same logic of randomness and frozen accident.

Of course, the working life of a physicist or mathematician is not continually a mystical experience. Most of the time it is a kind of game playing. But this is also a beautiful, if more prosaic, experience. All human beings play, indeed, so do all mammals. But there is something special about the games that mathematical scientists play. We play games that, when we win, answer questions about the world.

This is a separate experience from the comprehension of the truth of an equation. After you understand the Pythagorean theorem or Newton's laws, you use them to answer questions about what will happen in different circumstances. You pose a question: Suppose I throw a ball up in the air, at such and such a speed and at such and such an angle. Where will it come down? Then you solve an equation and you get the answer.

Certainly, whatever the ultimate fate of the mysticism of mathematics, there is something here that is simply true: the world is such that it is possible, at least in many cases, to invent games that teach us things about it. To do this one needs no pretension that there is an ultimate game or formula that perfectly mirrors the real world. Whatever the world is, in many of its aspects it behaves with a logic that is comprehensible. This means that it can, to a greater or a lessor extent, be comprehended by playing games that we invent to mimic that logic. Even in cases in which we suspect that the underlying reality is random, there are still games we can play—statistical games—to teach us what will happen.

This distinction between an equation that one believes to be a true mirror of nature and a game whose rules capture some observed regularity is often expressed by making a distinction between a theory and a model. The notion of a theory carries with it, in its ancient etymology and current practice, the mysticism of the desire to capture reality in symbolic expression. A model is just a game, meant to mimic some aspect of the world whose observed regularities can be posited in some simple rules.

There is no doubt that, whatever happens, scientists will continue to learn about the world by playing games. What is at stake is only the kinds of games we will play. In the first half of this book I have given two examples of contrasting claims of what kind of games might be most appropriate when we face the problem of constructing a theory of cosmology. At present the great question in theoretical physics is whether the desire to invent a beautiful equation that will capture the whole world will in the end succeed. Will there be a final game, and will it be of the kind that Newton, Maxwell, and Einstein played? Against this, we have the possibility I explored in Part Two and the last chapter: that many questions about the world might be answered by playing games more analogous to those played by biologists. Rather than finding the theory that uniquely predicts the parameters of the laws of nature, we imagine a game played by an ever-growing collection of universes that explore the consequences of different choices, with the rules designed so that those that are most successful at the game become the most probable.

To see if this might actually work we must try ourselves to play the game. This is a different kind of activity than has been traditional in theoretical physics, where the goal is to find solutions to an equation that has been posited to represent reality, and to then see if these solutions describe our world. In the case of this new theory, we play a kind of "what if" game in which we imagine changing the laws of physics in different ways and ask how these imagined worlds would differ from our own.

In Chapter 10 I described two games that have been invented to explain a particular astronomical phenomenon, the organization of the spiral galaxies. The first, the density wave theory, follows the traditional methodology for theoretical physics. One writes down certain nonlinear differential equations, which are supposed to capture the motion of matter and energy in the disk of the galaxy. One then tries to try to find solutions to these equations and compares them to the galaxies that are observed in the world.

The second kind of game is epitomized by the Gerola-Schulman-Seiden theory of galactic structure. Here one tried to capture the logic of the processes involved in star formation in a few simple rules. There is no equation to solve, only a rotating checker board and a few simple moves. To ask the game a question one must play it. As one must play out many moves to see the answer, one will typically program a computer to do the playing, but in a pinch a group of children would do just as well. One then literally sees whether or not one has won, for winning means that no matter how long one plays the game the pattern of pieces on the rotating board will look like a photograph of a spiral galaxy.

There are advantages and disadvantages to each kind of game. In many cases, including galaxies—each one captures some aspect of reality. Some galaxies, it seems, are more "density wave-ish", some are more "game of life-ish." Which kind of game will be best depends to a large extent on the kind of question one is asking.

For example, let us suppose one wants to ask where a ball thrown in the air will land. The game we play in Newtonian physics certainly works best. What is good in this case is that the logic of the situation is captured in a few simple relations which may be expressed as mathematical equations. These equations hold in an enormous number of different circumstances. To describe them one must find solutions to the equations. Because the relations hold true in an infinite number of different cases there are infinitely many solutions. This is not bad—it is good. It means that the theory captures something very common, perhaps even universal.

The hard part is often to find the particular solution one is interested in. These are usually distinguished by specifying the situation at some initial moment of time. For example, in the case of the equations that describe the motion of a ball, the solutions are labeled by where it is thrown from, and the direction and speed with which it is thrown. For each of these starting, or initial, conditions there is a solution to Newton's equations that tells us where it will go.

The beautiful thing about such a game is that the rules are deterministic. Given the initial conditions one can predict exactly what will happen. Let us call a game like this a Newtonian game. Electrodynamics, general relativity, and the standard model are all Newtonian games. They have infinite numbers of solutions; to answer a question one must find the solution that matches the initial conditions of the problem one is interested in.

We are interested in this book in the question of how to construct a theory of a whole universe. If this is our ambition, there are some particular advantages and disadvantages to the different kinds of games we should consider. The advantage of Newtonian games is certainly completeness. Given the initial conditions, every question about anything that might happen in a system a Newtonian game describes ought to, in principle, be answerable.

But there also lies the difficulty. For there is only one universe, but there are an infinite number of solutions to any Newtonian theory. If we write down an equation of this kind to represent the universe we have a problem, because only one of the infinite number of its solutions can represent the universe. This is very different from the case of the flight of a ball, because there are an enormous number of actual instances of balls being thrown in the air, and it is good to know that one equation describes them all. The freedom to choose initial conditions is necessary for any useful theory of a part of the universe; it gives us the freedom to comprehend the phenomena of many different parts with one equation.

But the cosmological case is much different. Each solution to the universal equation describes a whole world. But only one can have anything to do with reality. This means that any theory of the whole universe must, if it is a Newtonian kind of game, come with a supplement that tells us which of the infinite number of solutions describes the actual universe.

This is called the problem of the initial conditions. It is a problem for cosmology because it implies that there must be some reason why the universe started off in one state rather than another. But if this reason lies outside the universe, then it seems that the universe is not all that there is, which is absurd, for then it is not the universe.

There is thus a danger that the need for such a theory of initial conditions leaves the door open for a return of religion. Not the mysticism of the mathematical I have been speaking about, but the idea that there is a god who by conscious decision and choice made the world. Einstein is quoted as saying, "What I am really interested in is whether God had any choice in the creation of the world." If the theory of the whole universe turns out to be a Newtonian theory the answer must be yes, for the theory does not tell us how the initial conditions are chosen; if God made the world this way, then he did leave himself choice.

As often as I have heard this issue discussed I still have no idea how to make sense of it. Must all of our scientific understanding of the world really come down to a mythological story in which nothing exists before twenty billion years ago, save some disembodied intelligence who, desiring to start a world, chooses the initial conditions and then wills matter into being? I suspect that the attraction for such a story is at least partly fueled by the nostalgia for the religious conception of the world, and by a desire to see ourselves in the place of the creator of the world. And as such, the desire to see the world this way is a religious yearning and not an expression of any principle of scientific methodology.

The problem of the initial conditions in cosmology has not yet been solved. These days it is usually couched in the language of the quantum theory, where it becomes the problem of specifying the quantum state of the universe. From time to time someone has proposed that there should be a unique solution to the equations of quantum cosmology. But in each case, closer inspection revealed that there are many solutions to the equations, each of which describes a possible cosmology.

A different kind of solution to the problem would be possible were it to turn out that the present state of the world does not actually depend on the choice of an initial condition. If such a theory worked, then there would be no arbitrary choice in the construction of the world. The reason for everything that we see around us would then lie only in the world itself, and not in something outside it. For this reason the possibility of a universe that is self-organized and not chosen, would, if it can be achieved, lead to a more rational comprehension of the world. Can this be achieved by the second kind of game we have been describing, typified by cosmological natural selection or the Gerola-Schulman-Seiden theory? The answer is, at least partly, yes. Such games certainly go in the direction of a conception of the universe as a self-organized entity; when they succeed they develop structure and patterns without any need to choose initial conditions or tune parameters.

But there are certainly disadvantages to the proposal that the final explanation for the order of the world is to rest on a statistical process of self-organization, such as cosmological natural selection. One is that it then may not be the case that every question we ask can be answered to arbitrary precision by the theory. At some point statistical uncertainty must necessarily limit the accuracy of its predictions. This is in striking contrast to what we would expect if the parameters were set by some fundamental mathematical theory. In that case we should expect that, when we got down to the truly elementary level, we should see simple mathematical expressions emerge for the parameters. The reason for this is that if the fundamental theory is simple, the numbers it produces must be describable with a small amount of information. This means that there must be a reason for every one of the infinite numbers of digits in a decimal expansion of each parameter. On the other hand, if the universe is tuned by some historical and statistical process, such as the mechanism of cosmological natural selection, it will not be the case that every last digit in the decimal expansion of every parameter is meaningful. There will instead be a bit of roughness in the construction. Perhaps only a few digits of each parameter are really going to be meaningful; the rest may be essentially random.

The question is then whether fundamental physics and cosmology should, in their use of numbers, be more like pure mathematics or more like biology? In pure mathematics we work most of the time with exact numbers. Every digit in the decimal expansion of π matters: if we have a number whose decimal expansion is the same as π except for one digit in the tenth, or even the 10^{100}th place, it is not π. On the other hand, in sciences such as chemistry, geology, or biology we use mostly approximate numbers. This is because we often work with statistics which allows things to be defined only up to a certain precision. In these cases, usually only the first few digits of a number are meaningful.

If cosmological natural selection, or something like it, is to be the final explanation for the parameters in elementary particle physics, then this will also be the case for fundamental physics. It is likely that there will be no explanation for more than a few digits of each of the parameters of the standard model. We will be able to determine as many digits as we like by measurement, but there will be no reason for them, they will reflect only the fluctuations inherent in any statistical process.

This does not mean that there will be limits to the precision of physical theory, for we may still be able to discover relations between physical quantities that hold to precisions much greater than that to which we are able to predict the values of the fundamental constants. But it does mean that we will have to give up the idea that it will be possible to find a rational explanation for every last digit of every measured quantity.

Is this sufficient to satisfy the desire to comprehend the world through physical theory? I believe the answer is yes, but to explain why we must recall that the opposite supposition also has a price. When we signed on to the hypothesis that there would exist a deterministic physical theory that would allow us in principle to predict the future with perfect accuracy, we also gave up something precious. This was the belief in the possibility of novelty. If the whole world is just the working out of deterministic laws then the future is, in a real sense, always just a manifestation of the present. Nothing new can happen that is not already coded in the present.

In general, the Platonic conception of physical theory makes it difficult to believe in the possibility of novelty. To the extent that all structures in the world are reflections of ideal forms there can be nothing new; the forms are eternal. The biological world seems to belie this, as the history of natural selection is full of moments at which forms are invented that did not previously exist. It is very tempting to want to say that novelty is possible in biology. But if we believe that the fundamental laws are deterministic, are we really allowed to believe in the actuality of new, or must we always insist on the impossibility of novelty? At the very least, there seems to be a question here that is worth looking into: How is it possible for processes that are completely described by physical laws to create categories of things that did not exist at earlier times? And is the answer to this question changed if the laws of physics themselves are the result of processes of self-organization or natural selection?

Certainly we and all the other living things are enormously structured and highly improbable collections of atoms, that obey the very same laws as do atoms everywhere. At the level of atoms, it is clear that there is likely to be no possibility of novelty. But there really is no contradiction here. Reductionism of explanatory principles, which certainly works in this case, does not prevent us from conceptualizing structure and organization that can only be perceived on larger scales. Nor does it prevent us from discovering principles of self-organization that are to be comprehended in terms of these emergent structural categories.

A good thing about quantum physics is that it allows structure and information to be conceived as real things in the world. The existence of atoms allows us to count things in the world, which means that we may apply the logic of the natural numbers to real phenomena. This makes it appropriate to apply logic and information theory to naturally occurring structures. The fact that DNA can carry a code that can only be understood in terms of information, and only realized in structure, is in no way in contradiction with the idea that the only laws acting are those of physics.

So there is certainly not a scientific problem here. Natural processes acting over time can indeed create novelty. But there is a philosophical problem, or to put it better, a problem for philosophy. The process of natural selection is supposed to be simply the working out of logic and probability on processes involving structured molecules. And probability is supposed to be nothing but counting, and counting is also, according to the logicians, really nothing but logic. So in the end, natural selection is just the working out of the principles of logic on certain populations of structured molecules. But logic is supposed to be tautological. There is not supposed to be any actual information in tautology, because its meaning is those things which are true in all possible circumstances. But if something is true in all possible circumstances, it is true always. It can then never be new. So, how can that involves nothing but the working out of logic and probability generate novelty? This is why the problem of how novelty is possible is a problem for philosophy.

One possible answer to this is that there is in reality no novelty. The possibility of all species, and indeed of all possible mixes of species, exists as soon as one has the basic mechanisms of life. All possible sets of living things are coded in DNA, and the set of all possible DNA sequences, and all possible collections of them, may be said to exist timelessly, as possibilities. It might then be argued that natural selection does not create novelty, it merely selects from a list of possibilities that always exist.

But this is a completely uninteresting use of the idea of existence. Certainly we want to distinguish between the notion of the existence of actual things and the existence of names of things on lists of possibilities. It may indeed help to formalize some problem to imagine (for we never actually construct them) the collection of all possible species, of all possible histories of life, or even of all possible computer programs or chess games. But almost all the things on these imagined lists do not exist. On the other hand, a small number of them do exist in the world, and they are not timeless; they are created at definite times by definite processes. If the word existence is to have a useful meaning, then we must hold that the world does create new things, so novelty is a fact of reality. Thus, we are indeed faced with the problem of how processes that are explicable completely in terms of logic and counting can produce novelty.

It might be useful to approach this problem from another direction, having to do with the problem of what kind of knowledge the theory of natural selection represents. The philosopher Immanuel Kant argued that statements about the world could be classified in two ways. First, they are either analytic or synthetic. Analytic statements are those that are true only by virtue of the definitions of words, something like, "all unmarried men are bachelors." Synthetic statements are those that make assertions whose truth cannot be discovered by analyzing the definitions of the words involved, for example, "the writer of this book likes cats."

In addition, for Kant statements are also either *a priori* or *a posteriori*. A priori statements are those that must be true in any possible world, so that it is possible to ascertain their truth without doing any observation of the world. *A posteriori* statements are those whose truth must be demonstrated by observation.

Clearly analytic statements are *a priori*. The difficult question, which Kant struggled with—and which I want to raise in a different way here—is the status of certain synthetic statements. Of particular interest is the status of synthetic statements whose explanation is supposed to be based on a mechanism of natural selection or self-organization. For example, we may ask about the statement: there exist animals that fly. This is certainly a synthetic statement; we can easily imagine that a planet exists on which natural selection has not yet led to the discovery of flight. At the same time there are good reasons why flying creatures may have a selective advantage over those that walk or crawl. This makes it possible to argue, using the principles of natural selection, that it is quite probable that most inhabited planets with a suitable atmosphere would be home to some kind of flying creatures. The question to be asked then is whether the statement is *a priori* or *a posteriori*. It seems that the relevant argument from the theory of natural selection is one that would hold in any possible world in which there were animals. Thus, it seems that the explanation by natural selection makes the truth of the statement, "it is probable that on any planet in which there are animals there are some that fly," *a priori* rather than *a posteriori*.

The result seems to be that once we allow explanation of features of the world by natural selection, we admit a new category of synthetic *a priori* knowledge. And this is for us, as it was for Kant in a different circumstance, a problem. How can there be knowledge that is synthetic, that has content beyond tautology, that is also true in any possible world?

The key to this problem must lie in the element of time that is present in the formulation of the question, How is novelty possible? For our understanding of how natural processes can produce novel structures by processes that are explicable in terms of only logic and counting involves time as an essential element. On the other hand, the idea that logic and mathematics are tautological comes from viewing them as timeless systems of relations, in which anything that is true is true eternally.

The problems we have been discussing seem more puzzling than they really are only because time has been left out of the discussion. The existence of features of the world, which seem both synthetic and explicable by *a priori* arguments (true in any possible world), seems puzzling only if we think of the world as fixed, absolute, and static in time, so that anything that is ever true is true always. But natural selection must be understood as taking place in time, so that the properties of a species are time dependent statements that hold only during the necessarily limited time that its members are around. Whereas pure logic seems to have no power to create anything when viewed in the context of a static, Platonic world of propositions that are eternally either true or false, a process which acts over time to transform structures in the world, such as natural selection, may be both completely explicable in logical terms and truly capable of the invention of novelty.

Indeed, we may reverse the question and ask how it is that the theorems of mathematics can be understood to be timeless. How is it possible for us to discover any truth that is true always? The only reasonable answer to this question, which really just emphasizes Kant's point in a different way, is that mathematical and logical truths may be true for all time because they are not really about anything that exists. They are only about possible relations. Thus, it is a mistake—a kind of category error—to imagine that the theorems of mathematics are about some "other" or "Platonic" realm that exists outside of time. The theorems of mathematics are outside of time because they are not about the real. On the contrary, anything that exists must exist inside of time.

If we insist that existence means existence bounded by time, we can reverse the trap that the old metaphysics imposed on us, in which all that really exists—the true Being—exists only eternally, while those things that exist in time are only appearances, only faint reflections of what is really real. If existence requires time, then there is no need and no place for Being, for the absolute and transcendent Platonic world. That which exists is what we find in the world. And that which exists is bound by time, because to exist something must be created by processes that act in time to create the novel out of what existed before.

A few chapters back, I joked about the chicken-and-egg problem. But the point is really serious. The chicken-and-egg problem is, as most children discover, insolvable. But it remains so only as long as chickens and eggs are assumed to be eternal categories. As soon as time and evolution are allowed into the picture the problem dissolves, because clearly eggs came first. Thus, this simple joke suggests how the notion that structure in the world is formed through natural selection may allow us to escape from the prison that the Platonic view of the world imposes on epistemology, and in particular on the expectation that objective knowledge is necessarily knowledge that is eternally true. If all that is real is created in and bounded by time, then objective knowledge—knowledge of the real—is also bounded by time.

What about the conflict with determinism? In the case of biology, the statistical noise present at the molecular level at the finite temperatures at which biology operates makes the question of whether the laws of nature are actually deterministic or not irrelevant for the formation of biological structure. Thus, for all practical purposes there is no issue, even if physics is deterministic. Of course, quantum physics, as we presently understand it, is not deterministic, but what seems more important for coming to this conclusion is the fact that the existence of discrete quantum states allows information to be a genuine physical quantity.

Still, the question persists at least at a purely theoretical level: if the world is just the working out of preexisting mathematical law, how is novelty possible? The possibility that the laws may not be eternal, but may actually be constructed in time through physical processes, sheds new light on this dilemma, for if the parameters in the laws of physics are set by statistical processes of self-organization that occurred in real time, then novelty is possible down to the level of the fundamental laws themselves. If this is what we gain by giving up the possibility of a rational explanation for every last digit of the parameters of the laws of physics, it is perhaps a price worth paying.

This point of view may also shed light on some of the philosophical problems at the heart of mathematics. There are, for example, paradoxes in the foundations of mathematics associated with the possibility of self-reference. These arise from the possibility that a mathematical statement may refer to itself. The simplest of these is the liar's paradox: What are we to make of a person who says, "I am lying"? Translated appropriately into logic this becomes Godel's theorem, which results from the possibility that a statement about arithmetic can also assert about itself that it is not provable. From this theorem we learn that any mathematical system complicated enough to include arithmetic can be either consistent—meaning without contradiction—or complete—meaning that everything that is true about it can be proven—but not both.

To be bothered by this, we must think of mathematics as some timeless reality, such that anything that is true about it is true forever. If we stick to the view that logic and mathematics are about nothing, and that all that exists is bound in time, then these difficulties may be seen in a different light. If we construct a real system, say a computer or a living thing, that is capable of self-reference, then what we have done is to construct a feedback loop. Self-reference in a real entity must exactly be the possibility that its state at the next moment is a function of its state now. In a real system, which can have only one state at a time, self-reference must be understood as something that happens *in* time.

As we saw in the last few chapters, feedback is an essential element of any process of self-organization. And processes of self-organization are what gives our world structure. Thus, self-reference, which leads to paradox when we try to envision knowledge as timeless, leads instead to structure and organization when it is realized as a real process that acts over time in the real world.

At its root, the most elemental operation in logic is the making of a distinction. We do this whenever we separate one part of the world from another, or when we define a set that contains only certain things and not others. It is possible to make distinctions in the real world because we live in a world full of structure and variety. To the extent that this is the result of intrinsic processes of self-organization, it might be considered that the possibility of self-reference is what makes logic possible. Thus, if it is the case that the laws of nature have been constructed, over time, by processes of self-organization, then the old dream of reducing science to logic may to some extent be realized. But this will be possible, not in the timeless sense dreamed about by Platonists, but only in the sense that logic, expressed in time and complicated by the possibility of self-reference, is self-organization.

In the end, the two different kinds of mathematics on which fundamental physics might be based come down to different notions of form and how it may have arisen. Let us think, for example, of a flower and a dodecahedron. Both are beautiful and ordered, and the flower may seem no less symmetrical than the geometric construction. Their difference lies in the way that each may be constructed. The dodecahedron is an exact manifestation of a certain symmetry group, which may be written down in one line of symbols. If I can't make a perfect one, I can make a pretty good representation of it, either with paper, scissors, and glue; or with a page of computer code. A flower, by contrast, is not perfect. If we examine it closely we will see that while it may appear symmetrical it does not adhere precisely to any ideal form. From the coiling of the DNA in its trillions of cells to the arrangements of its petals, a flower's form often suggests symmetry, but it always fails to precisely realize it. But even given its imperfection, there is no way I could make a flower. It is the product of a vast system that extends far back in time. Its beauty is the result of billions of years of incremental evolution—the accumulated discoveries of blind, statistical processes; its meaning is its role in a much larger ecological system that involves many other organisms.

Neither the ancient Greeks nor the physicists who made the Copernican revolution knew about the possibility that structure could be formed through such a process. They had no alternative to explain the beauty and order of the world except through the dream that it represented a reflection of eternal Platonic mathematical form. The question we must face now is whether our physical theory will be limited to this conception, or whether we will take advantage of the new possibility for the construction of an ordered world that processes of self-organization, such as natural selection, make possible. The question, in the end, is whether the world is more like a dodecahedron or more like a flower.

In the history of science, there is no argument for the power of mathematics stronger than the story of Johannes Kepler's struggle, in the early seventeenth century, to discover order in the motions of the heavens. For him and others of his time, the cosmos consisted of the Sun, six planets, and the sphere of stars surrounding them. Like us, part of Kepler's problem was to understand why a certain list of parameters that governed the overall shape of the universe should take the values they do. For Kepler these parameters described the orbits of the six planets then known.

Kepler was an inventive soul with a powerful imagination, and among the products of his search were the basic laws of orbital motion. But to learn about these is just to touch the surface of this man's prodigious vision, for the same books which present what we now call Kepler's laws contain a startling variety of attempts to make order out of the cosmic motions. Among the things he imagined was that the six planets comprise an orchestra that sing out notes which are proportional to their orbital speeds. As they speed up and slow down during their journeys around the Sun they play an ever-changing symphony which he called the harmony of the spheres.

If we borrow just a shadow of his genius, we may imagine with him that the universe is indeed a great harmony, it is only that his scale was wrong. The instruments of the cosmic orchestra are not the planets; they are instead the fundamental particles and forces. The themes they sing are nuclei, atoms, stars, supernovas, galaxies; their symphony is the history of the universe. Everything we have learned since tells us that Kepler's vision was true. To make such a beautiful and varied sound, the instruments must all be tuned; if we seek to commission a universe but neglect to tune the orchestra, we will have only chaos and noise. The question now, as it was then, is whether we can make sense of all of this harmony without going outside the confines of explanation by physical causes and effects.

Kepler was in his heart a mystic, who one imagines would have preferred a universe that exactly mirrored some beautiful mathematical principle than a world tuned roughly by blind historical processes. In this sense, many contemporary theorists of cosmology are his descendants. But, in the end what made Kepler great was that he was willing to give up his preconceptions as he struggled over many years to listen to what his charts of the motions of the planets on the sky were trying to tell him. So, after almost ten years of struggle, he was able to hear the ellipses in the motions of the planets in the sky, when he and everyone before him had their ears tuned only to the sounds of circles. And in the end, if we listen closely enough to the harmony of the spheres, we will hear either the precise mathematical intervals that could only be the sign of a fundamental mathematical order behind the world, or we will hear all the remnants of roughness and disharmony that a blind and statistical process cannot erase.

The fundamental question of philosophy is thus precisely:
What is given to us?   —Stanley Rosen

The main problem philosophy faces at the present time is
how to have knowledge without faith.
                              —Paola Brancaleoni

FOURTEEN

PHILOSOPHY, RELIGION, *and* COSMOLOGY

Since at least the seventeenth century, the writings of Western philosophers and scientists have rung with the ambition to have complete knowledge about the world. We see in the writings of physicists from Copernicus and New-ton to Maxwell and Einstein the faith that the world is constructed on a rational basis, and that it is possible for human beings to represent this rationality in com-prehensible language. And in the ambitions of the great philosophical system-makers, from Descartes and Leibniz to Kant and Hegel, we see the ambition to dis-cover, from an exercise in pure thought, the whole meaning and shape of the world, at least as it is perceivable by us.

In the writings of most of these scientists and philosophers, we find also the belief that the world is rational and explicable because both it and our minds were made by a rational God. The ambition to comprehend the world is then the ambition to mentally take the place of God and see the world from the outside, as its creator did. For some, such as Newton, the religious underpinnings to the search for scientific knowledge are explicit, even celebrated. But even in Einstein, who denies belief in such an anthropomorphic god, one sees in so many writings and remarks his yearning to know the secrets of "the old one." And, indeed, in his autobiographical notes, one reads of a lonely adolescent who, after a profound disillusionment with religion, discovered in science a search for transcendence and identification with the absolute more acceptable to a young secular European of the Nineteenth-century fin de siècle.

It was indeed this promise of transcendence that I found in Einstein's writings that first captured me for physics. And, having started with a master and not with a textbook, I began reading the original writings of those who had invented the science I was struggling to make mine. Of course, to get anything out of the old books, with no background in general history, let alone the history of science, it was necessary to learn to skim, to read selectively, to take in what meant something to me and leave the rest uncomprehended.

As a secular child of a much different period, with more Marxism and mysticism in my upbringing than religion, I skipped over the references to God in the writings of Newton, Copernicus, Kepler, Descartes, and Einstein. Only later, preparing to teach about them, did I reread these founders and discover how much their search for truth was a search for God.

The references to God in the founders of my science made no sense; they seemed so quaint, so unnecessary. Can there be any doubt that science is a better road to truth about nature than any received dogma? But, if this is so clear to us now, when we live amidst a world created from the knowledge constructed by science, could they not have understood at least the promise of what they were beginning? Of course, there is the myth, and perhaps even the reality, of Galileo, who, with his lack of religiosity and his faith in the judgment of the individual mind, speaks to us like a brother over the centuries. But, as much as his battle with the Catholic church is now celebrated, he was alone among the great visionaries who made physics for his lack of interest in the mind of the creator. Almost every one of the founders of physics write as if their search and the search for God were one and the same. How many times, reading late at night, have I wished it were possible to confront Newton and the others with the contradiction between their irrational identification with God and the rationality they created.

The ambition to construct a scientific theory that could explain the world, as conceived from the seventeenth to the early twentieth centuries, shares a great deal with the search to know God. Both of them are a search for the absolute, for an understanding of the world that attributes its beauty and order to an eternal and transcendent reality "behind" the world. In diverse aspects of the production of European culture in these centuries—in the sciences, philosophy, theology, and art—one sees a striving to construct an absolute and objective view of the world that would ground the vicissitudes of our lives in an eternal and unchanging greater reality. Whether the talk is of God, or of an eternal and universal Law of Nature, the idea that dominates is that the rationality responsible for the coherence we see around us is not in the world, but behind it.

I believe that the transition that science is now undergoing is in part a necessary process of liberation from the influences of this essentially religious view of the world. What ties together general relativity, quantum theory, natural selection, and the new sciences of complex and self-organized systems is that in different ways they describe a world that is whole unto itself, without any need of an external intelligence to serve as its inventor, organizer, or external observer. These are all steps towards a more rational and more complete comprehension of the world based more on what we know and less on myths that have been passed down to us from past generations. Such a science will be able to satisfy two aims that have become, at least implicitly, the goal of much current research: to construct a cosmological science that has no need of reference to fixed, eternal frameworks outside of the dynamical system of the world, and to provide a physics and cosmology within which life has a natural and comprehensible place. Most importantly, as I have tried to argue in this book, there is good reason to hope that the realization of these two goals will involve the understanding that they are intimately related, so that a universe hospitable to our own existence will also be a complete world that can be rationally comprehended without any need to refer to external agency or intelligence.

We still have part of the story to tell, that concerned with the theory of space and time and the relationship between them and the quantum. But even with what we have seen so far, it is clear that, whatever the outcome of the many controversial questions we have discussed, the new physics and cosmology raise several philosophical questions which must be faced before we will be able to completely appreciate their significance. Some of these, which we have touched on before, are:

- To what extent is it possible to conceive of a world whose fundamental regularities arose as the result of a historical process, rather than being the manifestation of some fundamental and absolute law?
- Why is it easier to conceive of a world structured by law imposed from the outside, as has been imagined from Plato's cosmology to the current day, than it is to imagine that the regularities of the world are all the result of processes of self-organization that take place in real time in the world?
- Can we live with a scientific cosmology that posits, and draws verifi-

able predictions from, the existence of regions of the universe we can-not directly observe? What are we to make of cosmological theories that predict that we cannot in principle observe most of the universe?

- If the world has existed for only a finite time, what could it mean to conceive of things that might be true eternally, whether these are the laws of nature or mathematical theorems?

- What kind of knowledge does the theory of natural selection represent? Is that theory just an application of logic, or probability, or does it have some empirical content? Can we imagine a world in which there is biol-ogy but in which the theory of natural selection is not true?

- What do the incompleteness theorems of Godel and others mean for the Platonic vision that the order of the world is essentially mathe-matical?

- How, if the world is the working out of simple and universal law, is novelty possible?

I freely confess that I am not a professional philosopher and I bring to a discus-sion of such problems a spotty and autodidactical reading of the philosophical tradition. So my ambition here is only to raise philosophical questions, not to resolve them. Perhaps the best thing I can say to justify the following remarks is that there is a need for an examination of questions like these that is both philo-sophical and informed by the dilemmas we in science are facing. Those of us with hands dirty with the mess and detail of the work of science may not be in the best position to finish this job, but perhaps it requires something from us to help it get started.

There is, of course, a community of professional philosophers of science, some of whom are very well trained in physics and mathematics. However, while I have learned a lot from conversations with a few of them, I must say that I often go away with the feeling that they are too nice to us. For example, I sometimes get the impression that specialists in the philosophy of quantum theory see their role as cleaning up the way we speak about the theory, when many physicists suspect the problem is much deeper. It may be helpful to be able to sort out the really bad ideas about the interpretation of quantum mechanics, but what if, as seems likely, the problem in the end is that we have the wrong theory rather than just the wrong interpretation? If not for the philosophers, who is going to have the courage to tell the physicists when quantum theory, or another of our construc-tions, just cannot be made sense of? In the past, philosophers like Leibniz did not hesitate to tell physicists when they were speaking nonsense. Why now, when at least as much is at stake, are the philosophers so polite?

Certainly, there is a great deal of sorting out going on among philosophers, and perhaps in my ignorance I am being unfair if I say that little of that discussion, as important as it may be for philosophy, seems to touch the points of crisis in contemporary science. It may very well be that if more scientists were able to understand more of the difficult and, may I say, inelegant prose in which much of contemporary philosophy is coded that we would find answers for some of the questions that trouble us. I certainly do get the impression that some of the muddle in which philosophy finds itself has its origin in the same circumstance I was just discussing, which is that it is caught between its historical roots in the religions of its founders and the implications of new knowledge of our real circumstances that has come from twentieth century biology and cosmology.

It may then be that philosophy is in crisis for reasons not unrelated to the deep problems facing physics and cosmology. This is another reason to bring philosophy into a discussion of the future of cosmology. But, beyond this, there are other reasons that we must discuss philosophy if we are to find a way around the dilemmas raised by our current attempts to construct cosmological theories. Not the least of these is that philosophers such as Leibniz and Kant, have noticed some of these issues, and may have something useful to tell us. Beyond this, the fate of the metaphysical project is likely to be relevant for the current crisis in fundamental physics because, however divergent their methodologies, there is an affinity between the ambition of theoretical physics and the ambition of metaphysics. Both have often presumed that there is some absolute truth to be discovered about the world, which they conceive variously as the final, fundamental law, or the true essential—the true Being. Both have found inspiration in the Heraclitean doctrine that "Nature loves to hide," and thus see their highest purpose to be a search for a transcendent and timeless actuality beyond the appearances of the world. Running though both is a tradition that asserts that the world we see around us is not completely real, but is only a kind of movie constructed by our eyes. Beyond these appearances lies the true reality, the true existence, that both fundamental physics and metaphysics have endeavored, in different ways, to discover.

In European philosophy, the period of the great metaphysical philosophers coincided with the period, roughly comprising the eighteenth and nineteenth centuries, when people believed in the picture of nature invented by Newton. This is no coincidence, for there was much in Newton's concept of nature that is friendly to the desire for a world rooted in an absolute and transcendent Being. Even more, the idea that a human being had discovered the absolute and final laws of nature was both a puzzle and a challenge to philosophers. How could such an achievement have been possible, especially given the possibility of doubting the impressions of our senses, as Descartes, Berkeley, and Hume had taught us? But it had apparently happened, which meant that philosophers could dream that the mind of a human being might grasp truth beyond the merely empirical.

Our century, which began with the fall of Newtonian physics, has then not surprisingly also seen a great reaction against metaphysics. The attack on metaphysics was begun with the logical positivists, who taught that the only meaning a sentence can have is the conditions that could be given for its empirical verifica-tion. But perhaps its deepest expression is found in the work of Ludwig Wittgen-stein, who attempted to discover, first from logic, and then from language, the limits of thought and knowledge.

Certainly, the overthrow of Newtonian physics was not the only reason that the twentieth century saw the rise of these anti-metaphysical movements. The problem of the uncertainty of scientific knowledge is not new; that Newton was wrong could not have been a complete surprise to anyone who had read Berkeley and Hume, let alone Leibniz. One cause of the growing skepticism about what philosophy can accomplish must simply be that the metaphysical ambition has been tried, and the results are not encouraging. Although Hegel and the other grand metaphysicians continue to be read and studied, it seems clear beyond a doubt that their central project has failed. Human beings cannot by pure thought alone arrive at the truth about Being—about what, if anything, is behind the appearances.

In the history of philosophy, many have argued against the idea that science can lead to knowledge of the absolute reality behind the appearances. I do not want to begin this argument again. There is no way to climb the ladder of empiri-cal knowledge, or fly on the wings of logic, to ascend to the absolute world of what really is. But I think that the situation I've just described makes it possible to confront a different and more difficult question. This is whether there might not be something wrong with the whole conception of an absolute and timeless real-ity lying behind the appearances. If possible knowledge is knowledge of the world of appearances that we live in and interact with, why is it necessary—or even desirable—to believe that the reality of the world is somehow behind the appear-ances, in a permanent and transcendently absolute realm?

Is there any reason we might not conceive of the world as made up as a net-work of relationships, of which our appearances are true examples, rather than as made up of some imagined absolute existing things, of which our appearances are mere shadows? Why should there be any "things in themselves," besides the effects that all things have on each other? This is related to another question: If the laws of nature are only the working out of principles of logic and probability by processes of self-organization, must there still not be some fundamental parti-cles, on which those processes act? And must they not obey *some* universal laws? Perhaps a principle such as natural selection, self-organization, or random dynamics might explain why the parameters of the standard model come to be what they are, but just as biology requires molecules on whose combinations the principles of self-organization and natural selection can act, does not physics still require some fundamental substance for the laws to act on? Must not the world consist of something beyond organization and relations?

I do not know the answer to these questions. They are in the class of really hard questions, such as the problem of consciousness or the problem of why there is in the world anything at all, rather than nothing. What in the end is the reason the world is called into being? I do not see, really, how science, however much it progresses, could lead us to an understanding of these questions. In the end, perhaps there must remain a place for mysticism. But mysticism is not metaphysics, and it is only that I seek to eliminate. Wittgenstein said, in his Tractatus, "Not *how* the world is, is the mystical, but *that* it is." Perhaps in science, as in philosophy, by eschewing the metaphysical fantasy, the dream of an absolute being forever unknowable behind the veil of the appearances, we bring ourselves in closer proximity to the genuinely mysterious.

Let us turn away from these genuine philosophical questions, and come back to the question of the effect of philosophical preconceptions on our expectations about what a scientific theory should be. The radical atomist point of view has so captivated the minds of physicists and philosophers that the existence of a realm of things with absolute and timeless properties seems almost axiomatic. If we want to go beyond this point of view, we may try to understand why it has come to seem so necessary. Why has the highest goal of physics been almost universally felt to be the construction of a final theory that captures the logic of this eternal reality?

I want to suggest that perhaps the answer is that the belief in radical atomism—in the existence of a final and absolute theory that governs the behavior of the elementary particles—is as much a religious as it is a scientific aspiration. Although I will not try to develop the argument in a scholarly way, I suspect that the ambition to discover a final theory that is absolute and immutable, while remaining completely atomistic and reductionist, reveals its roots in the religious yearnings of European civilization of the last four centuries. As such, our continuing adherence to it reflects the general difficulty of completely adjusting our conception of the world to our growing knowledge of it, so that we still see in the shapes of our ambitions a *nostalgia for the absolute* that was lost when the Newtonian universe was overthrown.

Sometimes one finds physicists asserting that the belief in such a fundamental theory is an antidote to belief in a god who made the world. However, it can sometimes seem as if the belief in such a theory is more a replacement for than an antidote to a belief in a god. Like a god, a fundamental theory is something whose truth is believed to transcend any particular fact about the actual physical universe. And, like a god, a fundamental theory must in some sense exist prior to the universe, so that the things in the world may obey it from literally the first instant of the creation of the world. A universe made according to a fundamental theory is then very like a universe made by a god, in that it is made according to a rationality that exists prior to and independently of the actual universe. Such an agent, whether it is a fundamental law or a god, is something that acts on the universe, but is not in turn acted on or influenced by anything that actually happens in the history of the actual universe.

Thus, belief in a final theory shares with a belief in a god the idea that the ultimate cause of things in this world is something that does not live in the world but has an existence that, somehow, transcends it. This is why the belief in god and belief in the existence in a final theory are both related to the metaphysical idea that what is really true about the world is true about a timeless transcendent realm and not about the world of the things we see around us.

There is still another issue that arises if we aim to give up on the idea that the goal of physics is the discovery of a final theory, in which the properties of the elementary particles are fixed by first principles, independent of the history of the universe. For it might seem that if we give up on the idea that there is a single, final theory, we may also be giving up on the possibility of gaining a complete and objective description of the world. Is it possible then to have objective knowledge, if that knowledge does not tell us how the world of appearances is constructed out of what ultimately exists?

I would like to argue that the answer to this question is, in fact, yes. It is, to begin with, not really the case that the aspiration to discover the final theory, or apprehend the true Being, has really helped the project of gaining objective knowledge. It is true that it is often presumed that objective knowledge, to the extent that it is possible, is knowledge of some absolute reality that lies beyond the subjective appearances. But it seems to me that to equate the world of appearances with the subjective is to make a kind of category error. What we have given to us, from which we will deduce all possible knowledge, is nothing other than the appearances of the world. If objective knowledge exists at all, must it not be knowledge about the world of appearances? Must it not then be possible to construct or deduce any real knowledge from the appearances alone? Do we, as observers who live in the world, have any other choice?

The idea that objective knowledge must be about something other than the appearances carries with it a presumption that it is possible to imagine a view or a picture of the world that is somehow more true than the views of human observers. Such a view would not be limited to the incomplete and incompletely reliable views of observers present in the world. It might be a view of the world in its entirety, as it is.

But such a view cannot be the view of any real observer living in the world. It could only be the view of some imagined being who is outside the world. In this way the idea that there is a world behind the appearances, an absolute Being, a world as it is, carries with it, in every context in which it appears, the dream that there is a view of the world from outside of it. And if one subscribes to this dream, then it is clear that the ultimate justification for objective knowledge must lie not in any incomplete view from inside the world, but in this all encompassing view from the outside. Thus, if one believes in the possibility of this view from outside the world, one is led to identify objective knowledge with knowledge of the absolute world behind the appearances. All other knowledge is at best incomplete and tainted its subjectivity.

If such a view were possible, then we would certainly like to aspire to it, for we would all like to have a kind of knowledge which is liberated from our situation, just as, indeed, we would all like not to die. The questions is then, is such a view possible? Or, at least, is it conceivable?

I do not think that such a view can be achieved; we can learn this from both relativity theory and quantum theory. I think, however, that the idea of such a view is conceivable. This is how the view of God has usually been understood. For the Aristotelians, the universe had a boundary and God simply dwelled outside of it, looking down on all of his creation from the outside. But even in Newton's infinite cosmos, God had a view as if from the outside, as the entirety of infinite space itself was conceived both as his dwelling and his means of perception of everything. For certainly the view of physics that one gets from Newton's physics is a view in which every system that is described is seen from the outside, with the observer playing no role other than to be directly aware of all that is.

But there is a problem here for those of us who prefer a scientific to a religious understanding of the world. If there is a view of the whole universe that can only be read as a view from outside of it, it must be the view of God. Indeed, whose else could it be? This is then another reason to suspect that the idea of an absolute reality behind the appearances is a religious idea. Even in its modern forms that are not explicitly religious, the presumption that in physics we are constructing a picture of the world that could be read only as the view of an observer who is not part of the world, cannot be completely divorced from the presumption that there is a possible view of the world that apprehends some absolute reality. But this then, being impossible for us, implies at least the possibility, if not the existence, of a god.

Is there any alternative to this situation? Does faith in the possibility of a scientific understanding of the world necessarily involve the dream of the possibility of a view of the whole world which could be the view, not of any person, but only of a being apart from and outside of nature? Must science in this way always lead back to the religion of its inventors and first practitioners? Or is it possible to imagine a science that aspired to a complete and objective description of the universe while, at the same time, denying the possibility that that description could be read as the view of a being outside of the universe?

The answer to this question is yes. The purpose of the remaining parts of the book is to explain how the task of combining relativity with quantum theory both requires us to develop such a science and suggests how it might be accomplished. Recent developments that I will describe tell us that it is indeed possible to imagine a new approach to the weaving of an objective view of the universe that, by its very construction, denies the possibility of its being read as the view of an observer outside the world. For this reason, it is perhaps not too much an exaggeration to say that the present crisis of modern cosmology is also an opportunity for science to finally transcend the religious and metaphysical faiths of its founders.

But before doing this, it is necessary to first clear away one very influential reflection of the idea that the world was made for us, which is the anthropic principle.

FIFTEEN

BEYOND *the* ANTHROPIC PRINCIPLE

There is a certain kind of idea which is wrong, but which is also necessary at a certain stage of the development of a science. Examples of these are the earth-centered universe and the conception of elementary particles as points that take up no space. Such ideas are necessary because they allow people to express certain sets of observations in language that is then available to them, that otherwise could not be formulated meaningfully. Later, the new observations can become the basis for new theoretical frameworks, which will supersede the original ideas.

The anthropic principle is, I believe, such an idea. It is a vestige of the old meta-physics, whose prominence shows, more than anything else, the power of the nostalgia for the absolute I discussed in the last chapter. But, at the same time, the anthropic principle has up till now played a useful role in the development of cosmology. In this chapter I would like to argue that it is time we give it up, but let me emphasize that I urge this with a great deal of respect both for the inventors and developers of the principle and the role it has played to this point.

The anthropic principle arose as a response to the observation that much of the structure of the world depends on the parameters of physics and cosmology being finely tuned. This fact poses a challenge for science, as it necessarily points outside of the kinds of questions usually asked in elementary particle physics. The idea that the world was created for human beings is not new, so it was available to express these remarkable new observations. Like many such cases, in both science and ordinary life, the first idea that comes to mind as an explanation of some new fact is often only a temporary measure, which serves to distract us from the possibility that we are facing a true mystery long enough to collect that facts that we need to invent a better explanation.

Of course, this is a personal view, and it is self-interested as I believe I have a better explanation for the same set of facts. As such, it would not be fair to dismiss the anthropic principle so glibly, especially as it has been a central idea for many of the most thoughtful astronomers and physicists of the last decades. If the idea is insufficient, it should be possible to see why on its own terms. For this reason the anthropic principle deserves a full discussion as we attempt to address the broader philosophical issues raised by the problems of cosmology.

The anthropic principle takes several forms; I will discuss here only what is called the weak form of the theory. (The strong form is explicitly a religious rather than a scientific idea. It asserts that the world was created by a god with exactly the right laws so that intelligent life could exist.) The weak form, at least in the version that is easiest to defend, begins with a postulate that there are a large number of universes, or regions of the universe, of which ours is only one. This may seem a strange starting point, but as I've mentioned before, there are several hypotheses about cosmology that lead naturally to the idea that our universe is not unique. To this is added the postulate that the laws of physics, or at least their parameters, are different in these different universes. Given this, it is asserted that we, who are living, intelligent creatures, could only find ourselves in those members of the community of universes that are hospitable to our existence. This is then taken as the explanation for why we find ourselves in a universe with the improbable laws and conditions that are necessary for life to exist.

The point of the weak anthropic principle is that given these postulates, no other explanation is needed. All one needs is to postulate that there are a number of worlds, with a variety of properties, at least one of which is hospitable to our existence.

I do not think there is anything wrong with the logic of this argument. However, as it stands it leads nowhere, because it cannot produce a prediction that could be falsified by observation. This is because it is based only on the already (apparently) confirmed fact that there is intelligent life in the universe. It does postulate something beyond this, which is an unknown number of other universes with other properties. But the argument uses only one property of this collection, which is that it contains at least one world which, like ours, is hospitable to life.

It has sometimes been claimed that the anthropic principle has led to physical predictions, as in the case of Hoyle's discovery that the existence of carbon in the world would be inexplicable were a certain state of its nucleus not to exist. But, as I mentioned earlier, this is not a prediction based on the anthropic principle; for the prediction follows only from the assumption that lots of carbon exists in the world, which is a true fact.

The theory of cosmological natural selection is different from this, and does lead to predictions that could falsify it, because it dares to make much more specific assumptions about the collection of universes. It postulates a specific mechanism by which they are created, which leads to a specific prediction about a property that almost all of the members of the collection must share. Rather than requiring a specific, and already evident, property of one world, it makes assumptions which lead to a prediction that holds for almost all members of the collection. For this reason it has the possibility of explaining why there is life in our universe. As we discussed earlier, it seems that at least one way for a universe to make a lot of black holes requires that there be carbon and other organic elements, as well as stars that produce these elements in large quantities. The theory then predicts that our universe has these ingredients for life, not because life is special, but because they are typical of universes found in the collection.

This theory may certainly not be true. But as an example it shows what is wrong with the anthropic principle, and what we may ask of a theory that would really explain why the universe is hospitable to life. Like many first attempts to construct an explanation from a new set of facts, the anthropic principle is a view that forces us to take on faith certain important features about the world that might be explained by a real theory.

While the anthropic principle as I've stated it leads to no testable predictions, it has sometimes been extended by adjoining two more postulates, which do lead to predictions that can be tested. The first of these is that the existence of life is a rather rare property among universes, with the existence of intelligent life being even more rare. This suggests, but does not imply, the second hypothesis, which is that the more life a universe contains, the less common or probable it is.

In the literature on the anthropic principle one can find arguments for these assertions, but I must insist that as we cannot observe any other universe but our own, we are free to make any assumptions we like about them. There is no way then to deduce such hypotheses from the theory; in particular there is no way they can be deduced from the anthropic principle.

Once these postulates have been added, it follows that it is most probable that we find ourselves in a universe with the least possible amount of intelligent life. The reason for this is that if a universe that contains one such species occurs much more often in the set of all universes than a universe that contains zillions of such species, we are more likely to be in a universe that contains one such species.

This combination of hypotheses could then be falsified by the discovery that our galaxy contains a great many other planets with intelligent life. For this reason, some of the proponents of the anthropic principle have devoted no small amount of time and effort to attempts to demonstrate that we are the only intelligent species in the galaxy.

These arguments are worth examining. The argument most often mentioned may be called the "if they existed they would already be here" argument. The logic of this argument, which seems at first strong, is that if a large number of other planets in the galaxy are going to evolve intelligent life, it is extraordinarily likely that a fair number of them have already done so, it being very improbable that we are the first. Many of them must be at least tens, if not hundreds of millions of years, in advance of us, which means that they must already have evolved the ability to explore the galaxy. This being the case, it is very unlikely that they have not visited our solar system. If they did, they would have left some message telling us of their existence. Since we have not found any such messages they must never have existed.

It seems at first that there is a big hole in this argument. We do not really know that some other intelligent civilization would want to explore the galaxy. We certainly have no reason to think that they would want to explore it thoroughly enough to visit this particular solar system. But there is an answer to this worry. While it is true that there may be intelligent civilizations that have no wish to explore the universe, it is hard to imagine that there are none that do. And what the theory predicts is only that intelligent life is not common. If even a hundred intelligent civilizations arose in the galaxy during the last several billion years, it is hard to imagine that at least one of them would not want to explore the galaxy.

This turns out to be enough, because it is possible to argue that any civilization that developed the technology to explore the galaxy would quickly develop the technology to explore it thoroughly. This, it is claimed, is true for the same reason that there are now millions of Walkman tape players and millions of personal computers around: the cheapness of computer technology. Given the enormous time required to cross the galaxy, a civilization wishing to explore it would almost certainly rely on robots. But, as soon as one can build a few robots fit for this job one can also easily build enormous numbers of them. Given that robots could build other robots, the number of them that our curious aliens might construct is limited only by the available material. But this is also no limit, for as soon as they get off their planet they may turn the material of any number of other planets into robots.

Given the exponential rate at which the cost of computer technology is falling, it is sometimes claimed that all that is needed is one fifteen year old alien kid, who constructs a galaxy-exploring, reproducing robot from a science kit and sends it off. Thus, so the argument goes, a rather short time after any such civilization came to our technological level, they would be sending out enormous numbers of robots to explore the galaxy. So, why, if life is common, do we not see their robots here?

This may seem like a strong argument, but there are problems. Cute as the story may seem, the fact is that technology for travel does not get cheaper at the rate that computers have. A sailboat that could reliably cross the Atlantic (even a small one carrying only a computer) is still too expensive for almost all fifteen-year-olds, at least five centuries after it was first done. This is true in spite of the fact that many fifteen-year-olds could afford, and program, a computer that could navigate the boat across the ocean. So it is not obvious that a civilization that could easily make millions of robots could as easily make as many rockets to carry them to the stars.

A second problem emerges when we inquire into the time scales involved. The problem is that because there are more stars in the galaxy than there have been days since it was formed, it is difficult to think of a reason why such an intelligent race of living things would want to maintain a permanent presence in every solar system. It is much more likely that any given solar system would be visited from time to time to see what was up. If one tries to work out the time-scale for galactic tourism (which some authors do) it seems to come out to on the order of a hundred million years. Thus, we can only be assured by the argument that if other intelligent life exists in the galaxy, they have been here at least once in the last few hundred million years.

Thus, what we should look for to confirm the existence of other intelligent life in the galaxy is a message left for us sometime in the last several hundred million years. The question that must be asked is then not so trivial. Where could a message be left that would survive so long? Certainly if they could have traveled through the galaxy these creatures must know enough biology to understand that the life they may have found here a few hundred million years ago might one day bring forth a creature they could communicate with. But they must also have understood that the time-scale for this to happen is the same several hundred million years, so that they must seek a way to leave a message that would last that long. We can only assert that they have not been here if we are confident that we have looked for all possible messages that might be designed to last a hundred million years and not found them.

There are several different ways in which such messages might have been left. I will mention only one here. It is that our galactic sojourners might have left a message in the genetic code of some living creature. They could have inserted a message, coded in the language of nucleic acid bases, in the DNA of one or more living creatures, confident that the ability of living creatures to replicate DNA would keep the message relatively uncontaminated for time scales of this order. Although I am not about to propose a search for such a message, it is amusing to mention that there is a great deal of extra DNA in most species, that does not appear to play any biological role at all. The presence of this extra DNA is one of the puzzles of molecular biology. What makes it even more curious is that the amount of extra DNA varies enormously from species to species without apparent cause.

While it is amusing to contemplate that rather than being ignored by intergalactic travelers, we might have been used as a kind of cosmic answering machine, the point is that the possibility of messages left in this way deflates the argument for the non-existence of other intelligent life in the galaxy. And with that goes the only evidence I know of that has been proposed as supporting even an extended version of the anthropic principle.

Given the weakness of these and other arguments put forward in favor of the anthropic principle, we might ask why the idea has been so popular among scientists during recent years. I believe that one reason has been that attempts to realize cosmological theories based on atomistic reductionist precepts always end with a residual arbitrariness that must be resolved if one is going to assert that one has come to a final theory. In several different contexts, from string theories to the inflationary models, one is left at the end with an embarrassment of riches. A theory that started out attempting to explain the properties of our universe ends up either allowing or requiring the existence of a large number of alternative universes. In these cases the anthropic principle is often brought in to save the day, to explain how our universe is selected out of this large set. As I emphasized before, this is not logically wrong. The question is only whether it is possible to do better.

There is one more topic concerning the anthropic principle that ought to be mentioned in any such discussion. At least two of its advocates, George Ellis and Frank Tipler, have suggested connections between the anthropic principle and Christian theology. Some of this involves the strong anthropic principle, which, being explicitly theological, has little relevance to the discussion here. I have no desire to enter any discussion of theology or religious faith. However, it does seem that there are interesting analogies between the use of the weak anthropic principle I've discussed here and the use of religious, and particularly Christian, faith as an explanatory principle in pre-Galilean science. In both cases, our existence is made central to the logic of explanation employed in cosmology. In both cases, we play this role particularly as intelligent minds. The fact that we are alive is quite incidental (indeed this is made clear by some of the proponents of the anthropic principle who assert that a race of robots would serve at least as well, if not better, than us). Furthermore, in both cases our significance is heightened by our being the only intelligent life in the universe. I believe that these analogies are not accidental; and that what motivates some, although certainly not all, proponents of the anthropic principle is the nostalgia for a world in which there is both a god that stands eternally outside the world and the possibility of our transcendence by sharing, at least through comprehension of the world, something of his power and vision.

There is nothing wrong when a scientific idea shares themes with religious ideas. I mention the relationship between the anthropic principle and theology not to attack it, but to suggest that part of its appeal is that it is a scientific idea that makes us special, by making the answers to many questions about the natural world depend at least partly on our existence. The anthropic principle exemplifies a particular strategy for making our existence meaningful in the face of the apparent meaninglessness of the Newtonian picture of the cosmos. Not surprisingly, this strategy shares something with that favored by Newton himself, which is that it makes us special through our power to comprehend, which Newton, as a profoundly religious man, understood to be a poor reflection of God's infinite understanding. I am certainly not objecting to religion, or to its influence on people's thinking. I only want to suggest that when we are discussing such difficult questions which lie on the edge between science, philosophy, and religion, the discussion is more likely to succeed if we are cognizant and honest about the roots of our ideas.

There is also an interesting analogy between the weak anthropic principle and the Aristotelian view of the place of humanity in the universe, which is to be found in its formal logical structure. The anthropic principle is not a conventional scientific principle, to be applied in the context of a causal explanation. Instead, it functions like a teleological principle, as it has the logical form of an explanation in terms of final causes, because it takes the existence of intelligent life as the starting point from which aspects of the fundamental laws of nature are to be deduced.

One might ask whether it is legitimate to criticize the anthropic principle for eschewing conventional causality in favor of something analogous to a teleological principle? For it is not offered as the solution to a conventional scientific problem. Instead, the anthropic principle is offered as an answer to an apparently different kind of question: Why are the laws of nature as we find them?

The problem with trying to answer this question, while employing only conventional forms of explanation, such as causality, is that normally we do not think of a law of nature as something that may be true only at a particular time. But, causal explanations are necessarily connected with the notion of things changing in time. As a result, it might seem that if this question is going to be answered at all, it must be answered in some way that does not rely on the conventional scientific forms of explanation. If the laws of physics are timeless, if they are true everywhere and for all time, any explanation of them must lie in something that is not in the universe. It must instead rest either on some absolute principle, on faith or on appeal to an explanation by final causes.

Physicists who attempt to confront the question of why the laws of physics are as we find them and not otherwise, but who keep the conventional view that the laws of nature are timeless, are in the situation of biologists before Darwin, who wished to understand why the different species were as they were found to be. Before the discovery that the different species were created by the mechanism of natural selection, it was assumed that the species were timeless categories. Any attempt to understand why we find a particular set of species, and not others, living here on Earth had to rely on some principle outside of conventional scientific causality. They then had to search either for logical or structural principles, appeal to final causes, or appeal to God.

Indeed, the two problems are completely analogous. That is why the attempts to explain why the laws of physics are as we find them to be that also assume their timelessness have resorted to the same kinds of explanations as the biologists before Darwin. And, they have done, perhaps, about as well. For example, it is certainly the case that structural principles exist which limit the possible species. Elephants cannot fly. Mathematical consistency also limits the possibilities for the fundamental laws of physics. But neither of these principles suffices to explain why we find the particular species or the particular laws that we do. In each case the set of possibilities, while limited, is still large. Thus, in each case some recourse to either a God or to a teleological principle must be made.

The question why we find one set of species and not others became a scientific question when it was realized that the species are not timeless categories. They are created in a particular mechanistic and causal process, which occurred in time. Indeed, this first step, even before the process is identified, was radical enough, and it is fair to say that many of those who opposed the theory of evolution were attempting to preserve the old and mistaken idea that the species are eternal categories. Once one accepts the idea that the particular mix of species we find here on Earth is neither necessary nor permanent, the question becomes how new species can arise and why certain ones arise and not others. Once the question is put this way, there are not many options for an explanation—if one wants to stay within the conventional notions of causality. As many biologists have argued, random variation followed by selection is the only explanation that has been offered that is powerful enough to account for the enormous variety of species we find among living things on Earth.

For exactly the same reason, I believe that to make the question why the laws of physics are as we find them to be into a scientific question, we have to give up the idea of their timelessness. As in the case of the origin of the species, this is the only alternative to ultimately resting our comprehension of the world on either faith or an appeal to final causes. Once this step is taken, the only question that remains is to discover the right causal mechanism that acted in the past to produce the laws as we find them here and now. Once the laws of physics are posited to be time-bound rather than timeless, and once we therefore reject attempts to explain them on the basis of faith, mathematical consistency, or teleology, we make the question, why they are as we find them, solvable by more conventional approaches.

Perhaps it is time to summarize where these reflections have taken us. What I have been arguing is that the presumption that that which is timeless is somehow greater, more real, or more true than that which is time-bound is essentially a religious idea that, sooner or later, comes into conflict with our desire to have a rational comprehension of the world in which we find ourselves. It points always to a worship of an imagined world, and with it an imagined intelligence, that exists somehow outside of our world, free of the bounds of time that life and death impose on us. The opposing idea, which natural selection makes possible first of all for biology, is of a world that makes itself, that is self-organized.

But, if this can work for biology, why could it not also work for the universe as a whole? If the laws of physics exist eternally and independently of the universe, then the reasons for them are beyond our rational comprehension. But if they are created by natural processes acting in time, it becomes possible that we may come to understand why they, and ultimately our world, are as we find them to be.

Thus, no less than for any other regularities we observe in the world, the extent to which we bring the laws of physics inside of time is the extent to which we make them amenable to rational understanding.

Time is then the key for the aspiration to construct a theory of the whole universe. As long as the eternal only means for a very long time, longer than the time over which we have been observing the world, we can pretend that science is the search for the eternal. But, this is an illusion; as soon as we confront the problem of making a theory of cosmology we realize that to the extent to which it is also a scientific theory, it cannot be a theory of the eternal. Time is thus the most central and most difficult problem we must face as we attempt to construct a theory of a whole universe.

PART FOUR

EINSTEIN'S

LEGACY

What are space and time?

In Buddhism, the concept of linear time, of time as a kind of container, is not accepted. Time itself, I think, is something quite weak-it depends on some physical basis, some specific thing. Apart from that thing, it is difficult to pinpoint—to see time. Time is understood or conceived only in relation to a phenomenon or a process.

—The Dalai Lama

SIXTEEN

SPACE *and* TIME
in the NEW
COSMOLOGY

When people speak of political change, they often speak of a rearrangement of the relationship between the individual and society. This is a euphemism, for society is an abstract concept that refers only to those human beings that are alive in one time and place. This is not to say that there are not hierarchies of organization in human society, but each interaction I have with any level of this hierarchy is really only an interaction with one or more people, even if the exchanges may be increasingly scripted as the hierarchy is ascended. What is then rearranged when society evolves is nothing other than the myriad of relationships between individual human beings. We speak of society because, our social instincts having been formed through millions of years of evolution, during which we lived in groups of never more than a few dozen, we have difficulty envisioning directly the fantastic complexity of the human relationships that tie together the world we have made.

By the same token, abstract conceptions of space and time came into use when human beings first began to perceive the immensity and complexity of the universe. As long as the universe contained only those things that could be seen with our eyes, completely concrete notions of space and time (as we find, for example, in Aristotle) sufficed. It is only after the telescope had exploded the stellar sphere and we had to conceive of a world containing uncountable suns at unfathomable distances that it became useful to talk of space as something absolute, distinct from that which it contained. And similarly, it was only when people began to grasp that six thousand years might be but the briefest hesitation in the life of the cosmos that a concept of time as something distinct from change became common.

But, as I hope to convince the reader in these next chapters, space and time, like society, are in the end also empty conceptions. They have meaning only to the extent that they stand for the complexity of the relationships between the things that happen in the world. And, just as we have learned in this century the cost of too much abstraction in our understanding of politics, we are learning also now that talk of space and time as abstract entities may hide complexities and structures that we need to understand if we are to have a theory of a whole, complex cosmos such as our own.

In this fourth part of the book we will learn that complexity and variety are not only needed if the world is to be interesting enough to contain galaxies or stars. We will see that complexity is not an option, it is required of a world constructed according to the principles that underlie our modern understanding of space and time.

We may begin very simply, by asking how we talk about where things are. One way is to describe their position relative to me: my left shoe is on my foot, my computer is in front of me, my guitar is on my favorite chair which is ten feet to my left, my cat is on my head. This suffices for most purposes, but it seems not completely satisfactory, for where am I?

I am sitting in my apartment, on Eleventh Street in New York, on the planet earth, which is a bit past the winter solstice in its orbit around the sun, and so on. What I am doing now is telling you where I am relative to other objects in the universe. As long as you are familiar with those objects you will be able to find me. But this is still not a complete answer, for where are the earth and sun?

The sun is in a particular orbit, near a particular spiral arm of a certain galaxy, which is in the local group, which is. . . .

Perhaps this still seems not entirely satisfactory, but a moment's thought will, I think, convince the reader that this is what we always do when we talk of location: we give an address, or a room number, or directions for how to get to the place through the woods. All ordinary talk of place is talk of relative position.

It is then natural to ask whether there might be something more to where we are than relative position. Is there some way to tell, absolutely and without reference to anything else, where something actually is?

In the history of physics and philosophy there are two great traditions about the nature of space that stem from the two profound answers that may be given to this question, which are yes and no. These lead to two views about the nature of space, which are called the absolute view and the relational view.

Newton was the great advocate of the absolute position. In the introduction to his great *Principia*, which was the culmination of the Copernican revolution, he could not have been more direct: "Absolute space, in its own nature, without relation to anything external, remains always similar and immovable ."

Many people at the time found this absurd, and many people still do. Leibniz, among Newton's critics, saw most deeply why Newton's conception could not, ultimately succeed. In a debate with Clark, one of Newton's followers, he argued: "These gentlemen maintain, then, that space is a real absolute being; but this leads them into great difficulties

. . . As for me, I have more than once stated that I held space to be something purely relative. . .space being an order of coexistences. . . ."

The argument Leibniz makes for his relational point of view is one of the most important in the whole history of philosophizing about nature. I cannot do better than to reproduce his own words here.

> I am granted this important principle, that nothing happens without a sufficient reason why it should be thus rather than otherwise. . .I say then that if space were an absolute being, there would happen something for which it would be impossible that there should be a sufficient reason, and this is contrary to our axiom. This is how I prove it. . .if we suppose that space is something in itself, other than the order of bodies among themselves, it is impossible that there should be a reason why God, preserving the same positions for bodies among themselves, should have arranged bodies in space thus, and not otherwise, and why everything was not put the other way round (for instance) by changing east and west.

I expect that even the most skeptical reader will concede that Leibniz has a point. We have already seen the power of this *principle of sufficient reason,* as the philosophical idea behind the gauge principle, on which the standard model of particle physics was built. We will see shortly that it is also the idea behind relativity and quantum theory. It is hard to think of any argument in the history of science that echoes more loudly today than Leibniz's dissent from Newton's physics.

More than perhaps any other philosopher of the Western tradition, Leibniz seems to have believed that it must be possible to give a specific and explicit reason for everything that happens in the universe. As we see in this passage, for him this meant that it must be possible to give an answer to any question one could pose about why something occurs in the way it does, *and not otherwise.*

This principle of sufficient reason expresses a supreme faith in the rationality of the world. Why did Leibniz believe in it so strongly? I believe the reason is that he had thought hard, perhaps harder than anyone before or since, about what would be required to construct a theory of the whole universe. This thinking led him to a conclude that any cosmological theory must satisfy this principle because, unlike a theory that describes only a part of the world, there is no cause outside its domain.

I suspect that the reader will agree that it is impossible to think of a reason why the universe might not have been created, in its entirety, two feet to the left. This being so, it makes no sense to talk about where the universe, as a whole, is. Moving the entire universe two feet to the left is not going to have any imaginable effect on our perceptions, or on the future behavior of the things in the universe.

If it is not going to make any different whether the universe is as it is, or two feet to the left, does it still make any sense to distinguish the two? This question is exactly what separates the relational from the absolute view of space. Newton, as an absolutist said yes. Leibniz said no.

Exactly the same arguments can be given with respect to the question of the nature of time. What does it mean to say that something happened at a particular time? What do I mean when I ask what time it will be (for you, *was*) when I type the question mark at the end of this sentence? The necessary qualification in this sentence illustrates the point: all talk of future, past, and present is relative to the moment of time of the person who is speaking. All ordinary talk of time is relational.

When we use a clock or a calendar to locate an event in time, we are giving its time relative to a system that has been set up by human beings. Although that system is arbitrary, its use is necessary. Without such a system, we would be lost, for we have no access to any absolute notion of when something happens. Of course, we can attempt to tie our system of measuring time to something larger. We may tie the calendar to the motion of the Earth around the Sun; if we like we could tie it further to the motion of the Sun around the galaxy, and so forth.

Ultimately, we come to the same question we confronted in the case of space. Is there in the end some absolute notion of time, which our clocks only imperfectly measure? Or, in the end, must all talk of time remain on the level of relations? Must time always be told with respect to some arbitrarily chosen clock?

Unsurprisingly, Newton, as described in his *Principia*, chose to believe in an absolute time: "Absolute, true and mathematical time, of itself, and from its own nature, flows equably without relation to anything external."

Leibniz answered along lines similar to his discussion of space:

Suppose someone asks why God did not create everything a year sooner; and that the same person wants to infer from that that God did something for which there cannot possibly be a reason why he did it thus rather than otherwise, we should reply that his inference would be sound if time were something apart from temporal things, for it would be impossible that there should be reasons why things should have been applied to certain instants rather than others, when their succession remains the same. But this itself proves that instants apart from time are nothing, and that they only consist of the successive order of things.

The logic, as we see, is exactly the same as before. Applying the principle of sufficient reason, Leibniz cannot believe in absolute time because he cannot believe any rational answer can be given to the question of why the universe was created when it was, and not a year earlier. Since no answer can be given, what we want is a conception of time that does not allow us to ask the question. A conception of time that satisfies this is called relational: it is based only on relations between things that happen in the world.

This argument between the absolute and the relational views of space and time has been waged passionately since the time of Newton and Leibniz. The argument is important because the side of it we take colors the whole of our cosmological theory. How we conceive of the universe as a whole depends on what one thinks space and time are.

For example, elementary particles that live in absolute space and time are different kinds of things than are particles that live in a world in which space and time are only relations. Against a background of absolute space and time it makes sense to speak of a universe with only one particle in it. That particle's motion is defined against the background in exactly the same way no matter what else is in the world. But if space and time arise only from the relations among the particles, a universe with one particle in it could not even be described using words like position, space, and point. A universe conceived as a collection of particles moving in absolute space and time is thus a very different thing than a universe of relations that define what is meant by space and time.

Because of this, the problem of constructing a unified theory of elementary particles is tied to the problem of constructing a new theory of space and time. A shift in the conception of space and time, from a Newtonian to a Leibnizian framework, cannot leave untouched the Newtonian conception of what a particle is.

Throughout the history of physics, there has been a tension between atomism and the relational conception of space and time. In its naive form, atomism teaches that each particle is endowed with properties independently of whatever else may exist in the universe. This implies an absolute notion of space and time. The most basic properties a particle can have are where it is, and how it moves. If these are to be defined independently of anything else, they can only be defined with respect to some absolute space and time that do not depend on relationships among things.

On the other hand, if there is no absolute space then the position of a particle cannot even be spoken of without bringing in its relationship with the rest of nature. For this reason, atomists have tended to be suspicious of relational approaches to space and time, while relationalists have tended to be suspicious of the radical atomist idea that the properties of each particle are defined independently of the others.

In the twentieth century we seem to live in a world composed of elementary particles that move in the relational space-time of Einstein's theory of general relativity. There is certainly good reason for both atomists and relationalists to see twentieth-century physics as their particular triumph. But, twentieth-century physics is not finished, and the old tension between the relational and atomistic philosophy confronts anyone who attempts to construct a theory that brings together these disparate elements. For example, the standard model of particle physics is relational in its use of the gauge principle, but atomistic in its description of the elementary particles. It is likely that the ultimate success of twentieth-century physics will rest on how this tension between the atomist and relational views may be resolved.

Atomism compels us to postulate that the world is essentially simple, while relationalism pulls the opposite way, towards a vision of the world as a complex system. What is at stake is not the question of whether there are fundamental particles, but where they get their properties. According to the atomistic view, particles simply have the properties they have, regardless of context. There is no reason for a world composed of atoms with fixed properties to be complex. The relational view requires more; it cannot make sense unless the universe is sufficiently complicated.

Recall another of Leibniz's principles, the *identity of the indiscernible,* which requires that any two particles which have the same relationships with the other things in the universe must be in fact the same. For if things are only distinguished by their relations, then there is no way to tell them apart. A world constructed according to these principles must be complex enough to allow observers to distinguish each particle uniquely, by talking about their relationships with the other particles in the universe.

This is the case even when we are simply talking about where things are. Suppose we are able to send a satellite on a journey out of the galaxy. We would like to put a message on it to let any intelligent creature who finds it know from where it came. We have no way of knowing how far it will travel before it is picked up, nor will whoever picks it up be able to reliably tell from what direction it came. There is also, of course, the problem that we cannot use language to describe our location. Furthermore, as we can assume no common landmarks or reference points, we cannot draw a map, for the satellite's discoverers will not know how to orient it. Is there a way we can draw a diagram so that, assuming only they have sufficient information about the universe, they will be able to find us?

In addition, it is likely that our intergalactic letter in a bottle will travel many millions of years before it is discovered. Its finders will want to know when it was sent. Without any common language, is there any way we can tell them?

There is an answer to both of these questions. Rather than trying to draw a map of the universe with an arrow pointing to where we are, we must do the opposite, which is to draw a picture of the universe, as we see it. We simply draw a picture of the sky, showing the brightest stars, galaxies and groups of galaxies, as seen from earth. If our neighbors in the cosmos have sufficient knowledge of the universe, they will be able to find us by searching for a place whose view of the universe is identical to ours. In order for this to work, the universe must have a certain complexity. It must be possible to draw a picture of our sky in sufficient detail that there will be no other place in the universe from which the sky looks exactly the same.

In fact, our universe is complex enough that it should be sufficient to give a picture of the brightest stars and galaxies, as seen against our sky. Although there are perhaps a trillion stars in our galaxy, we can be sure that from no two does the pattern of bright stars on the sky look identical. Similarly, we can locate our galaxy relative to all the others by giving the pattern of the brightest galaxies on our sky. If we give enough detail, this pattern should differ enough among each of the trillion or so galaxies to make us unique.

Furthermore, because our universe is evolving in time, this should also suffice to tell the finders of our bottle when it was sent. There is only one period in the history of the Earth when the sky is exactly as we see it, and it would be an extraordinary coincidence were our sky to be reproduced by the view from any other planet at any other time.

What if the universe were too simple for this to work? What if there were many planets whose people saw exactly the same sky? This is the question that divides the Newtonian from the Leibnizian conception of the universe. For Newton this would be no problem. It would not matter for him, even if the looked the same from every point. For Leibniz, on the other hand, this would be impossible. If we accept his principle of the *identity of the indiscernible*, any two places in the universe with exactly the same sky must be the same place. A completely symmetric universe, in which the world looks exactly the same from all points, must, according to Leibniz, consist really of only one point. A completely homogeneous three-dimensional universe, such as has often been used to model our cosmos, is not, if this point of view is right, something that it even makes sense to talk about.

How differentiated does the universe have to be, according to Leibniz's principles, in order to speak meaningfully of the universe as a three-dimensional space that exists in time? To use a word favored by Leibniz, the universe must have so much *variety* that no two observers experience the same thing, and no moment ever repeats itself.

Before, we learned that life could not exist in a universe unless that universe was sufficiently complex. We have just found that the universe must be complex for a completely different reason, which is so that it is possible to describe where we are in the world in terms of our relationships to real things. What this means is that the questions of why the universe has structure and why there is life in the universe are profoundly connected to the question of what space and time are.

The common view, which we have inherited from Newtonian science, is that we live in a universe composed from a great many identical parts. The parts—the elementary particles—are each very simple, and each is identical to every other of its kind. Their arrangement happens to be very complex, but this is in no way necessary—it is just our good luck. The opposing picture, posited by Leibniz and realized by Einstein, is of a world made by a great many particles, each of which is different. While each proton has the same charge and mass as every other, each is different, because each occupies a different place. Each elementary particle has a unique relation to the whole. The world they make is necessarily complex because a certain minimal complexity is required if each proton is to be distinguished from all the others by its relationships to the rest. We may say that where something is determined by its view of the rest, which is to say by its relationship to the others. If each of a vast number of particles is to have a unique view of the rest, the world must have a fantastic variety of views.

To make sense of this, we need a notion of the complexity of a system that is based on the idea that in a complex system each part has a unique relation to the whole. Such a measure of complexity does exist, and it was directly inspired by Leibniz's philosophy. It is called, the *variety* of a system, and it is defined in terms of how much information is required to distinguish each part of the system from the others, by describing how it interacts with the rest of the system. Formally, given any elementary part of the system, we may call its neighborhood those things that are nearest to it, or that it interacts most directly with. The variety of the whole system is then defined so that a system has more variety the less information is needed to distinguish each part from all the others by describing its neighborhood.

The notion of variety provides a way to think of what it means for a system to be complex that can be applied as easily to a biological system as it can to space and time. A universe with a great deal of variety is one in which it is easy to tell where you are just by looking around. A biochemical system with a lot of variety is one in which each kind of molecule is easily described by naming the other molecules it interacts with. Thus, using the notion of variety we may make a connection between the kind of organization required for life and the kind required for a relational notion of space to make sense.

Such a notion of complexity may also be more useful than others that are based on entropy, because highly symmetric systems like a crystal and highly organized systems like a living cell both have low entropy compared to thermodynamic equilibrium. But a regular crystal has zero variety because it is impossible to distinguish the atoms in terms of what is around them while a living cell, on the other hand, exhibits a high level of variety, as each of its many different kinds of enzymes will interact with the others in a distinct way.

By contrast, a universe in thermal equilibrium has a low level of variety. It may have more variety than a completely symmetric universe, but it still will have much less than our universe, which is far from equilibrium. In an equilibrium universe, it is not even clear that Leibniz's principle of the identity of indiscernible can be realized; to distinguish each atom from all the others would take a stupendous amount of information. If the amount of information required becomes so large that it could not be stored in the universe itself, then it may be more proper to say that the principle is simply not realized. If this is the case then the possibility of a universe in equilibrium is precluded in any relational theory of space and time that agrees with Leibniz's principles.

What is at stake in the conflict between the absolute and the relational views of space and time is then much more than the academic question of how space is to be represented in physics. This question goes to the roots of the whole of the scientific conception of the universe. Does the world consist of a large number of independently autonomous atoms, the properties of each owing nothing to the others? Or, instead, is the world a vast, interconnected system of relations, in which even the properties of a single elementary particle or the identity of a point in space requires and reflects the whole rest of the universe? The two views of space and time underlie and imply two very different views of what it means to speak of a property, of identity, or of individuality. Consequently, the transition from a cosmology based on an absolute notion of space and time to one based on a relational notion—a transition that we are now in the midst of—must have profound implications for our understanding of the place of complexity and life in the universe.

SEVENTEEN

THE ROAD
from NEWTON
to EINSTEIN

Just as the world, in Leibniz's vision, consists of a network of relations, so anyone who attempts to think about the world discovers that their thoughts are imbedded in a network of the thoughts of other people, both past and present, so that most of the ideas that they may for a moment have mistakenly taken for their own were in fact only passing through, having traveled from mind to mind from some origin far in the past. In my case, I always considered myself a follower of Einstein, having gone into physics after an encounter with his writings, but I was largely unaware of the roots of his way of looking at the world until a year or so after I finished my university studies.

I was spending the summer in Oxford, at the invitation of Roger Penrose, when a friend told me that there was someone called Julian Barbour who had thought very deeply about the nature of space and time. I wanted right away to meet him, but this didn't happen for some time, as he lived a quiet life in the English countryside and came into Oxford only from occasionally to talk with people. Finally, my friend arranged a meeting at his farm, and after a lovely journey through the countryside, we found ourselves at the door of a seventeenth-century farmhouse. After introductions and a walk around the village, we settled down to sherry around the fireplace and began to talk physics and philosophy. He, playing the role of the English gentleman to his brash young American guests, asked first what we had been working on. I told him of my efforts to construct a theory to unify quantum theory with space and time. He listened politely then asked if I had ever read Leibniz. When I replied no, he said, "Well, perhaps you ought to" and began to explain to me how Leibniz's philosophy could provide the starting point for a theory of cosmology and how he had tried to accomplish this in his own work. Ever since then, Julian has been a friend and a colleague. Among the things we have explored together is the notion of variety I described at the end of the last chapter. More than this, he has been for me a kind of a philosophical guru, as much of the thinking that led to this book had its beginnings in conversations between us. I have since also found that Julian is not the only deep thinker in science who considers himself a follower of Leibniz. So also do many others such as David Finkelstein, Louis Kauffman and John Wheeler.

There is, however, danger as well as comfort in the company of admired thinkers. In science, as in politics or love, one can have all the good arguments and still be in the wrong. When it comes down to it, what matters is not whose story is more logical or beautiful, but which leads to the greatest effect.

The relational philosophy of space and time was, for the two centuries following Newton, exactly such a case. It certainly seems that, in his debates with Newton and his followers, Leibniz had the stronger arguments. After hearing his arguments against absolute space and time, I think that many people will agree that one can continue to believe in them only by an act of something very like religious faith. But then we must ask why, if Leibniz had the better argument, did Newton choose to take the absolute point of view? Certainly he was not dumb, many consider him to have been the greatest physicist and mathematician of all time. Is it possible that he was just a bad philosopher?

We must also ask why, if it was based on the worse philosophy, was it Newton's physics, and not Leibniz's, that triumphed and became the basis of science for the next two centuries. Certainly Leibniz was also not dumb. One philosophy professor I know calls him the smartest person who ever lived. If Leibniz had the better philosophy, why could he not invent the better physics to go along with it? Or, to put the point slightly differently, if it is so easy to criticize Newton's idea of absolute space and time, why was it possible to construct a theory based on them that was so successful for so long?

Although I have posed them somewhat whimsically, these are not just rhetorical questions. From their answers we can learn several important lessons.

The first of these is that Newton's view of space and time was initially more successful than Leibniz's because it is much easier to construct a description of motion based on an absolute notion of space and time than on a relational notion. The problem is that any relational description is necessarily complicated, because to tell where something is relationally one must bring the rest of the universe into the picture. Any relational theory of motion must treat the universe as a complex system involving many particles. This is hard to do.

By contrast, an absolute theory of motion can be constructed one particle at a time, as the motion of each individual particle is described with respect to the fixed structure of absolute space and time. The laws that any one particle obeys are not affected by what the others are doing.

By thus paying a certain price, an absolute theory of space and time allows us to realize the idea that a fundamental theory of physics should be simple. This price is the postulation of an absolute entity, which exists to give meaning to properties and tie the whole thing together: absolute space and time. Newton understood this well, and was willing to pay the price, because he saw that none of his contemporaries had been able to construct a useful theory of motion based on relational ideas. Einstein also understood this, and has been quoted by John Archibald Wheeler as praising Newton's "courage and judgment" for going ahead to construct a workable theory against the better philosophy. He may also have been thinking of this when he wrote to Slovene, "Unless one sins against logic, one generally gets nowhere; . . . one cannot build a house or construct a bridge without using a scaffold which is really not one of its basic parts."

But the main reason for Newton's success lies in the way he actually used absolute space and time to give a theory of motion. To understand this, we should ask: What good does the notion of absolute space actually do for Newton's theory of motion? After all, absolute space is completely unobservable. How can something that is unobservable play a role in a scientific theory?

Furthermore, even if one believes in absolute space, this belief gives one no help when one attempts to actually locate one's position. If one wants to know where one is, the only way is to mark one's position with respect to some real physical objects. So, what did Newton have in mind when he introduced the idea of absolute space?

The answer to these questions is that it is not quite true that empty space has no observable properties. It is just that in order to see them, it is necessary to ask about something more than just where things are. To see them we must ask questions about how things move.

For example, there is an intrinsic difference between rotating and non-rotating. As anyone who has ever ridden a roller coaster knows, when you rotate you feel it in your stomach. If I ask whether you are rotating now at this moment, you do not need to think to answer no. If the rotation is fast enough, it will be evident, without any need to refer to anything else.

Let us imagine that we are kidnapped by some philosophical terrorists. We wake to find ourselves in a room with no windows or doors. Like the characters in Sartre's play *No Exit*, we have no way to see outside or to contact the outside world. To pass the time and also to plot our escape, we may try to learn about our true situation. One thing we may want to know is to what extent we can, trapped in this room, learn something about where we are.

We may begin by asking about our position. Are we able to detect the position of the room without being able to look out? A few moments thought I suspect will convince the reader that the answer is no. This reflects the fact that the only notion of position that can be given a concrete meaning is position relative to something else.

Good enough! But surely if we cannot determine where our room is, at least we ought to be able to find out if it is moving or not. For example, as I just described, will be able to agree about whether the room is rotating or not. But can we detect all motions? Without being able to look outside, can we tell in all circumstances whether we are moving or not?

The answer to this question is no. Even with all the most sophisticated equipment, there is nothing we can do to determine whether our room is at rest or moving, as long as it is moving at a constant speed, without rotating or accelerating.

This is now common experience, due to travel in airplanes, trains, and cars. But in the early seventeenth century, travel was not so common, and people had to be convinced that they were in fact moving. One of the classics of science writing is Galileo's description of this, in his book, *Dialogues Concerning the Two Chief World Systems*.

> Shut yourself up with some friend in the main cabin below decks on some large ship, and have with you there some flies, butterflies, and other small flying animals. Have a large bowl of water with some fish in it; hang up a bottle that empties drop by drop into a wide vessel beneath it. With the ship standing still, observe carefully how the little animals fly with equal speeds to all sides of the cabin. The fish swim indifferently in all directions; the drops fall into the vessel beneath; and, in throwing something to your friend, you need throw it no more strongly in one direction than another, the distances being equal; jumping with your feet together, you pass equal spaces in every direction. When you have observed these things carefully (though there is no doubt that when the ship is standing still everything must happen this way), have the ship proceed with any speed you like, so long as the motion is uniform and not fluctuating this way and that. You will discover not the least change in the effects named, nor could you tell from any of them whether the ship was moving or standing still.

In jumping you will pass on the floor the same spaces as before, nor will you make larger jumps toward the stern than toward the prow even though the ship is moving quite rapidly, despite the fact that during the time you are in the air the floor under you will be going in a direction opposite to your jump. In throwing something to your companion you will need no more force to get it to him whether he is in the direction of the bow or stern, with yourself situated opposite. The droplets will fall as before into the vessel beneath without dropping towards the stern, although while the drops are in the air the ship runs many spans. The fish in their water will swim toward the front of their bowl with no more effort than toward the back, and will go with equal ease to bait placed anywhere around the edges of the bowl. Finally, the butterflies and flies will continue their flights indifferently toward every side, nor will it ever happen that they are concentrated toward the stern as if tired out from keeping up with the course of the ship, from which they will have been separated during long intervals by keeping themselves in the air.

But what about the dizziness of the amusement park ride, what about the way my stomach feels when the pilot announces we are passing through what they like to call "mild turbulence?" The point is that it is possible to detect any form of motion that involves either a change in direction or a change in speed. If our room spins around, this is a change in the direction of motion and we can feel it. If it suddenly accelerates, we can feel it as we, along with the birds, the fish and the butterflies will be thrown against the back wall. But, we cannot detect uniform motion, which is motion in a straight line at a constant speed.

The realization that uniform motion cannot be detected played a crucial role in the Copernican revolution, because Galileo had to explain how it is that we don't feel the motion of the earth. At the same time, it posed a trap for Galileo, because his adversaries could respond: "So, you claim that the earth is moving, but you also claim that we can't feel this motion. Is there any way, then, to prove that we are moving? If motion is just a relative concept, perhaps it would have been prudent (especially in a time of religious war and social upheaval) to have stuck to doctrine and continued to teach that it is the Earth and not the Sun that moved."

The Jesuits were no slouches, and they indeed caught Galileo up on this point. The result (to greatly simplify a story that is still controversial) was that Galileo was imprisoned, perhaps tortured and had to recant his views on the motion of the earth before the inquisition, after which he spent the rest of his life under house arrest.

Galileo could have answered his critics by pointing out that the Earth's rotation and revolution are accelerated motions. While we don't feel them in our stomachs, they can be detected by more subtle means. Instead, on this point he gave an erroneous argument, backed up by a lot of rhetoric, that his enemies were easily able to see through. Why he did not do better is one of the mysteries of his story, but this is not the first time that the inventor of a new idea was blind to some of its consequences.

It was left for Newton to realize the importance of accelerated motion, and put the whole picture together. But, in doing so, he faced a problem. It is clear that there is a physical difference between rotating and non-rotating motion. But, what is the origin of this difference? If, in the end, all motion is relative, is there anything we are rotating relative to when we feel dizzy?

We may ask the same question about acceleration. It is clear that there is a physical difference between accelerating and moving at a constant speed—we feel it when we push on the accelerator. So there is something special, and apparently intrinsic, about the notion of motion at a constant speed.

The problem is that to move at a constant speed means to cover equal distances in equal intervals of time. But this implies some standard that distance and time is being measured relative to. When we say equal distances, we imply that there exists some standard of distance, some ruler, that we are moving relative to. But how do we know that the ruler itself is not in motion?

The same problem arises with respect to time. To speak of uniform motion, we need a clock. But which clock should we use? How do we know that the clock we are using is not itself speeding up and slowing down?

Newton's predecessors and contemporaries, such as Descartes, Huygens, and Leibniz, had struggled with these issues. If all motion is relative, then to what do we attribute the difference between rotating and not-rotating motion? Newton introduced his notions of absolute space and time to resolve this conundrum. This made everything much simpler. Newton had to suppose only that at each point of space there abides an intrinsic sense of rotating and non-rotating. This sense is, according to him, a property of absolute space that is fixed once and for all. Rotation, and acceleration in general, are then defined directly with respect to absolute space and time.

By making use of absolute space in this way, Newton was able to formulate apparently meaningful laws of nature. The most important example of this is his law of inertia (also called Newton's First Law) which asserts that bodies with no forces acting on them move in straight lines at constant speeds. We might worry that such a statement is not meaningful, because it leaves unspecified how "straight line" and "constant speed" are defined. But for Newton the law is meaningful because these are meant not relative to any particular physical reference point, but with respect to absolute space itself.

Of course, saying something is so does not make it so. There is still a problem to be faced, for it remains true that absolute position and absolute time are absolutely unobservable. How are we, who live in the real world, to apply Newton's laws? All we can perceive is motion with respect to real bodies. The only time we can measure is the time counted by real physical clocks.

God can solve this problem because He knows where everything is in absolute space. But we are not gods, and we have no direct access to the absolute. How then are we, who perceive and measure only relational space and time, to apply laws that are expressed in terms of their absolute counterparts?

This is not unlike a problem some religions face: how are we to know what God's word is if it can be heard only through the voices of other human beings? Both the genuinely pious and the genuinely schizophrenic can claim to speak the thoughts of the divine. That this dilemma is unavoidable in principle has not for one moment stopped the exercise of religion, to both good and bad ends. What happens of course is that the believers in a religion find ways to exercise their best judgments to decide which claims are most likely to represent the real word of the absolute. And who can say that some of the time they do not choose wisely? One does not have to be a Buddhist to admire the wisdom of the person whose words opened the last chapter.

It is the same with respect to the problem of connecting Newton's absolute space and time with space and time as they are measured by real observers. Any experimentalist who wants to test Newton's laws must make a judgment about which clocks and which rulers are most likely to measure the absolute quantities. The success of the theory, over several centuries, tells us that whatever confusion there may have been about the logical foundations of the subject, it was not difficult for experimentalists to actually design situations in which Newton's laws could be used to good effect.

No wonder physics is so hard to learn. As it is usually taught to students, Newton's physics does not make complete sense, for they are almost never told the whole story. Position, velocity, and acceleration are usually introduced as if they have simple and obvious meanings, but they do not. Even more difficult are the concepts of force and mass; the definitions given of them in textbooks are almost always circular. The students are seldom told that if they are puzzled it may be for good reason, or that the things that confuse them have been debated for centuries. Some figure it out for themselves. Many go away with an unjustified sense that they cannot learn science.

After reading an introductory physics textbook, one might worry whether there is any completely coherent or logical way to understand Newton's physics. In fact there is, but it was not known to Newton, as smart as he was. It was only worked out in the nineteenth century. The problem that had to be solved is the same as the religious one: whose observations do you trust? What was needed was a way to replace the informal judgments of working experimentalists, astronomers, and engineers with a formal distinction that would allow anyone to objectively distinguish those observers whose measurements should be trusted to mirror the passage of absolute space and time. This problem was eventually solved by the introduction of a very useful conceptual device, which was to have a dramatic effect on the development of physics.

The step that was needed to make complete sense out of Newton was revolutionary, for it was no less than the explicit introduction of the observer into the description of the laws of motion. This made it possible to clarify and to formalize the sense in which statements about motion are meaningful only when expressed relative to a particular set of observers.

According to the theory, an observer is someone or something that carries around rulers, by which they can locate the position of a body in motion with respect to themselves. Typically, an observer is pictured as carrying three perpendicular rulers, which serve as explicit physical coordinate axes which allow them to give the position of anything in motion. The observer also carries along a clock, which may be used to note the time of various events.

The whole apparatus-the observer, rulers and clocks is called a *frame of reference*. What Galileo was describing in the passage I quoted is exactly a frame of reference: the walls of the ship's stateroom provide exactly a set of reference points against which the sailors can measure the changes of position of the various animate and inanimate objects in the room.

This may not seem very useful, because all that is observed is the relative motion between the objects and the frame of reference. If the frame of reference can move arbitrarily—if it can accelerate or rotate in any way imaginable—it will be difficult to sort out the causes of the relative motion. We are back to the dilemma of relative motion that led Newton to posit the existence of absolute space. All that we have done is to make the problem explicit, by bringing the observer into the story. But, now, because we can speak of the observer, there is a way out. We can make use of the fact that not all observers are the same. We can use the fact that, as we said before, some observers can feel that they are moving, while others cannot.

The observers who feel no effects of their motion—who are, as we said, not accelerating—may be put in a special class. They are called *inertial observers*. What is good about the inertial (non-accelerating) observers is that the motions of things seem especially simple when described relative to one of them. For example, Newton's first law of motion may be stated in full in the following way: A particle with no forces acting on it travels in a straight line at a constant speed, *when seen by an inertial observer*. Thus, there is no need to ever speak about absolute space or time. One may express Newton's physics entirely in terms of what inertial observers see.

There is one point about the inertial observers which is at first confusing, but then turns out to be the key to the whole subject of motion: there are many inertial observers, and they are all moving with respect to each other. For example, it is not hard to see that any observer that moves in a straight line at a constant speed with respect to one inertial observer will also be an inertial observer. This is because if the first is not accelerating or rotating, neither is the second. Thus, while they will differ about the speed and direction of the motion of a particle, any two inertial observers will agree that it is moving in a straight line at a constant speed.

This whole game works because none of these different inertial observers can feel any effect of their relative motion. Because what is felt intrinsically is only acceleration; each observer is insensitive to any effect that might be ascribed to their own motion. Because of this, each of the two inertial observers can equally well declare that, as they feel no effects of motion, they are at rest, and it is the other that is moving. Because each of them may equally well declare this, with equal validity, we see that the notion of "at rest" can only be a relative concept— one can only say that one is at rest with respect to a particular observer. Not only is there nothing observable in the world that might correlate with an absolute notion of place, there is nothing observable in the world that might correlate with an absolute notion of at rest. Whether something is at rest, or moving with a constant velocity, depends on what frame of reference one is using to describe the motion.

This is the meaning of what, since Einstein, has come to be called the relativity of motion. However, it is clear from what I have been saying that the fact that motion is relative was discovered by Galileo and Newton; Einstein realized the centrality of the idea and took it further.

Let's pause for a moment to reflect on all of this, because what we have been talking about is the most fundamental discovery, and the deepest mystery, in the whole history of physical science. There is no objective meaning that can be given to something's position. There is no objective meaning that can be given to the speed or the direction of its motion. There *is* an apparently objective meaning that can be given to acceleration. The distinction between uniform motion and accelerating motion is absolutely central to our understanding of motion; it lies at the heart of the modern conception of the world that has been with us since the seventeenth century. I think it may even be said, without exaggeration, that anyone who does not understand this, anyone who has not confronted, in some quiet moment, the relativity of motion and the meaninglessness of any objective notion of their being at rest, is living—at least as far as their conception of the physical world is concerned—in medieval times.

The relativity of motion is both a fact and a mystery. But, if nature is rational, no mystery should be impenetrable. The next step must be to frame a question that will take us to the heart of the mystery. In the case of the relativity of motion, one way to ask the question is to wonder: What is the cause of this distinction between velocity and acceleration?

For Newton the answer is simple: Acceleration is defined with respect to absolute space and time. Unfortunately, it seems that Leibniz never addressed this point. This was a crucial failure. Because one can feel the effects of acceleration in one's stomach, Newton could counter Leibniz arguments by saying simply: Absolute space and time exist—here is their effect on the world. It was certainly Leibniz's failure on this point that led to the relational view of space and time being reduced to a philosophical curiosity for more than two centuries afterwards.

But, could Leibniz have answered Newton on this point? We know that he could have, because we know the answer he should have given. This was the answer given by Ernest Mach, a Viennese physicist and philosopher, in the 1880s.

As a philosopher, Mach was one of the earliest and strongest proponents of a view called positivism, which took as its main principle the idea that in science one can only speak meaningfully of those things that are directly observed. In its extreme form, this philosophy leads to absurdity; indeed, Mach strenuously rejected the idea that atoms might exist, apparently because they were not directly observable. But it sometimes happens that a philosophical idea that is nonsense when taken literally as an absolute principle leads to important insights when taken only as a strategy. This was certainly the case for Mach's rejection of the use of unobservable elements in science.

Mach believed that in every case in which something observable is explained by reference to something unobservable, it will turn out that the unobservability of the cause is really a lie. If one looks, there will always be something real in the world that is the actual cause one is looking for. In these cases, the effect of the hypothesized unobservable thing is really a stand-in for the effect of the real thing. Mach was a realist and an optimist; he believed that we liberate ourselves from the myths of the past by replacing fictional, unobservable things by real, observable things.

Rotation has real physical effects which are experienced directly. Mach believed that there must be some cause for this, based on something observable in the world. Newton's explanation for is that one feels acceleration because one is rotating with respect to absolute space. But absolute space is not observable, and there is nothing we can do to affect or change it; therefore, it cannot be real. There must then be something in the world that one is rotating relative to, when one spins around, that is the real cause of one's upset stomach.

Having asked the question in this way, Mach found an answer: What we are rotating with respect to when we feel dizzy is the whole rest of the universe. Imagine going out on a starry night and looking up at the sky. There is a curious coincidence that few people notice without being told: Relative to our own sense of rotating and not-rotating, the universe as a whole is not rotating.

Of course, the stars do rotate overhead; once every twenty-four hours. But we know this to be due to the rotation of the Earth. Once that is removed, one is left with a remarkable fact. The galaxies move in great velocities relative to us (on the order of several hundreds of kilometers per second.) It is perfectly conceivable that on the whole, the universe might be rotating. After all, this is the case for the planets in the solar system and the stars in the galaxy. There is nothing in Newtonian mechanics to rule out the possibility that the universe would be rotating, relative to us. But it is not. If you believe in Newton's absolute space, there can be no explanation for this fact. It can only be a coincidence.

On the other hand, if you believe with Leibniz and Mach that all motion must be relative, this cannot be a coincidence. Ultimately there must be no meaning to our rotation, except the relative rotation between us and the rest of the universe.

It's normally rude to answer a question with another question, but sometimes this is the only thing to do. So also in science; sometimes the right thing to do is not to answer a question, but to find a new question that expresses it in a different way. This is what Ernest Mach did when faced with the fact that the universe is not rotating. He rephrased the question this way: We feel an effect in our stomachs when we rotate with respect to the distant galaxies. But, we may ask, what would be the effect were we to stand still while the whole universe was made to rotate around us? Would we still feel the same effect in our stomachs?

For Mach, who believed that reality consisted of nothing but that which was observable, the only possible answer to this question was yes. The relative motion between us and the galaxies is the same whether it is us or the whole universe that is rotating. If I feel dizzy whenever I rotate with respect to the distant galaxies, I must also feel dizzy if the galaxies rotate about me.

Newton would have given the opposite answer to this question. For him the two cases are completely distinct, in the first case, I am rotating with respect to absolute space—and the galaxies are not—and in the second case, I am at rest, and it is the galaxies that are rotating with respect to absolute space. Thus, for Newton, only in the first case do we feel dizzy, because that effect is caused by my rotation with respect to absolute space and has nothing whatsoever to do with the galaxies.

Unfortunately, we cannot spin the universe around and see who is right. But the discussion is still useful because what Mach has given us is a principle that any theory of motion has to satisfy, if it is to be in accord with Leibniz's relational view of space and time: *If we feel an effect when we rotate and the rest of the universe remains fixed, we must feel exactly the same effect when the stars are rotated the other way around while we are fixed.* This has since come to be known as *Mach's principle.*

Newton's theory of motion does not satisfy this principle. It was immediately clear to those who read Mach that it must be replaced by a new theory of motion that could satisfy it. One young student, reading Mach during the many hours of freedom he gained from skipping classes, went on to invent that theory. The student was Albert Einstein.

It is a matter of fact that Leibniz applied his principle successfully to the problem of motion and that he arrived at a relativity of motion on logical grounds. . . . The famous correspondence between Leibniz and Clarke . . . reads as though Leibniz had taken his arguments from expositions of Einstein's theory.

—Hans Reichenbach, "The philosophical significance of relativity"

EIGHTEEN

THE MEANING *of* EINSTEIN'S THEORY *of* RELATIVITY

I t is, I suspect, a peculiarity of the twentieth century that we like to believe that what is deep must also be difficult. Those considered the greatest writers of the century—Joyce, Beckett, and others—produce masterpieces like *Finnegan's Wake* that no one but the experts can read. Then there is the standing joke about how hard it is for the uninitiated to appreciate abstract art—let alone penetrate the intellectual contortions behind postmodern installations and performance pieces. And in physics, we have the general theory of relativity which, although it captures our furthest progress in understanding our universe, is usually expressed in terms of mathematics that is not even taught in the undergraduate physics curriculum.

Of course, it is not really true that the deep and the difficult are connected in this way. The stories of Joyce and the plays of Beckett require nothing but an openness to the pathos of our situation to respond to them. Certainly they produced other, less accessible works, but I suspect they would not matter so much to us were that all they had done. Similarly, abstract expressionist painting and the other great art of our century is packed with feeling that enters directly through the eyes, with no need of theory.

It is the same with general relativity. The whole theory is based on a very simple idea, which can be explained to a teenager. One must only imagine the experience of falling and consider that those who fall have no sensation of weight. In the hands of Einstein, this everyday fact became the opening to a profound shift in our way of understanding the world, just as Cezanne changed our way of seeing by painting simple pictures of everyday objects.

It is also true that behind the direct emotionality of the best of our century's art there is often something profound, which is only indirectly expressed, and which we may call the artist's confrontation with our situation as human beings at this strange point in history. Despite its simplicity, general relativity also reflects a profound understanding of our situation in the world, as it is the first physical theory to express Leibniz's relational conception of space and time.

Of course, if one really wants to learn the general theory of relativity, as a physicist, and be able to use it, there is a certain amount of mathematics that must be learned. But all of that can be ignored here; as we are only interested in the key ideas that bear on the general problem of what it means to construct a theory of cosmology.

In the last chapters we took two steps towards the construction of a theory that could realize a relational conception of space and time. Leibniz taught us to reject any reference to a priori and immutable structures, such as Newton's absolute space and time. But he did not tell us what to replace them with. Mach did, for he showed us that every use of such an absolute entity hides an implicit reference to something real and tangible that has so far been left out of the picture. What we feel pushing against us when we accelerate cannot be absolute space, for there is no such thing. It must somehow be the whole of the matter of the universe.

We are now ready to take with Einstein a third step in the transformation from an absolute to a relational conception of space and time. In this step the absolute elements, identified by Mach as stand-ins for the distant galaxies, are tied into an interwoven, dynamical cosmos. The final result is that the geometry of space and time—which was for Newton absolute and eternal—becomes dynamical, contingent, and lawful.

To do this we must confront the reason why Newton originally rejected the relational point of view. This was to express the apparent fact that, while position and velocity seem to be only defined with respect to particular observers, acceleration seems to have an absolute meaning. To have a relational theory of space and time, we must then relativize the notion of acceleration. We must discover that the distinction between who is accelerating and who is not accelerating is not fixed once and for all, but is due to some contingent circumstances. We must identify and name the entity that causes us to feel dizzy when we rotate, and we must find out what laws it satisfies.

Generally, to bring something from the imagined world of the absolute into the real world of the contingent and relational is to bring it under the domain of natural law. We must make it changeable, make it a dynamic actor that can be influenced as well as influence. General relativity, as I will now explain, is exactly the theory that emerges when we make the distinction between accelerating and not accelerating contingent and dynamical.

This may seem a bit abstract. This distinction between who is accelerating and who is not may not seem to be something tangible. Like space, time and gravity, it is invisible. It is felt in every experience of motion, of carrying, of lifting. At any moment and in every place, there seems to be a ghostly entity that makes an accelerated motion different from a uniform one. This is the thing that we want to speak about.

To make this thing easier to conceptualize, it will be helpful to give it a name. It is called the *metric of spacetime*. The word metric means something that measures, and the entity that distinguishes accelerating from not accelerating is called that because it is necessary to any measure of motion.

In Newton's physics the distinction between accelerating and non-accelerating motion is assumed to be fixed everywhere for all time. Likewise, for Newton, the metric of space-time is assumed to be likewise fixed once and for all. This means if Newton's physics is right, the metric of spacetime is something that acts on everything, but it is in turn not acted on by anything. In a relational world, this is unacceptable. We cannot have things that act, but are not acted on. Following Mach, we must search for a concrete reason why the distinction between accelerating and not accelerating is made in a particular way at a particular place and time. We must be able to uncover a relationship between our sense of rotation on Earth and the motion of the distant galaxies.

How are we to do this? The key is a simple fact, which is that we cannot mess with the definition of acceleration without confronting the phenomenon of gravity. Because of certain features of the gravitational force that I will now explain, to make the definition of acceleration changeable is *precisely* to construct a theory of gravity.

There is no phenomenon in nature that is simpler than gravity. It is simple because, unlike the other forces in nature, its effects are universal. This means not only that all things fall under the influence of gravity. It means that all things fall in exactly the same way.

Go to the nearest window and drop any two objects. As far as the effects of gravity are concerned, they will fall with exactly the same speed, hitting the ground at exactly the same time. This is true, irrespective of their mass, what they are made of, whether they are alive or not, or anything else.

Of course, if you actually do the experiment, you will see that much of the time the two objects do not hit the ground at the same time. This is due to the effect of the friction of the air, which resists their acceleration. But if you were to do the experiment in the absence of air, say on the moon or in space, the motion of all falling bodies would be identical.

This applies to any motion in which gravity is the only force that is acting. If you throw any two objects, their motions will be exactly the same as long as you throw them so that at the moment they leave your hand they are traveling in the same direction at the same speed. This is the origin of the phenomena of weight-lessness. The space shuttle or the Soyuz space station, together with their inhabitants and everything they contain, are falling freely under the influence of gravity. But, because of the universality of gravity, they all fall together. The inhabitants of the space station cannot sense their motion because everything inside keeps always the same relations among them.

Weightlessness can be described in the following very interesting way: someone who is falling feels no effect of gravity! This simple thought is the whole basis for the general theory of relativity. Einstein called it the "happiest thought of my life."

This led him quickly to a thought experiment which captures the essential idea of the theory he then made. Let us go back to our room without windows that we visited in the last chapter. Let us now imagine that, while we are sleeping, someone puts our room in a very tall elevator shaft, from which all the air has been taken out, so there is no friction to impede any motion. They then cut the cable and let us fall. We wake up and feel that we are falling, we feel weightless. But can we be sure? Is there any way, without looking outside the room, to be certain that we are actually falling in the gravitational field of the Earth?

It seems that we know we are falling from the fact that we are weightless and no longer feel the force of gravity pulling us to the ground. But not so fast. Are we really sure we are falling? How do we know that instead of putting us in an elevator shaft, our manipulative friends have not moved our room way out into space, far from any star or planet, and thus any effect of gravity? Can we tell, without looking outside the room, the difference between free motion far from any gravitating body and free fall in a gravitational field?

The answer is that we cannot. I hope that the reader who has not heard this before will pause a moment and reflect on this fact: those who fall in a gravitational field have no way to detect the presence of that field.

At least for a short time. Because, of course, if our actual situation is to be trapped in an elevator shaft, eventually we will come to the bottom and then we will know the difference. So let us imagine that our friends (who really do love us) have attached a rocket to the bottom of the room, so that by firing it they are able to gently slow our motion and bring us safely to rest at the bottom of the shaft. Certainly then we know that we have been on Earth all along, because now we feel the effects of gravity. Or do we?

We do know that we have been accelerated, because we can feel it. But do we truly know that in the end we are settled on earth? Is it not possible, instead, that we are actually far from home, in deep space, and that the rocket motor under the room has simply settled into a steady acceleration equal to that of the Earth's gravitational field? In that circumstance we would feel pushed down to the floor, as we do in the earth's gravitational field. If we drop anything, say this book, it would seem to accelerate towards the floor. Of course, it is not accelerating; it is really just sitting there. It is instead we who are accelerating up to meet it.

Einstein saw that there is a deep principle here, a principle that is the key to all the mystery of the phenomena of gravity. He called it the *equivalence principle*. It asserts that it is impossible to tell whether one is in a room which is freely falling in a gravitational field or in a room moving uniformly in deep space. It also asserts that the room, which is accelerating steadily in deep space, with the same acceleration as falling bodies at the surface of the earth have, is indistinguishable from the room sitting on the surface of the Earth.

Many beautiful effects follow from this simple principle, such as the bending of light and the slowing down of clocks in a gravitational field. If this were a textbook, I would describe them. Instead I want to return to the main argument and bring out the implications of what we have been saying for the task we set ourselves, which was to see how to make the fixed structure of Newtonian spacetime dynamical. For the equivalence principle has a beautiful interpretation in terms of the distinction between accelerating and non-accelerating observers.

Look around at the room in which you are sitting, and imagine different observers all around like ghosts, with different motions, each noting and describing the motions of the things around them. Some are sitting still next to you, some remain in one place but spin around, some are moving through the room with a constant speed, some accelerating, some falling, and so forth.

Now, ask which are the non-accelerating observers? Which are the ones that correspond to those we called inertial observers? By the definition we gave before, these must be those for whom free particles (those with no forces on them) sit still or move with constant speed and direction.

It is natural to respond, I am sitting still, so I am one of the inertial observers. But there are two problems with this. The first is that you can find no free particles with which to confirm your statement. Anything that moves, falls. As gravity is universal, the idea of a free particle that hovers in space when everything around it falls is a fantasy that does not correspond to anything real.

The second problem is that this would violate the equivalence principle. For we, sitting in our chairs in our room, are in a situation that corresponds exactly to that of people sitting in a room in deep space that is being steadily accelerated.

This is perhaps a bit confusing. To straighten it out, we will need to keep our heads clear, which can best be done if we stick to things we are sure of. In the absence of gravity we know which are the accelerating observers and which are the non-accelerating, inertial ones. What is confusing is how this distinction is to be made in the presence of gravity, because the effects of gravity and acceleration are in fact indistinguishable. The only way not to be confused is to stop trying to fight this indistinguishability, and instead to use it. This means that we must rely on the equivalence principle to tell us, in the presence of gravity, who is accelerating and who is not.

If we do this, then we will want to say that, here on earth, the inertial observers are those that correspond, via the equivalence principle, to the inertial observers out in space. And those are the ones that we are used to think of as falling!

Take a moment to meditate on this, as when you understand it you will have experienced one of those famous paradigm shifts that philosophers and literary theorists love to write about. What I want to assert is that if we consider those observers who are falling to be inertial observers, we can understand easily all the phenomena of gravity. The fact that I feel pushed down into my chair, or that things I drop fall to the floor, are now to be understood as consequences of the fact that the room I am sitting in is accelerated.

We may now tie this together with our earlier discussion. The whole point of the last chapter was that the distinction between who is accelerating and who is not may be thought of as part of the intrinsic structure of space and time. For Newton, it was absolute. We, following Mach and Einstein, want to make it dynamical. Now we can. To make it dynamical means that the distinction can be made differently at different places and at different times. All we need do now is to assert that the effect of a massive body, like the Earth, is to cause the distinction to be made so that the inertial observers are those that seem (from the point of view of someone standing on the Earth) to be accelerating towards the center.

We may thus identify gravity completely as the effect of making the distinction between inertial and accelerating observers dynamical. Somehow, mass has the property of influencing how this distinction is made, so that at each point near a large mass, the inertial observers are those that fall towards its center.

We do not need to go into detail about how this is actually done. There is an equation, called Einstein's Equation, that tells how the distribution of matter and energy influences how the definition of acceleration is chosen. If you solve these equations, you discover all the phenomena we have been describing.

This is the essence of Einstein's general theory of relativity. We fall and do not feel it, but how we fall constitutes the whole phenomenon of gravity.

It is a measure of how much Einstein's vision has triumphed that to the astronomer the general theory of relativity has become commonplace. The theory has by now been confirmed in many examples. I will mention only two. The theory predicts that the paths of light rays will seem to be bent when passing near a star or galaxy, just as if a lens surrounded it. This is due to the equivalence principle, as it requires that light (as well as matter) must fall towards massive objects. This effect caused a sensation when it was first observed in 1919, now it is used routinely by astronomers to measure the distribution of mass in the universe.

Another prediction of the theory is that waves can propagate carrying energy through the gravitational field. These waves can be excited by two objects circling each other. The result is that two stars in close orbit around each other will slowly give off their energy into gravitational waves, and slowly fall towards each other, eventually colliding. This effect has been seen in observations of pairs of neutron stars. These stars rotate rapidly, which produces radio waves that may be received on Earth. As the stars move around each other, the frequencies of the pulses we receive vary as a result of their traveling through their changing gravitational fields. Each such system provides us a laboratory in which we see the metric of spacetime and the orbits of the stars evolve in real time. In this way several effects of the theory, including the loss of energy by the production of gravitational waves, are observed. The changes of the pulses over time match the predictions of Einstein's theory to unprecedented accuracy, more than ten decimal places.

There have been, in the history of science, a few experiments that go beyond the confirmation of particular theories to the confirmation of general philosophical principles. Tycho Brahe's observations of the orbits of the planets were among these, as Kepler found there was no reasonable interpretation of the results that did not accord to the Sun a role and a centrality that could not have been accommodated by the principles of Aristotelian science.

Similarly, the observations of the binary pulsars do more than confirm the general theory of relativity—although they do that exceedingly well. They signal the irreversibility of the replacement of the absolute, static conception of spacetime with Leibniz's relational and dynamical conception. The relational universe I have described is our world. There can be no going back.

The most basic properties we may imagine an object to possess are its position in space and its existence in time. After the triumph of Einstein's theory of general relativity, these must be seen as meaningful only in the context of the relations of that body to the rest of the universe. It can no longer be maintained that the properties of any one thing in the universe are independent of the existence or non-existence of everything else. It is, at last, no longer sensible to speak of a universe with only one thing in it.

*All these fifty years of conscious brooding have brought
me no nearer to the answer to the question, 'What are
light quanta?' Nowadays every Tom, Dick and Harry
thinks he knows it, but he is mistaken.*

— Albert Einstein, in a letter to Michael Besso

NINETEEN

THE MEANING *of* *the* QUANTUM

By dissolving Newton's absolute space and time into a network of relationships, general relativity takes a first step away from the notion that the coherence of the world lies outside of it. But general relativity does not by itself represent a complete transformation. For this reason, both physicists and philosophers have, in my view, yet to completely absorb the implications of the theory. Even if we know better, it is still very common among physicists to think of a spacetime as some new kind of object that can be seen from the outside. On our blackboards and in our notebooks, we draw pictures of spacetimes as if such were the case.

General relativity radically challenges the classical notion of an absolute reality behind the world, not because it says space and time are "curved," but because it asserts that the structures that are absolute in Newton's conception of space and time have become dynamical, and that all properties that have to do with space and time must be constructed from relations between things in the world.

But we are in the world. So how, if we forbid ourselves from entertaining the fantasy of a view from outside the world, are we to try to conceive of the whole universe? Is it at all possible that living in the world we can have complete and objective knowledge of the whole universe? Can we contemplate the possibility of constructing an objective description of a whole cosmos without telling from whose point of view that description is understood to have come? These are some of the questions that general relativity raises and that, as we are about to see, the quantum theory also forces us to confront.

The notion that a true view of the universe could be thought of as the view of a single, outside, observer is closely connected to the notion of the absolute that we discussed in earlier chapters. The notion of the absolute includes the idea that the rationality of the world arises from a source that is outside of it. Once we have accepted the possibility of such a transcendent rationality, that is independent of any contingent facts of the world, it is possible to conceive of an intelligence that could both stand outside the universe and be cognizant of what goes on in it. This allows us to imagine how the world might be seen by an observer who was not of it but had complete knowledge of everything in the universe.

It is easy to imagine this. What I want to question is only that to the extent that we do so, we end up with a notion of objectivity whose roots are as much in religion and mythology as they are in science.

Let me emphasize that I am not questioning the possibility of a complete and objective description of the universe, but only the idea that this could be read as the point of view of a single observer. To make the distinction clear I will call this notion of objectivity *single-observer objectivity*. The question that drawing this distinction implies is then: Can there possibly be a true and complete description of the world that is not readable as the view of a single observer?

General relativity, by telling us that the world must be conceived solely as a network of relationships, goes part way towards answering this question. But it is also true that general relativity was the invention of a person who longed to construct an absolute, objective picture of the world. In Einstein's work and writings we see a continual tension between his adherence to the principles of Mach and Leibniz, who had taught him that space and time should describe only relations, and his yearning to construct an objective picture of the world, and thereby capture something of the eternal transcendent reality behind nature. Indeed, it is often the case, in science as well as art, that the greatest creative acts are driven by a tension or a contradiction in the world view of the creator. There is often an ironic aspect to the work of such people, for their work does not succeed in resolving the conflict that drives them. But they do open up the way for people who come later. Einstein is truly one of these cases. For this, he can truly be thought of as the Copernicus of the twentieth century.

But the deepest irony of Einstein's contribution is that it was he who first realized that a new quantum physics would be needed to encompass the worlds of radiation and the atom. It was he who first established that light has both a wave and a particle nature. By his own admission, Einstein spent "a hundred times more effort" on his attempt to comprehend the quantum than he did on relativity. But this effort was unsuccessful, and when a complete theory of quantum phenomena was finally constructed, by other, younger, people, he opposed it because it failed to provide the absolute view of reality that he yearned for.

It has been said that the slaughter of World War I shattered forever the nineteenth century European faith in the power of rationality to lead to continual progress and ever greater human happiness. Certainly one has only to look at the writings of philosophers or the work of artists just before and after the war to see how that catastrophe tore the comfortable world of the nineteenth century off its firm foundations and set the twentieth century in motion. Of course, in the art of the early years of the century, one sees that something was already happening to the reliance on a single point of view in the construction of a picture of reality. But the complete break with the idea that the purpose of art is representation came only after the war.

Is it then not surprising that general relativity, constructed during the war that was to change Europe forever, has elements that are both classical and revolutionary? And is it not also surprising that Einstein, educated, as he was, in the Germany and Switzerland of the nineteenth century must, in spite of his primacy in bringing the quantum to light, be regarded as much as the last great architect of classical physics as he is the prophet of a new physics? For, as important as general relativity is for our story, the really decisive break with the notion of reality in classical physics came not with it, but with quantum mechanics, born of a very different generation raised during the tumult of the first world war, and indeed, perhaps the last great development in physics to be initiated solely by Europeans.

So we come finally to quantum mechanics. Absolutely the first thing that must be said about it is that the discussions and arguments begun in the 1920s about the meaning of the quantum theory remain unresolved. Many, apparently equally viable, interpretations of the theory have been proposed, and there is now as much contention among the experts as there has ever been.

The debate over the interpretation of quantum mechanics does not prevent the theory from being useful. Nor does it keep it from being fun to use. Although many people have called the quantum theory counterintuitive, once one gets used to it, quantum mechanics provides a very simple and intuitive framework for talking about nature. I suspect that this is because, while it does not give us an "objective picture" of what is happening, in its logical structure it corresponds more closely to what we actually do when we do physics.

When a human being does a scientific experiment, he or she interacts with nature in a carefully scripted way that experience has shown leads most surely to reliable results. In each experiment a system must be prepared; it is then manipulated in some way, after which some property of it is then measured. The basic mathematical elements of quantum theory correspond to exactly these three stages of an experiment. It may be said that while Newtonian physics is a language of description, dominated by nouns, the language of quantum mechanics is primarily one of verbs. When we talk quantum, the focus is on what we do and what we observe, rather than what is.

While it is true that quantum mechanics doesn't provide an objective description of the world (of the sort we have become used to from classical physics), this may actually be the great virtue of the theory. It frees us from the fiction of the absolute observer, looking at everything from outside the world. We may miss the "picture" of the world given to us by Newton, but we ought to understand by now that the idea of representing the whole universe as a collection of classical trajectories reflects a fictitious ideal that corresponds only very approximately to what is real.

At the same time, while the overthrow of Newtonian physics is certainly irreversible, we should avoid the temptation to take quantum mechanics, in its present formulation, too seriously. The theory is genuinely puzzling, and, while it works, it clearly does not make complete sense. Part of the unfinished business of twentieth-century physics must then be to find a better understanding of quantum physics. For this reason, when we speak of the quantum we must be very careful to separate talk of quantum phenomena from talk of the quantum theory. Quantum phenomena are real, and are genuinely puzzling. But not all of the idealizations and postulates of the quantum theory may actually correspond to nature. We must then seek to separate those parts of the quantum theory that really correspond to nature from those parts that only represent an incomplete attempt to comprehend nature.

Such a separation requires a point of view, and I would like to propose one here. This is to focus on the problem of extending quantum theory to encompass gravity and cosmology. We will try to keep those parts of the theory that seem consistent with the goal of constructing a theory of the whole universe, and seek to modify those parts that give us trouble when we try to extend its reach to cosmology. As we do this we will try to read its difficulties more as opportunities for the invention of new physics than as incentives for the invention of new philosophy. Put in this way, as a problem for physics and not for philosophy, we may hope to progress.

If we keep the focus on the attempt to unify relativity and quantum theory, then we are continually impressed by the fact that each of these are transitional theories. Each radically challenges the Newtonian conception of the universe, but only in part. Each holds unchanged a certain, but different, part of the classical picture. So the situation is genuinely confused. However, underlying both theories is clearly the move from an absolute, Newtonian picture of nature to a relational, Leibnizian conception. As I will try to argue here, this is the ground on which we may hope to find their ultimate reconciliation and unification.

The essential core of what quantum mechanics has to tell us about nature is its insistence that the world can only be described completely if it is described as an entangled whole. As such, it is definitely in disagreement with the view I have been calling radical atomism, according to which the properties of any one particle are supposed to be independent of everything else there is in the world.

To explain this, I need to sketch briefly how our knowledge of the world is organized in the framework of quantum theory. The first and most important thing that must be said is that in quantum theory we deal not with a description of the system itself, but with what we know about it. Thus, at the very beginning, as quantum physicists we envision a situation in which the world is divided into two parts. On one side of the divide is the particular system under study. On the other side are ourselves, as the observers, whatever tools or instruments we intend to use in the study, and the rest of the universe. This is very different from the description of the world in Newtonian physics in which we are invited to imagine that the mathematics gives a picture of reality, in which the observer need not be glimpsed.

The information that we, as observers, have about a quantum system is coded into a construction that is called *the quantum state of the system*. This is a necessarily abstract concept as it is the name, not for something in nature, but for a mathematical entity that is invented to keep track of the information that one part of the universe can have about another part. The quantum state is then not a property of the system it describes. It is a property of the boundary or interface that separates that system from the rest of the universe, including the observer who studies it.

I must pause here to warn the reader that we have now set foot on the part of quantum theory which is controversial. The point of view I am expounding here is not shared by all my colleagues. Many physicists and philosophers believe that there is more to the quantum state than a representation of the information the observer has about the system. But many, including Einstein, have believed that as the quantum state changes when we make a measurement, it is nothing but an encoding of what we know. I choose to adopt this point of view because it is the only one I know that definitely makes sense in a cosmological context.

Regardless of what system we are interested in, the quantum state that contains what we know about it is most accurately represented as a point in an abstract mathematical space. This space, which is also an invented construction, is called the state space of the system. It represents the different possible outcomes of experiments that could be made on that system. This space is constructed so that it has one dimension for each possible outcome of an experiment that we might make on the system. This space typically has an infinite number of dimensions because many things we would like to measure, such as the position of a particle, have an infinite number of possible answers.

I will not go further into details, but the important thing to appreciate is that this state space, where the quantum states live, is very different from the three-dimensional space in which we seem to live.

The quantum state is sometimes pictured as a wave moving in ordinary three-dimensional space. While useful for some purposes, this is a very misleading representation, because it is only appropriate if one is studying a system that contains only one particle. In that case, and only in that case, the quantum state can be pictured as a wave that represents the probability for the particle to be found in different places. However, if there are two particles, then the quantum state cannot be described as two waves moving in three- dimensional space. It can only be pictured as an object in some higher-dimensional space.

But why is this necessary? Why doesn't quantum theory allow us a simple and intuitive picture of phenomena, as things happening in ordinary space? The reason is that it seems intrinsically impossible for one observer to have all of the information that would be necessary to give a complete description of any system in nature. It seems that, whatever part of the world we are studying, we are allowed to know at most half the information that would be necessary to fully describe it. This limitation is called the Heisenberg Uncertainty Principle, and it is this, above everything else, that makes the real quantum world different from our Newtonian imaginings.

I must stress that I do not know why Heisenberg's Uncertainty Principle is true. Neither, as far as I have been able to tell, does anyone else. It may have something to do with the fact that we are in the world, irreducibly, and that any attempt to form an objective description cannot be complete, because it must leave us out of the picture. I certainly have the impression that Bohr and others believed this to be the root of the trouble, and I can imagine that they were right. But I have never seen a real argument that starts with this idea and arrives at the uncertainty principle.

What we do know is that if we take the uncertainty principle as a postulate, we get a theory that works quite well to describe all known atomic phenomena. So while we may look forward to someday doing better, for the moment we can only accept the principle and go on.

As a result of being unable to gain all the information we would like about the quantum world, it seems useless to mimic classical physics and try to draw a complete picture of the system we are studying. Instead, it is more fruitful to face the situation directly and talk only about the information that we do have. This is why we use the abstract language of quantum states.

Quantum mechanics may seem abstract or remote, because it is usually used to describe things that are very small. But size, in reality, has nothing to do with it. If quantum theory is right, then anything simple behaves in a way that challenges our powers of imagination.

Let us imagine that we have a friend who lives in Quantumland, where big things can be simple. To tempt us with the beauty of her world, she has sent us a present. We go to the airport to receive it and are given a sealed box with a door on each end. On the top is written, "QUANTUM PET CARRIER—OPEN ONLY ONE END AT A TIME." In Quantumland no one would have to be told that you can only open one end of a box at a time, but we will see that we have reason to be grateful for this advice.

Bringing the box home, we hurriedly open it to see what is inside. Opening one end, we see the head of a cat peer out! Lovely, but the cat stays inside. It seems that one property of pets in Quantumland is that they can never come out of their box, one can only interact with them by opening one of the doors. OK, we can live with this, but we become curious at least to know the sex of our cat. Well, we can use the door on the other end for this. We try to open it, but we find it is closed tight. Remembering what is written on top, we close the first door. Immediately the back door comes open. By looking in, we are able to ascertain that our pet is indeed a boy.

This done, we go back to the first door to play with our cat. To do this we must first close the back door. We then open the front door to find a jolly looking puppy gazing out at us!

After some trials and examinations, we discover that we are in an interesting situation. When we open the first door we discover that our quantum pet is either a cat or a dog. If we open the second door, we discover that our pet is either male or female. However, by the peculiar properties of the box, our vision is obscured so we cannot be sure, when gazing in the front, what sex our cat or dog is. And when we look in the back door we can ascertain the sex, but we cannot judge reliably whether it is a dog or a cat.

We cannot have both doors open at once, so we can never be sure of both the species of our pet and its sex. Ascertaining one destroys the knowledge of the other. If all we do is look at the front end, then once we have seen a cat there, we will always see a cat. If we wish, we may at any time close that door and peer in the other side to learn our pet's sex. Whether it is a cat or a dog, we will discover that there is a fifty percent probability that it is male and a fifty percent probability that it is female. But, once we have done that, if we go back to the front ,we will not necessarily find a cat, as we did before. For once we have ascertained the sex, the species again gets scrambled, and half of the time our pet will be a cat, and half the time it will be a dog.

What we are experiencing is exactly the Heisenberg Uncertainty Principle. It is happening because a complete description of our quantum pet would include its species and its sex. According to classical science, we ought to be able to take the animal out of the box and see what it is. But a quantum pet can never be removed from the box and, for reasons that are perhaps mysterious, we can only observe one aspect at a time.

Perhaps the reader thinks I am being facetious, or teasing. But no, I am describing what we believe is the general situation we are in when we observe any physical system. The Heisenberg Uncertainty Principle limits the information we can have about any system to always exactly half of the information we need to have a complete description. We always have some choice of which information we would like to have. But, try as we may, we cannot exceed this limit.

Of course, we do know both the species and the sex of our ordinary pets. But, this does not mean that the Heisenberg uncertainty principle does not apply to them. It is just that ordinary pets are much more complicated beasts than quantum pets, and we know this information only at the expense of other information we might like to have, such as knowing *exactly* where all the atoms at the ends of its whiskers are. By contrast, quantum pets are much simpler, and there are only these two things to know about them.

The uncertainty principle seems to prevent us from glimpsing the reality of the quantum world, from knowing—independently of our involvement with them—what our quantum pets are. But, this is only the beginning of the story. The real strangeness of quantum mechanics emerges when we apply it to systems that contain more than one thing.

The next week we receive another quantum pet. And this time it comes with a letter that advises us to wait for a phone call before opening the box. As we are worrying over what to feed our new pets, and what to call them, the phone rings. It is my brother, who is not happy at all. It seems that the same friend has also sent him and his family a quantum pet. He has also received a letter, which contains instructions about how we may jointly enjoy our two pets.

What the letter tells us is that the two pets have been created in a certain quantum state that has certain very puzzling properties. Both pets, taken separately, are completely random with respect to both species and sex. If either of us opens either door, we will see either possibility with fifty-percent probability. But our friend in Quantumland has tied the quantum pets together so that what I see is correlated in a certain way with what my brother observes.

These correlations involve what will happen in the case that my brother and I both open the same door. Suppose we both look to learn the species of our pet. Each of us may get either answer, with equal probability. But our friend in quantum land guarantees that whichever answer I get, he will get the same. If my brother received a cat then so will I, with one hundred-percent probability. This is the case even though, to begin with, there is for each of us an equal probability of seeing a cat or a dog.

Each of us are equally free to check instead the sex of the pet. If we do, then each of us has an equal chance of seeing a male or a female. But, again, this is not the whole story, because what each of us sees is correlated. If we both check the sex then, whichever one I get, my brother will get the same one.

How is this possible? What is true about the knowledge we may have about a quantum system is also true about how the system may be prepared. When our friends in quantumland prepared the two packages, they were free to fix exactly half the information that would be necessary to completely describe their contents.

I said before that I have some freedom to choose which half of the information about a system I may know. This applies also to preparing a system: our friend is free to prepare the two pets in different ways, so long as no more than half of the information is determined. Now we come to the really interesting part. The limitation applies to the whole combined system as well as to each pet individually. This allows us to do something quite elegant, which is to describe the system in such a way that we give half the information about the whole, while giving no information at all about each individual pet. This is possible if we describe only how the properties of the pets are related to each other, while not saying what properties each has. For example, we may specify that both species are the same. This gives us half the information we would need to determine both species, but it leaves us completely ignorant about the species of either one, taken separately. The same goes for sex. It is exactly this possibility that our friends have taken advantage of in preparing the presents.

It is easy to see that to know the existence of a correlation is to know exactly half the information we may have about the sexes of the two animals. There are four possibilities: male-male; male-female; female-male; and female-female. If we know they are correlated so that both have the same sex, then these four possibilities are reduced to two.

The same is true of the species. Thus, in preparing the two pets, our quantum friends have chosen to make definite the properties that the two pets share, while making their individual properties completely indeterminate.

One might think that this is a very unusual or contrived situation. But actually what I have been describing is the general case. In the quantum world—which I must insist is, as far as we have been able to determine, our world—whenever two systems have interacted it is more common to find them sharing properties in this way than to find them in states such that each have definite individual properties.

This plays havoc with the traditional idea that things that are isolated have definite properties. For, once created in this state, our pair of pets may be separated as far apart as we like before their properties are measured. Until we choose which doors we open, my brother and I possess together a system whose wholeness has in no way been compromised, whatever distance may separate its two parts. Quantum theory says quite generally that whenever two systems have interacted, their description is tied together in this way no matter how far apart they may be. This is called the *wholeness* or the *entanglement* of the quantum description.

This aspect of the quantum world is so strange that it was missed at first. For the first few years after quantum theory began to take shape, people argued about the uncertainty principle and wave-particle duality. These were strange enough. Moreover, most of the early applications of quantum theory were to systems with only one particle, which did not reveal the possibility that properties could be shared in this way between two widely separated things.

At the center of the discussion about quantum theory was a great debate between Einstein and Bohr. Nearly every time they saw each other, from their first meeting until the death of Einstein, they argued about the quantum theory. Their debate is of central importance for the issues of this book, because what they were arguing about was exactly the extent to which the view I have called single-viewpoint objectively can be maintained. Einstein believed that the goal of physics was to construct a description of the world, as it would be in our absence. Bohr believed this to be impossible. For him, physics was an extension of common language, which people use to tell each other the results of their observations of nature.

The fact that quantum systems are usually entangled like our two pets, no matter how far apart they may be separated, first came out in the course of this debate. It emerged as a trick that Einstein discovered in the course of trying to invent an argument that he could use to show that quantum mechanics could not give a complete description of nature. This argument was first presented in a paper Einstein wrote in 1935 with two young colleagues, Boris Podolsky and Nathan Rosen. It has since been called the Einstein-Podolsky-Rosen experiment.

I don't think anyone can have a useful discussion about quantum theory that does not put this experiment at the center of attention. But it is not difficult to explain it, given what I have said so far about quantum pets.

When we left the story, I was sitting on the floor talking to my brother on the phone as I wonder what to do with my new quantum pet carrier . My first problem is that I must make a choice whether I wish to know the species or the sex of my pet. According to how our friend has prepared the two packages, I know that I have no way to know which answer I will get to either question. But, whichever I choose, and whichever answer I get, I will know immediately something about my brother's pet.

Suppose I choose to see if I have a dog or a cat. I open the front door of the box and find that I have a cat. Then, from the instructions, I know right away that if my brother opens the front door of his box he will also find a cat. I tell him this on the phone, and he opens his box and confirms that I am correct.

This is very interesting, because I have been able to learn something definite about my brother's pet, without in any way interacting with it or disturbing it.

There is no question of my measurement somehow affecting his pet, for I have not touched it. But I can tell my brother that if he looks, he will certainly find he has a cat. This led Einstein and his friends to conclude, quite reasonably, that his pet must have been a cat all along. Because I can find out what my brother's pet is without looking at it, its catness must be something real—it cannot be created by the act of observing.

But this is not the end of the story, because I might have made the other choice. Had I looked instead at the sex, I would have gotten a definite answer, let us say a male. Then I would know, from the way our friends prepared the gifts, that if my brother looked not at the species but at the sex of his pet, he would see the same. I could tell him this on the phone, and when he looked he would tell me I was right.

So the whole discussion we just had about the catness of my brother's pet would now apply instead to the sex. By looking at the sex of my pet I would definitely learn the sex of his pet. But I wouldn't have looked at his pet, it is very far away. So it must be that its being a male is something real; it must have been a male even before my brother looked.

But now we can see that there is something wrong, because the same argument applied to both the species and the sex. In either case, if I look at which my pet is, I learn definitely which my brother's is. This means that both the sex and the species of my brother's pet must somehow exist, before I looked. The conclusion must be that before we looked, both the species and the sex of my brother's cat were determined.

But this implies that quantum mechanics does not give a complete description of reality. For quantum theory asserts that we can only know definitely either the sex or the species, but not both. This means that there are things that are true about the world that quantum mechanics cannot represent.

Since it was invented in 1935, this argument has convinced many people that quantum mechanics cannot be the final theory of nature. However, it did not convince Bohr. The reason for this is extremely interesting. Bohr did not believe this argument because he could not accept Einstein's assumption that when I look at my pet I do not affect my brother's pet, even if it is very far away. While he conceded that there is no possibility of a physical interaction with my brother's pet, he insisted that there is still an effect *on the conditions that make the information about its sex or species meaningful or useful.*

This is a strange thing to say, and it is one part of the story that strains the analogy I am using here. I have to admit that the original discussion was about position and velocity, and not about species and sex. But let me go on, if the reader will allow a little whimsy the point can be made.

Suppose that both the sex and the species of quantum cats really only reflect aspects of their relationship with each other, so that the properties of any one pet are not meaningful in isolation. Were there one pet in the universe, it would not be interesting or useful to know if it were a cat or a dog, or if it were male or female. Of course this is not true of real dogs and cats, but they are complicated creatures, are made of enormous numbers of parts. The discussion we are having concerns things that are so simple that they cannot be said to be made of parts. According to both general relativity and Leibniz's principle of sufficient reason, the properties of elementary things, that are not made of parts, can only be defined in the context of their relationships with other things. To the extent that its context does not suffice to determine some property, it cannot be otherwise defined.

What Bohr is doing is using this principle in a new way to defend the quantum theory. His point is that the sex of my brother's quantum pet is only meaningfully or usefully defined when we know the sex of my pet. If I choose to determine the species of my pet, then I lose any possible information about its sex. But, according to Bohr's relational theory of properties, this means that I have destroyed any possibility of giving meaning to the sex of my brother's pet. So in spite of the fact that I have not physically interacted with my brother's pet, *what I have done nevertheless has the effect of making its sex meaningless.* Because of this, Bohr cannot accept Einstein's argument that there is something real about both the sex and the species of my brother's pet. Only one can be real at a time, because to be real a context must exist in which that property is meaningfully defined by its relationships with the rest of the universe.

Not surprisingly, neither Bohr nor Einstein were won over by the other's arguments. Einstein's view of nature was classical; he accepted the truth of what I called radical atomism and single viewer objectivity. Bohr argued from assumptions that began with the denial of these principles. Instead, he believed in the principle that properties are only defined relationally, and that physics is an aspect of our relations with the world.

What is most disturbing to us about the example of the two pets is that properties seem to be shared by two things which may be very far from each other. We would be more comfortable if it were possible to describe the properties of each of them completely, independently of whatever we know about the other. Let us call a theory in which this is the case a *local theory,* because it implies that one can describe something completely by speaking only about what is nearby it.

Quantum mechanics is not a local theory. As I have described it, it is radically *non-local.* A very interesting question, for those of us who feel uncomfortable with the quantum theory, is whether it could be replaced by any theory that is local.

The answer is no. We know this because of a remarkable piece of work by an Irish physicist named John Bell in the early 1960s. What Bell did was to find a way to test directly the principle of locality. What Bell found was that in certain cases, in which the pets have three properties each, the predictions of any local theory must satisfy certain constraints, which we call the Bell inequalities. Quantum theory, being non-local, must violate these constraints.

It turned out that it was not simple to test Bell's inequalities experimentally. A definitive answer was only gotten in the late 1970s by a group led by the French physicist Alain Aspect. As may be expected, they didn't work with quantum pets, but rather with photons of light. They studied situations in which two photons are prepared in such a way that observations of them must be correlated in exactly the way our pets were. They then let the photons fly very far apart—about thirteen meters—before observing them. This may not seem so far, but let me note that it is 10^{11} times bigger than the atomic domain. Indeed, the experiment represented a test of the predictions of quantum theory on a scale which was many orders of magnitude larger than it had been previously.

The results of the experiment performed by Aspect and his colleagues demonstrated that locality is not a principle that is respected by nature. This means that the entangled nature of the quantum state reflects something essential in the world. It also means that the principle of radical atomism—that the properties of one particle are independent of the properties of other particles—is wrong, disproved by experiment. This makes it one of those rare cases in which an experiment can be interpreted as a test of a philosophical principle.

I should hasten to point out that this non-locality coexists peacefully with other senses in which influence can only be transferred locally, between nearby particles. It is still true, and completely consistent with all of this, that it is impossible for energy or information to travel faster than light. As a result, one cannot actually use the quantum entanglement to transmit any information.

Still, the situation is very surprising. Once any two photons, pets, or anything else have interacted, one cannot separate any description of the properties of one from the properties of the other. Given any one electron, its properties are entangled with those of every particle it has interacted with, from the moment of its creation, indeed quite possibly from the moment of the creation of our universe.

I remember very well the day I first learned about all this. It was during the Spring semester of my first year in college. I had a great physics teacher, Herbert Bernstein, who insisted that if quantum mechanics was fundamental it should be taught first, and so our first year physics was quantum mechanics, with the culmination of the course being the Einstein-Podolsky-Rosen experiment and the Bell inequalities. I was sitting outside on the grass studying the papers of Einstein, Podolsky, and Rosen, Bohr and Bell, and when I understood what they meant, I went back to my room and lay on my bed for a long time, staring up at the corner where the ceiling met the two walls. I was convinced immediately that there was only one possible conclusion: space is an illusion, so the coherence of the world must be behind and outside of space. While space may be a useful construct for certain purposes, a fundamental theory cannot be about particles moving in space. Space must only emerge as a kind of statistical or averaged description, like temperature. I recall also being very struck that there were atoms in my body that were entangled inextricably with atoms in the bodies of every person I had ever touched.

To my mind, this experimental disproof of the principles of locality and atomistic reductionism brings to a close any possibility of going back to the world of classical physics. This does not mean that there may not be a possibility—even a necessity—of replacing quantum mechanics with a theory that gives us still deeper insight into the coherence of the world. It is still possible to imagine that there may in the future be a theory that can do what quantum mechanics cannot do, which is to give, at least in principle, a definite prediction for the outcome of any experiment we do.

But what this situation does mean is that any such theory must be explicitly and radically non-local. It must involve additional variables, beyond those known to either Newtonian or quantum physics, which describe relations between any two particles that have ever interacted. That is, if there is to be a more complete description of an electron than that given by quantum mechanics, the additional information provided by that theory will not consist of more detail about the internal structure of the electron; it will instead involve more detail about the relations between that electron and the rest of the universe.

If such a theory is possible, then it will not allow us to give a complete description of any single particle unless we give a complete description of the whole universe. In other words, any physical theory from this point on that represents progress beyond quantum mechanics must be an explicitly cosmological theory.

Might quantum mechanics turn out to be an approximation to such a non-local cosmological theory? All we can say at present is that this is not impossible. A few examples of such a theory have been invented. Unfortunately, none of them have been terribly elegant or convincing. At best they stand as prototypes for theories that might yet be invented.

But whatever the future brings, we have already learned an irreversible lesson from quantum mechanics. The naive idea that each particle in the world can be described completely independently of all the others fails. We can even say as a result of Aspect's experiments, that the principle of radical atomism has failed empirically. Instead, if we want to give a complete description of an elementary particle we must include in the description every particle it may have interacted with in the past. This means that we can only give a complete description of any part of the universe to the extent that we describe the whole universe.

We, who live in the universe, and aspire to understand it, are then inextricably part of the same entangled system. If we observe some part of the world, we become entangled with it in the same way that any two particles that interact become entangled, so that a complete description of ourselves is impossible without incorporating the other.

The aspiration for an objective description of the world—by which we mean a complete description of the world as it would be independently of whether we were here or not—is then apparently in conflict with the results of experimental physics. Thus, the question of what space and time are is intricately connected to the final of my five questions: *How can we, who live inside the universe, construct a complete and objective description of the universe as a whole?*

This is the key question that confronts us as we attempt to combine quantum mechanics with relativity and cosmology to construct a unified cosmological theory.

PART FIVE

EINSTEIN'S

REVENGE

How can we, who live in the world, construct a complete and objective description of the universe as a whole?

It is always an I who says we.

    —Jacques Derrida

TWENTY

COSMOLOGY *and* *the* QUANTUM

At the end of the last century human beings did not know that the stars are organized into galaxies, and they had not imagined that gravity could be merely an aspect of how space and time are arranged. They did not know how atoms or stars work, and they had heard of neither the quantum nor of the atomic nucleus. Neither did they know that the continents move, or that genetic information is stored in DNA, and they had only the faintest notions of the history of life on Earth. Beyond this, the idea that the universe itself has a history would, had they heard it, have seemed almost inconceivable.

It is fantastic indeed what has been achieved in this last century. At the same time, when we think about what we would still like to know, what, indeed, we still must know if we are to make sense of what we have so recently learned, can there be any doubt that this extraordinary period of transformation in our knowledge of the world has not yet come to a climax? What we do not yet know sometimes seems so daunting, so almost impossible. We do not know why the proton, neutron, electron or any other particle has the mass it does. We do not know what happens inside a black hole, or why our universe was created to exhibit the extraordinarily improbable combination of uniformity and structure it does. We do not know how the galaxies formed, or how life began here on earth. And we do not yet have a way of thinking that allows us to understand our world as a single entity that unites the quantum with the discovery that the structure of space and time are dynamical and contingent.

This last circumstance alone means that we are living through a period of scientific crisis. We have made great progress, but the fact that general relativity and the quantum are not yet united means that we have no single picture of what the world is that we can believe in. When a child asks, What is the world, we literally have nothing to tell her.

As so many examples from the history of this century attest to, human beings have a remarkable ability to live with crisis, to live even with insupportable contradictions. And once we accommodate to something, and become used to it, it is often extremely difficult to imagine things could be any other way. This is perhaps the most difficult thing about any attempt to transform the world on any scale. I wonder sometimes how many diplomats, historians, or even pacifists ever really imagine a world without war as a normal aspect of the intercourse of nations, no matter how clear it may be that this must happen in the next century if there is still to be humanity? How many cancer specialists or AIDS activists wake up every morning with the conviction that in only a few more years these diseases will be in retreat? Or, despite everything we know, how often do we clearly perceive that the genetic differences that lead to different skin color are as trivial as those that lead to different hair color or body types, so that beyond what is in people's imaginations, the problem of what we call "race relations" is no different from the problem of relations between people of different histories, a problem which has been solved, many times, in the United States and other countries. At every level, as the political theorist Roberto Unger tells us, the greatest obstacle to transforming the world is that we lack the clarity and imagination to conceive that it could be different.

It is no different in science, and I often think that the greatest obstacle to the construction of a theory that will answer the questions I mentioned above is that we who work professionally on them have become too used to having them around. It indeed sometimes seems that we spend more time studying the problems than we do trying to imagine the solutions. But the wonderful thing about ideas is the extent to which they have a life of their own, so that while contradictory conceptions may live side by side for years in many people's minds, one day, in someone's imagination, a new thought, a new combination, leads to a new viewpoint from which it all suddenly fits, differently.

I have no doubt that there will someday soon be a quantum theory of gravity that, which by combining relativity and the quantum into a single cosmological theory, will answer many of the mysteries about our world that now seem so impossible. And while I cannot be sure what that theory will be, I have tried, in this book, to put together clues and to imagine what the world may be like, once we have discovered and understood it. If for no other reason, I have done this to keep alive, at least in myself, the sense that quantum gravity must be the name, not of an activity, or a field of research, but of a theory whose construction will transform our understanding of the world in which we live.

In this book I have discussed many different questions, ranging from the history of the cosmos to the properties of the elementary particles to the definition of life. But I have been talking all the time only about the quantum theory of gravity. If I have not yet come to describe what we know about how space, time, and gravity may be conceived quantum mechanically, it is because I wanted to first make it clear the extent to which the problems in all of the areas we have discussed are illuminated when seen in the light of a single point of view. For, as I will explain in these last chapters, this same point of view must also be the basis of the theory that will combine quantum theory and relativity into a single, beautiful, and rational understanding of our universe.

To put it most briefly, the point of view I have been championing comes from taking very seriously the idea that all properties of things in the world are only aspects of relations among real things, so that they may be described without reference to any absolute or fixed background structures. I have described in the last several chapters how this point of view, which arose first clearly in Leibniz's criticisms of Newton's formulation of physics, lies behind the most central principles—and most surprising predictions—of both relativity and quantum theory. Before this, in Part One, I described how this same point of view led to the greatest achievement of elementary particle theory—the gauge theories—and underlies also its greatest aspirations, as we see represented in string theory. In Parts Two and Three I showed that we are now in a position to radically extend the reach of this principle, by making possible the conception of a world whose structures and regularities are the result of contingent and historical processes of self-organization, rather than being imposed by absolute and pre-existing law. Now, in these last chapters, it remains for me only to bring all of this together and to explain how these same principles drive and illuminate the search for a quantum theory of gravity, to tell the story of what we have so far learned in this search, and explain how we are trying now to conceive of the resolution of the puzzles that remain.

In the last chapters I have argued that we must go beyond two ideas that, despite their attractiveness, are holding up progress because they are incompatible with the conception of a theory of a whole universe. I gave each of them a special name, which were radical atomism and single-observer objectivity. The reason for these perhaps awkward, constructions was to make it clear that I am not against reductionism in general. I am only against the much more radical idea that the world is made solely of fundamental entities with fixed immutable properties that are defined with respect, not to each other, but only to an unalterable and absolute background. Similarly, I am in favor of, not against, the idea that there is an objective world, and I believe that the goal of science is its comprehension. But I think this goal cannot be reached unless we give up the idea that this comprehension can be put in a form that could be read as the view or picture of the world that could be seen by a single observer, as this must necessarily be an observer outside of the universe.

Against these two ideas, whose power over us, I have tried to argue, is only historical accident having to do with the religious and philosophical roots of science as it originally developed—I have proposed three principles. First, a cosmological science must be based only on relations between things in the world and cannot depend on any properties whose meaning requires the postulation of fixed, background structures. Second, a world in which such a relational notion of properties makes sense must be a complex world, with sufficient variety to distinguish completely the different things in it from each other. The fact that we find ourselves in a world full of a variety of phenomena and structures must be essential and not accidental. Third, to uncover the real relations of things, we must think clearly and carefully about the correspondence between what the theory describes and what is actually observed in the world.

Let us keep these principles in mind as I now turn to describe what we know about combining quantum theory, cosmology and relativity. There are three key questions which I want to focus on in this and the chapters that follow.

Quantum theory seems to accord a special role to an observer who is outside of the system under study. But if we hope to elevate it to a theory of the whole universe, there can be no place outside it for an observer to stand. The first question is, What are we to do about this? The second question is how we describe the geometry of space in quantum mechanical language. Guided in part by Leibniz's vision of a relational universe, we have come to some conclusions about this, which I will describe in the next chapter. And then I will end with the question of time.

The bold step that Bohr and Heisenberg took in their formulation of the interpretation of quantum mechanics was their insistence that to give the theory meaning the world must be split into two parts. The observer, together with a clock and the measuring instruments, is on one side of the divide; the system to be studied is on the other side. What, above everything else, is in contention in the search for a quantum theory of cosmology is the extent to which this division of the world is essential.

The divide serves several functions. As a result of the uncertainty principle we cannot have all the information about any system that would be necessary for its complete description. As a result, when we study a quantum system we must choose which quantities we would like to measure. If we later make a different choice, then our information will change, immediately and discontinuously, for some of the old information that we used to have may no longer be relevant.

As a result, when doing quantum physics we must always keep in mind that we never have a complete representation of the system we are studying. This means that the abstract-state space we use to represent the information we have is not necessarily something that is in complete correspondence with the system. For example, if we ask a new question and gain new information, we will represent it by a new state. This abrupt change in the state represents the abrupt change in our knowledge of the system. There may or may not be any corresponding change in the reality of the actual system. The divide helps us keep track of these distinctions because it circumscribes the part of the world that we are representing incompletely by a quantum state.

Further, because of the probabilistic nature of quantum theory, we will not be able to predict exactly what the result of the new observation will be, so the new information we gain cannot be predicted on the basis of the old information we had. This is the basis of the argument of Einstein, Podolsky, and Rosen that the quantum state corresponds not to reality, but only to the information we have about a subsystem of the universe.

Most of the time we have enormously less information about a system than the fifty percent that the uncertainty principle allows. Any macroscopic object is made of at least 10^{22} atoms, so there are at least this many things to know about it. We take advantage of this all the time in our use of clocks and measuring instruments. Because we require that they only give us one number, we can remain ignorant about almost everything else about them. Only in the systems we study quantum mechanically do we try to get close to the limit of knowing half the information. This is another reason the divide is useful: it delineates a part of the world, which we would like to get to know as well as possible, from the rest. The delineation is important because it is only because we are happy to remain ignorant about most of the details of our clocks and measuring instruments that we can use them to get close to the limit for the quantum system.

Quantum mechanics works very well in every context in which it has so far been tested. The insistence on the division of the world into two parts may trouble us philosophically, but it normally presents no real barrier to science, as we are normally interested in studying only a part of the universe. It is only in the desire to extend the quantum theory to the whole universe that it presents an obstacle that must be overcome.

From here everything I say must be controversial, as there is no settled opinion about how to extend quantum theory to cosmology. I would like to describe several different points of view that are being pursued, not all of which are my own.

The most natural response to this situation is to say that the quantum theory simply cannot be extended to a theory of the whole universe. If, according to quantum theory, the observer must necessarily be outside the system being described, and the observer is in the world, then quantum mechanics must necessarily describe only a part of the universe. So it is very reasonable to suppose that quantum mechanics is just an approximation to another theory, and that it is that deeper theory, and not quantum theory, that can be extended to describe the whole universe.

This conclusion dovetails nicely with the conclusion we reached in the last chapter, which is that if there is to be a deeper theory beyond quantum theory, that theory will have to be a cosmological theory. It is not easy to invent a new theory, especially in the absence of a direct contradiction between theory and experiment. But this has not prevented some courageous people, such as Roger Penrose, from pursuing this direction. The aim of such a theory must be to uncover a description that does not require a division of the world into two parts, with the observer on one side and the system on the other. One way to do this, which is being pursued by Penrose and several others is to hypothesize that the collapse of the quantum state—which is the immediate change in the state as a result of a measurement—is not just a reflection of the fact that our information about the system has changed. Instead, they suggest it may be a real physical process that happens from time to time whether an observer is present or not. To agree with the predictions of quantum theory they must presume that this happens very seldom for little things like atoms but very often for big things like a human brain. The challenge is then to understand how nature understands the difference between big and small. One natural possibility, which is being pursued by Penrose, is that this is done by the gravitational field, as that is the force that counts the total mass of a system.

Another way to try to make such a theory is through a non-local theory that goes beyond quantum mechanics, as I described in the last chapter. A number of people have attempted to do this, including myself. The important thing in any development that aims to replace quantum theory by another theory is to try to make contact with experimental physics. These theories must predict something different than would be predicted by the usual quantum theory. If they are true, there must be some scale, in size, energy or perhaps complexity, that measures the boundary between the quantum world and the new physics.

A second path is to leave the quantum theory more or less as it is, but to try to find an interpretation of the theory which eliminates the special role played by observers and measurements. If this can be done then perhaps quantum theory can be seen to be more like Newtonian mechanics, which claims to be able to give an objective description of a whole world, simply as it is. Of course, the theme of the chapters 16 through 18 was that Newtonian physics did in fact rely on the notion of an observer outside of the system, as we see in the implicit dependence of the theory on a special class of non-accelerating observers. Still, this does not mean that the goal of an observer-independent interpretation cannot be reached for a quantum theory of cosmology. This is then a possibility that must be explored.

The project of reinterpreting quantum theory so that an observer is not explicitly required has a long tradition. Many good physicists have worked on it in the last forty years; it is presently the direction with the most adherents. Hence, this is a story worth telling, whatever the outcome.

In the 1950s, a graduate student named Hugh Everett made an astounding suggestion, which was taken up and championed by two of the great pioneers of quantum cosmology, Bryce DeWitt and John Wheeler. They proposed that it is the quantum state, and not what we see around us, that should be taken as coinciding with reality. To the immediate objection that the quantum state coincides only probabilistically with what we observe, they replied that the world we see around us must be only a tiny part of reality. Instead, they imagined the universe to be multiplied into an infinite number of copies or "parallel branches," so that there is a branch for each of the possible outcomes of every experiment and observation.

How is the theory then going to make any predictions? It may, because within each "branch" of reality there must be consistency, for example, different observers who make the same measurement must agree about what they see. While the theory cannot predict what will happen in any particular branch, it can predict the extent to which observers in a particular branch will agree about what they observe.

This proposal is called the "many worlds interpretation of quantum mechanics." I must confess immediately that I do not like it, but that many of my friends and colleagues do. The fact that it has been championed by several of the people whom I most respect has left me deeply puzzled. First of all, the theory seems a bit extravagant and excessive. As an artist friend once remarked when I described it to him, to postulate an infinite multiplication of the world because one is unable to resolve a problem of interpretation in physical theory is a bit like moving and purchasing a new set of kitchenware each time one doesn't feel like doing the dishes.

The many worlds interpretation does succeed in letting the observer be part of the universe, but at the cost of saying that the world that we observe is only a small part of reality. At first this sounds like progress, but I must confess I have never been able to understand what its adherents mean by the claim that reality has more to do with the quantum state than with what I observe. Do they mean that I, who live in the universe, cannot observe reality? It seems so. According to this interpretation, only an observer who lived outside the universe, who had somehow the same relation to the whole universe that we may have towards some atoms of gas in a container, could observe this quantum state of the universe. According to the many-worlds interpretation, it is only such an observer who could know all of reality.

It thus seems to me that the many-worlds interpretation can be understood as an attempt to preserve, in a quantum theory of the whole universe, the notion of single-observer objectivity. I know of no other way to understand the desire to posit that the quantum state of the universe corresponds to reality, in spite of the fact that there can be no observer inside the universe (as we ordinarily mean it) who can observe it. This formulation preserves the idea that there is a single objective view of reality, by the extreme means of making that the view of an observer who does not live in the world.

It seems to me that the only possible name for such an observer is God, and that the theory is to be criticized as being unlikely, on these grounds. I say this not because I have something against God, but because I suspect that a theory that asserts that only such an outside observer could know the objective reality of the universe must lack the kind of logical coherence that we would like a theory of the whole universe to have. To put it most directly, I suspect that the many-worlds interpretation is another shadow of the metaphysical roots of modern physics; it is another manifestation of the nostalgia for the absolute. It diminishes the real in favor of the ideal and so reveals the ambition to imagine a view of the world from outside of it and the desire to identify with a being who could have that view.

Whatever one may feel about these issues of metaphysics and motives, there would be little that one could say against the many-worlds interpretation if it did provide a completely coherent interpretation of the quantum theory. However, it does not in the end do that. It does accomplish certain things, such as providing an interpretation in which the quantum state never changes discontinuously after a measurement. But it does not remove the necessity of adding an element to the theory that corresponds to the choice of what language we use to describe the world, or what questions we choose to know the answers to.

The problem is that there is an infinitude of different ways in which we might divide the quantum state into different branches. In almost all of these there is nothing remotely like the classical world. In only very few of them do the ordinary observables that characterize our world, like the positions of particles and the values of fields, have definite meaning. The many-worlds interpretation only makes predictions that correspond to the world we live in when we choose to divide up the branches according to the quantities that make sense in Newtonian physics. If the theory is to stand as a description of reality, and not just of the information we have about it, we must discover a reason why the branches should be divided up so that the quantities of Newtonian physics are meaningful.

Over the last ten or so years a point of view has emerged about how to do this which is called "decoherence." According to it, a quantum universe has to be very special if it is to have any interpretation that could be formulated in the language of classical physics. When this happens the universe is said to "decohere." In these cases alone, there can be observers who think they live in a classical universe that contains objects with fixed properties.

There are anthropologists who say that what is special about human beings is that we tell each other stories. From the ancient cosmologies to our friends' love affairs, we frame what we know as narrations of what happened. The idea of decoherence falls in this tradition; it is above all a proposal for how we may extract stories that tell the history of the world from the information that could be coded into a quantum state of the universe.

In quantum physics a story can be constructed by asking a series of questions about what happened at different times. Each set of questions evokes a sequence of answers; each possible set of answers is then said to comprise a history of the world. The problem is that most such histories in quantum physics cannot correspond to a classical world in which things have fixed properties. If we try to do this we find that the probabilities for the different alternative histories do not add up to one. (This is familiar from the double-slit experiment: the probability that the particle goes through either the top or the bottom slit is not the sum of the probabilities that it go through either one.)

There are special sequences of questions that can be chosen so that the probabilities for the different stories created by their answers do add up to one. When this happens the stories are said to be "consistent." In these cases, and in these cases alone, one can tell stories about quantum mechanical systems that make logical sense. This is the basis of an approach to the interpretation of the quantum theory that aspires to do without observers, called the *consistent-histories interpretation*. Its main developers have been Murray Gell-Mann, James Griffiths, James Hartle and Roland Omnes. Their formulation has made it possible to give a precise meaning to the idea of a quantum world "decohering" into a collection of classical worlds.

This has clarified the extent to which the quantum world allows the existence of observers like us who perceive the world as if it were classical. But, while it succeeds to this extent, there is still a problem. Whenever a universe allows a consistent set of histories that describe the world in terms of classical, stable properties, it allows also many other, equally consistent sets of questions and answers. These other sets tell the story of the world in language that cannot be described at all in our familiar classical terms.

This is not unlike the problems that afflicted the original many-worlds interpretation. To the extent that the theory allows the classical description as a possible reality, it also allows other equally consistent descriptions of reality. The definition of reality that the theory allows encompasses too much. Unless we tell it that we are only interested in one particular set of stories—for example, the ones that make sense in classical language—the theory is unable to predict anything definite at all.

This aspect of the theory was discovered by two young British scientists, Fay Dowker and Adrian Kent. It has been hotly debated at physics conferences over the last two years, and two kinds of responses have evolved. One is that the theory is simply incomplete. If decoherence or consistent histories are to be the basis of quantum cosmology, then we must find some criterion which distinguishes the description in terms of classical quantities from all others. In the absence of this, there is no way to explain why our world is so well described by classical physics.

The other response is to embrace the apparently damning conclusion of the argument, and assert that the reality of the world is indeed a many-faceted thing, which may be viewed from many points of view besides our own. According to this point of view, the different possible sets of consistent histories describe equally real tellings of the story of the world. The only question then is why they are constructed in such a way that we perceive the world in terms of one kind of story rather than another.

This is the view of Murray Gell-Mann and James Hartle, who make the following proposal. They say that the reason creatures like us see a "classical world" made up a stable objects that move in predictable ways is that we have evolved by natural selection to be sensitive to this way of perceiving reality. They hold that the other alternative ways of telling the story of the world are equally consistent and equally real, but a creature that adopted such a way of seeing the world could not have counted so regularly on having lunch. We have evolved sensory organs that process information about where things are and how they are moving because that is what we must do to survive.

What is remarkable, if this point of view is correct, is that the complexity of the world is a precondition for their strategy to work. According to this view, it is possible to perceive the world in classical terms because it is highly organized. This means that we cannot just believe in a world made of simple laws that happens to have evolved complexity. There must be something essential about the fact that the world is complex.

Even if this view is not completely right in the end, the discussion has come to a remarkable conclusion. In the last chapters we saw that general relativity requires that the world be complex if the world is to have a sensible interpretation. Here we arrive again at the same conclusion: one cannot have an interpretation of quantum theory that can be applied to the whole world unless that world is sufficiently complex. As we go on we will see that there is something robust about this conclusion, for every attempt I know which aims to give meaning to a quantum theory of the whole universe succeeds to the extent that it embraces the complexity and variety of the world.

TWENTY ONE

A PLURALISTIC
UNIVERSE

A nyone who has ever had a dispute with a friend or lover knows that there must be something funny about the notion that a single observer can have a complete and objective description of a whole universe. In life, as in quantum cosmology, the reason for this may not be in the difficulty of gaining knowledge of others; the root of the problem may lie in the impossibility of seeing ourselves completely and objectively.

There is a simple argument that can be made for the postulate that no observer can have complete and objective knowledge of him or herself. I learned it from an article by a young philosopher called Thomas Breuer. The core of the argument is simple. Whatever an observer observes must be recorded inside itself as a memory. Thus, the act of observing necessarily changes the observer, because he now must contain within his memory a record of what he saw. This means that just after an observer attempts to make a completer observation of himself, his memory contains a complete description of himself the moment before. But now his state has changed—dramatically, because by definition this recording of a complete copy of himself cannot be a small alteration of his state. So what is in his memory a representation of his present state. He may try again, but clearly there is a problem of an infinite regress. (Even assuming the observer has an infinite potential memory.)

If we cannot make a complete observation of ourselves, then our point of view cannot include a complete description of the whole universe. This means that no single view of the universe can exist that is both complete and objective, unless such a view would belong to some entity outside of the universe. If we rule out speculation about what fictional beings outside of the world could know, we must conclude that single observer objectivity is impossible in a quantum theory of the whole universe.

I should hasten to say that this argument does not preclude the possibility that we may gain a great deal of information about ourselves. Like all macroscopic systems, it would require a stupendous magnitude of information to describe ourselves completely. So there is only a problem for science to the extent that we aspire to construct a theory that could give a complete and objective description of the whole universe. But the problem is serious; it implies that such a description could not be constructed from the information available to any single observer in the universe.

Each of the two approaches to quantum cosmology that I described in the last chapter agree with this. In a non-local, hidden-variable theory much of the information necessary to describe the universe completely resides in correlations between things that are far from each other. No observer limited to a small portion of the universe will be able to gain this information. In the many worlds and consistent histories approaches, it is axiomatic that there is more information in the quantum state than can be recovered by any single observer in the world.

However, both approaches allow the invention of a fictional language in which we can speak of an observer outside the universe, who could know all the information needed to describe it completely. This seems somehow wrong. We should expect more of a cosmological theory than that it allow a partial interpretation in terms of observers in the world, while at the same time leaving open the fantasy of a more complete interpretation in terms of fictional beings outside of the world. The concept of an observer outside of the world is based on an elementary contradiction, for then there is a second world, larger than the first, that encompasses both what we called the universe and its fictional observer. In a truly fundamental theory that aspired to describe the whole universe, it should not be possible to make such a logical error.

To avoid this, I believe that we should ask more of a cosmological quantum theory than that it simply allow the possibility of an interpretation in terms of observers inside the world. We should require that the theory logically forbid the possibility of an interpretation in terms of an observer outside of the universe. We may make a principle of this; *the principle of the logical exclusion of the possibility of an interpretation of cosmological theory in terms of an observer outside of the world.*

This means that a quantum theory of cosmology cannot be achieved by simply extending the formalism of quantum mechanics to the universe. Whatever other interpretation we give to it, that formalism will always allow an interpretation in terms of an observer outside of the system, *because that is what it was invented to do.* To make a quantum theory of cosmology, we must invent a mathematical formalism that would make no sense were it applied to any subsystem of the universe.

How are we to do this? One possible approach is to simply give up the idea that a complete description of the universe can be composed of the knowledge available to a single observer. Instead, we may posit that a complete and objective description of the world might be found in the collection of views held by a community of observers, each of whom can never have more than partial information about the world. To put it most simply, I accept that *I* cannot know everything. Quantum mechanics seems to leave no way around this. But perhaps, at least in principle, *we* can know everything.

I would like to describe one approach to doing this which has evolved over many years in conversations with several friends. Like so many of the ideas I have described here, the inspiration for this proposal comes partly from the writings of Leibniz. In *The Monadology,* Leibniz posits a world that is constituted by a large number of entities, which he called monads. These monads do not live in space; rather, space is an aspect of their relations. Nor can there be any observer of the universe who is not one of the monads. Instead, in Leibniz's vision, reality is contained in the views, or perceptions, that each monad has about the others. As there is nothing but the monads, what one of them can perceive about the others is only relations. A complete description of Leibniz's universe cannot be obtained from the outside, and there is no monad who sees completely the whole of reality. Rather, reality is contained in the sum of the views of all the monads. Leibniz expresses this vision beautifully in the passage that opened this book.

Many philosophers describe *The Monadology* as a metaphysical work. Perhaps they are right, I am not a scholar. There are certainly parts of Leibniz's writings that I have difficulty understanding, and it is possible that I would disagree with them if I did understand them (Indeed, I've found in philosophy that this often happens, if I have difficulty understanding what someone is saying it is because of an unstated assumption that I don't share.) But the impression I get from those parts of Leibniz that I do understand is that he was a physicist struggling with the same broad problems that we face now as we attempt to make a quantum theory of cosmology. I believe he was led to the vision he describes in his writings by wrestling with the question of how to construct a physical theory that could describe a whole universe, without the postulation of absolute space and time, such as his rival Newton required. As I described in previous chapters, this means a theory that satisfies his principle of sufficient reason, something no physical theory was able to even partially achieve until Einstein's general theory of relativity. Now, I do not want to say that Leibniz solved this problem. Had he, he would have invented relativity, and having only just invented calculus, even he could not possibly have done that. But I do believe that, because he struggled with this problem more than any other philosopher, his work contains insights that we may learn from now.

A few friends and I have been thinking about whether it is possible to use Leibniz's idea as a model for how to make a quantum theory of a whole universe. Roughly speaking, we would like to make a theory in which there are many observers, each of whose incomplete views of the rest is represented by a quantum state. Such a solution to the problem of extending quantum mechanics to the universe would be, in a sense, opposite to the many-worlds interpretation. Rather than having many universes but one quantum state, we would have one universe whose description would require many quantum states. Each quantum state would be a description of a part of the universe—the part excluding a particular observer.

For this to work, the different quantum states which contain the information known by the different observers must be consistent with each other. However, this does not require that all observers have the same information. For example, suppose that two scientists, Louis and Carlo, want to know what the weather is. Louis calls the phone number that gives recorded weather information. Afterwards, his knowledge of the world includes something about the temperature, wind conditions, and so forth. Afterwards, he says nothing. What then does Carlo know? He knows as yet nothing definite about the weather. But, as a result of his observation of Louis at the telephone, he can conclude that his friend knows what the weather service says the weather is.

Rather than ask Louis, Carlo chooses to observe the weather by going into the next room and looking out the window. Then the situation is symmetric. Louis knows Carlo knows something about the weather outside, but he does not know the content of that knowledge. Later I come in and, looking at them, I know instantly that each of them has learned something about the other. Carlo tells me, "These crazy Americans, you ask them should you wear a coat today and they pick up the telephone." Louis responds, "These crazy Italians, they think that if they just look out the window they can know more than the weather service about whether it will rain this afternoon."

Suppose that each observer describes what they know about the world in the language of quantum theory. Then Carlo has a quantum state that describes his view of the world, which includes Louis, the weather and the weather service. In his quantum state will be found the information that there is a correlation between what the weather service knows about the weather and what Louis knows. But, just as there could be a state that described a correlation between the species of two pets, without specifying whether each is a dog or a cat, there can be a quantum state that describes the fact that Louis and the weather service will agree about the weather, but does not contain the information about just what it is that Louis and the weather service both know.

Similarly, Louis has a quantum state that describes his knowledge of the world, which includes something about the weather service, the weather, and Carlo. The state that describes the total of his information about the world may include the fact that Carlo knows something about the weather, without specifying what it is that Carlo knows. Then I also have a quantum state, that describes Carlo, Louis, the weather service and the weather. But the only information I may have is that Carlo and Louis each know something about the weather, and about each other, without knowing the content of any of this information.

From a Newtonian point of view, it is possible to imagine that each of these observers might in time gain a complete picture of the whole situation, what the weather is, and who knows what about it. But in quantum theory, any one observer can have at most only half the information about the rest of the world that would be required to give a complete picture. Imagine, for example, that the world were simple enough that the temperature and the precipitation were the only two measurements that could be made about the weather. In this case quantum mechanics would require that no one could know simultaneously whether it were raining and what the thermometer reads. In this case, each of the two observers could only know one or the other. Carlo might look outside to see if it is raining, while Louis hears from the weather service what the temperature is. And it is consistent with quantum mechanics that I could know that each of these knows the answer to one question or the other, as long as I do not know what the answers are.

One of the really wonderful things about doing science is the way one's arguments and conversations with friends continue over many years, in many settings. Beginning with this first meeting in New Haven, and continuing over many phone calls and encounters, Louis Crane, Carlo Rovelli, and I have been arguing about whether it is possible to formulate a description of the universe in terms of a large set of quantum states, each one of which describes the partial knowledge or point of view that an observer has about the other things in the universe. Given the fact that in quantum theory no observer has complete information, the possibilities for how the different views may relate to each other can get somewhat intricate, as in my story about the weather. If there is to be a sensible theory that takes into account all the possible views, it must rely on some principles that constrain how the different views may differ, while still being partial views of the same world.

To put it in terms that Leibniz used, what is needed is a mathematics that describes how it is that the views of all the monads are in harmony. Furthermore, this mathematical structure must be different than the conventional quantum theory, as it must degenerate into triviality or contradiction if one attempted to deduce from it a description of the whole universe as it would be seen by an outside observer.

This may all seem a bit crazy. So let me hasten to mention that the mathematical structures needed to make such a theory work are already under development. They have grown out of the study of a very special class of quantum theories, which are called *topological quantum field theories*. These were first invented by Michael Atiyah and Edward Witten to study the topology of three-and four-dimensional spaces. Some years ago Louis Crane realized that they provide exactly the kind of framework we need to construct a quantum theory of gravity and cosmology that would automatically forbid an interpretation in terms of an observer outside of the universe.

A topological quantum field theory is a model of a closed world, like a sphere, that has no boundary and no exterior. However, there are no quantum states associated with the whole system. Nor are there fields associated with points of space, as we imagine the electric or magnetic fields. Instead, the description of a world in a topological field theory begins by drawing a boundary that divides it into two parts.

Given a choice of such a boundary, one then constructs a space of quantum states. The idea is that these may be interpreted as containing the information that an observer, who may be thought of as living on one side of it, may have about the part of the universe contained on the other side.

There are many ways to draw a boundary to divide a world into two halves. In topological quantum field theory we do not pick one; instead, the theory treats all of them equally. One associates a quantum state to each possible choice of a division of the world into two. Each contains the partial information that an observer on one side of a boundary might have about the world on the other side.

In this language it turns out to be possible to formulate precise conditions that constrain the ways in which the information associated with the different boundaries are related to each other. It turns out that the laws that describe the dynamics of the universe are coded in the relationships between the different views, rather than in any single one.

At present, work is underway to underway to elaborate these topological quantum field theories into structures rich enough to describe our four dimensional universe. One of the most interesting aspects of this approach is that these theories have their own way of talking about the properties of things in the world, which is in terms of a language called *category theory*. This is a language that Leibniz would have embraced, had he known of it, for—as the mathematician John Baez explained to me recently—it resolves one of the difficulties of his philosophy. The principle of the identity of indiscernibles says that any two things that are identical are the same; it is impossible to speak of two distinct things that are identical. This sounds like good philosophy, but there is a problem: it makes it hard to do any mathematics. For what mathematicians do all day is to write equations in which the quantity on the right side of the equals sign is set equal to the thing on the left. But according to Leibniz, if two quantities are equal, they are identical. So what could an equation like $1+1=2$ actually mean? One answer to this is that all such statements are merely tautologies. According to standard logic, the only meaning a statement like $1+1=2$ can have is that $1+1$ is only another name for 2. But this seems silly. If mathematics encompasses only tautology, why do we have the impression it teaches us so much?

Category theory solves this by describing a mathematical world purely in terms of relationships between distinct things. In the language of category theory one simply cannot say $1+1=2$. What one can do is to postulate a relationship between two distinct things, one of which is a pair of 1's, and the other of which is 2. This relationship may be described by saying that there is a certain transformation, which we call addition, that acts on the first thing and turns it into the second.

This may seem pedantic, but it is the difference between a mathematics in which one can say things that are both true and novel, and a mathematics that is only elaborate tautology. It is also precisely the fact that category theory does not allow one to say things that appear to have content but are in the end empty that makes it impossible for topological quantum field theory to speak of the state of a whole universe as it would be seen by something outside, but unrelated to, it.

Topological quantum field theories have another attractive feature, which is that they describe worlds that are necessarily simpler than those imagined in previous theories. In ordinary quantum physics, there are an infinite number of possible quantum states of a system. This is necessary if quantum theory is to describe the fact that a particle might be at an infinite number of different positions. The topological quantum field theories seem impoverished compared to this; according to them, any observer in the universe, associated with a given surface, can have only a finite amount of information.

This is the origin of their name: they were thought originally to be good only for describing topology, which requires only a finite amount of information. For this reason, it might seem that the topological quantum field theories cannot describe our real world, as that would require an infinite amount of information.

However, there is good reason to believe that this is wrong, and that in fact no observer in the world can have more than a finite amount of information about the rest of the universe. This conclusion comes from another line of attack on the problem of quantum gravity, in which people have tried to understand how black holes might be described in terms of quantum theory.

Perhaps the greatest achievement, so far, in the search for quantum gravity was the realization, in the mid-1970s, that black holes are thermodynamic systems. This means that, as discovered by Jacob Bekenstein and Steven Hawking, each black hole has a temperature and an entropy. The entropy of a system is a measure of the maximum amount of information it may contain. What is remarkable about a black hole is that its entropy is proportional to the area of its horizon.

This discovery was made by Jacob Bekenstein, when he was a graduate student studying with John Wheeler. Since then he has been pursuing a radical consequence of his discovery, which is that there is a limit to how much information may be contained inside of a given region of space. The limit is given by the entropy of the largest black hole that would fit into the region. Since the entropy of a black hole is proportional to its area, the maximum amount of information any system can contain is proportional to the area of its boundary.

In outline, the argument for this goes as follows. According to the laws of thermodynamics, no process is allowed that can decrease the entropy of a system. Suppose that inside some boundary, a system exists that contains more entropy than any black hole, which fit inside the boundary, could contain. It turns out that in this case one can always add energy to the system until it is so dense that it must collapse to a black hole. But then the entropy goes *down* to that of a black hole that could be contained within the boundary. This is impossible, so there must be something wrong with the assumption of the argument, which was that a system can have more entropy than the largest black hole that fits into the same region. So the entropy of any system contained within a finite region is bounded. But then the information it can contain is also bounded, as entropy is a measure of information.

The existence of a bound on the amount of information that can be contained in a region of space opposes directly one of the basic principles of twentieth-century physics, which is that the world is made of fields. A field is something that can vary independently at the different points of space. Whether described by classical or quantum physics, a field can thus contain an infinite amount of information. Even if one thought that there was an ultimate limit to how small things could get, so that one could not infinitely divide space, a field theorist would still expect that the amount of information that could be contained in a region grew proportionately to its volume, not to the area of its surface. Bekenstein's bound, as we call it, tells us that any quantum theory that can describe black holes cannot be such a theory of fields.

How are we then to describe physics, if not in terms of fields at points? Physicists like to joke that no idea about quantum gravity has a chance to succeed unless it is crazy enough. Each of the few times I met Richard Feynman, as a young field theorist, and had a chance to tell him of my work, he dressed me down for insufficient craziness. Well, here is an idea that perhaps would have passed this test: it is called the *holographic hypothesis*. It holds that the Bekenstein bound allows no description of the world in terms of what is happening at points in space. Instead, the world can only be described if we draw boundaries around regions and describe what is inside in terms of information that is associated with each boundary. This information tells us what we would see were we to look through the boundary at the region inside. When we look through a hologram, we see a reconstructed image of the world on the other side. This hypothesis says this is all the world is; there is nothing to three dimensional space but images reconstructed from information that lives on two dimensional surfaces.

The holographic hypothesis was invented by Gerard 't Hooft and Leonard Susskind, both of whom have many important and imaginative contributions to their credit. Still, it is an idea that is genuinely hard to take seriously. One needs to know: Which surfaces are the holograms? Are these holograms associated with all surfaces, or only certain ones? Each observer can be surrounded by a surface. Is there then a hologram associated with each of us that codes what we see when we look around? If so, then when you and I look up at the sky, what principle guarantees that we see the same sky? If there is no reality but that coded on the surfaces we look through, these questions must have answers.

It is possible that topological quantum field theory might provide possible answers to these questions. This is because it tells us how to construct a theory in which there is a quantum description associated to every possible surface that separates a system from the rest of the world. Thus, the two lines of thought dovetail nicely and support each other. From them may emerge a theory in which the principles of physics are expressed as relationships that constrain how the information different observers hold about the universe may be related to each other.

I must warn the reader that the ideas I've been describing are very new. There is every chance that they will not succeed. Suffice it to say that something is happening which seems to address simultaneously the deepest issues in the foundations of physics, mathematics, and logic. If it succeeds, we may look back and say that a whole area of mathematics and physics was hidden in what Leibniz left out, when he said simply that the perceptions of the different monads are "in harmony, one with the other."

TWENTY TWO

THE WORLD *as a* NETWORK *of* RELATIONS

The word quantum aptly describes the physics of the very small, because there are many aspects of the microworld that are discrete in the sense that physical quantities are counted. In Newtonian physics, the situation is different, as almost all quantities of interest are continuous. According to the classical conception, the mass or charge of a particle, its energy, position, momentum, and angular momentum are all quantities that can vary continuously. In quantum mechanics, as it has been developed so far, some of these quantities become discrete. Electric charge, and angular momentum, as well as the energy levels of bound systems like atoms, are all counted in terms of discrete units.

But not every quantity in quantum physics is quantized. The position of a particle moving freely in space still varies continuously, as does its momentum and energy. This has led a number of thinkers to propose that our present quantum theory must be only a step towards a truly fundamental theory in which all physical quantities would be discrete. One reason for this is that a fundamental theory should be as simple as possible, and no mathematics is simpler than counting. Another reason is the desire that a truly fundamental theory be immune from the kinds of logical paradoxes that can arise as soon as one has sets with uncountable numbers of elements. The logic needed to simply define what is meant by a real number involves a subtle limit involving infinite numbers of sets, each containing infinite numbers of elements. It seems unlikely that the world at its most fundamental should require such contorted thinking for its description.

All the quantities that are still continuous in quantum theory have to do ultimately with motion in space. This has led many people to the idea that the continuous quantities would be eliminated from the quantum description were it discovered that fundamentally space and time themselves are discrete. As we see from the quotations at the beginning of this chapter, even Einstein wondered about this. Since him, there has emerged a tradition of speculation, which argues that underlying the description of spacetime in general relativity is some essentially discrete structure waiting to be discovered.

My initial encounter with this tradition took place at the first physics conference I attended, as a freshman student of physics. On the last day of the conference, I found myself sitting next to an impressive figure dressed in black, who possessed a calm presence, like an Old Testament prophet, and a beard to match. I struck up a conversation and asked him what he was working on. He replied that his approach to physics was to imagine how God might have made the world, and then to try to emulate Him. Having come to the conclusion that "God cannot integrate, but most likely he can count," he had constructed a game which described an electron moving in a discrete world. I went away perplexed, without getting his name. It was only many years later that I realized I had met David Finkelstein, a man whose diverse contributions to physics include the discovery of the meaning of black-hole horizons and the notion of solitons in statistical physics. I think I have never met a purer spirit in my science work, and his lifelong search for a description of a world simple enough for God to have made has been an inspiration to many seekers in the field of quantum gravity.

The tradition of searching for a discrete structure underlying space and time includes many of the deepest thinkers in theoretical physics. Besides Einstein, Finkelstein and Penrose, many physicists have contributed to it including Jacob Bekenstein, Richard Feynman, Chris Isham, Ted Jacobson, Alexander Migdal, Holger Nielson, Tulio Regge, Rafael Sorkin, Gerard 't Hooft and John Wheeler. These people and others have invented beautiful structures, which they have proposed as the underpinnings of space and time. There has been, however, a problem with all these approaches: they do not easily connect with the picture of continuous space and time that we know must be true at some level of approximation. For this reason, many less courageous people have followed a less heroic route to quantum gravity, which is to try to apply the rules of quantum mechanics directly to Einstein's theory. We (I have been among them) hoped by doing so we would at least find clues as to how to describe space and time at the quantum level.

The story I want to tell in this chapter is how one of these searches has turned out. Against our expectations, one line of work has succeeded in arriving at a picture of quantum geometry by starting with Einstein's description of space and time and applying the rules of quantum mechanics to it. To our surprise, the picture that emerged is one of a discrete quantum geometry that in many ways confirms the expectations of the prophets of our field. The picture we find is in fact not very far from a simple game that was born thirty years ago in the notebooks of Roger Penrose, which he called *spin networks.* Moreover, these structures emerged by following the Leibnizian philosophy and insisting, at each step, of speaking only in terms of meaningful quantities that are defined in terms of relationships.

The question we have sought to answer is how is the geometry of space to be described at the Planck scale, where quantum mechanical concepts must hold sway. When we apply quantum mechanics to space itself, we expect that there may also be some discrete set of quantum states, analogous to the orbitals of an atom. But how are we to describe the quantum states of space? They cannot be anything that can be seen as living in space, for what we are elevating to the quantum level of description is exactly space itself. Furthermore, we know from our basic principles that this description must be entirely relational; there must not be any fixed or absolute structure of space.

With all due caution, I think that it is possible to assert that over the last ten years a solution to this problem has been found. The result is easy to describe: *the possible quantum states of space may be labeled by the different ways to tie knots in pieces of string.*

Take a piece of string, tie a knot in it, then tie up the ends. Take another piece of string, tie another knot, taking care to make it different from the first, and again tie up the ends. Try it yet again. If you have never done this, you may be amazed to discover that it is possible to continue to tie knots that are different, one from the other, as long as one likes.

The problem you will run into as you continue this is not inventing more knots. It is telling each knot from the others. For very quickly one reaches a level of complexity at which it is very difficult to tell just by looking if two knots are the same or not.

The game can be made even more interesting if one takes two or more pieces of string and links them together while tying up each into a knot. Again, one sees that very quickly one reaches a point where the knots are too complicated to easily distinguish one from the other.

Mathematicians have been playing games with knots for about two centuries. Interestingly enough, the original motivation for studying knots came from Leibniz, who had speculated on the need for a new branch of mathematics "... which deals directly with position, as algebra deals with magnitudes." A few mathematicians of the eighteenth century, attempting to follow this suggestion, began to investigate knots. These early efforts eventually developed into the subject of topology. But the subject of knot theory really took off in the late nineteenth century when several physicists conceived of the idea that atoms were knotted tubes of electric field. They suggested that different kinds of knots could correspond to different types of atoms, which led them to begin to make lists and classifications of the different possible knots.

Rather quickly the idea of atoms as knots faded, but by then mathematicians were hooked, because knots are an ideal mathematical subject in which questions that can be asked of children turn out to hide enormous varieties of intricate structure, and remain unsolved for centuries. Indeed, despite many attempts, mathematicians have still not found a method that suffices to completely describe and classify knots. They have not even solved the problem of whether there is a simple procedure that, applied to any two knots, could tell if they are the same or not. But besides this, what is so beautiful about knots is that they embody in a pure form the principle that structure may arise out of a world of pure relations.

In retrospect, it seems that someone might have proposed for just these reasons that knots represent the solution to the problem of giving a quantum mechanical description of space. They are discrete, there are an infinite number of them, so they can represent an infinitude of possibilities. Further, they rely on no information about space except that it has three dimensions, and that it is possible to move things around smoothly; as such, they embody a world of pure relations. I am not aware that anyone made such a proposal, although at least a few of the knot theorists I know have been driven by an intuition that such beautiful structures must have something to do with how the world is made. What we have discovered is that just as in the case of an atom, the different possible quantum states of space can be classified. For every different way to tie a knot, there corresponds a different quantum state of space.

Actually the picture is a bit more complicated than this. It turns out that the pieces of string may meet each other and run along together for some distance before again going their separate ways. Thus, what we have to deal with are networks of bundles of string. Each network consists of some edges, along which a definite number of strings run. The edges meet at vertices, at which the pieces of string are routed from one edge to another. To get the complete picture of quantum geometry, one has to now imagine such a network in which the edges are linked and knotted with each other.

When he was a graduate student in mathematics, Roger Penrose began to play with networks just like these. His original motivation was to solve the four-color problem, which is the problem of showing that one needs no more than four colors for a map so that no two adjoining countries share the same color. But he found another application for the networks, as diagrams of quantum mechanical experiments involving angular momentum. In quantum mechanics angular momentum comes in discrete units, which are called spins. Penrose found that he could read these networks as pictures of processes in which particles with various spins travel and interact. The number of pieces of string in each edge then corresponds to the quantum mechanical spin carried by the particles, and the routing through the vertices tells the story of what happened at these intersections. He then imagined that these spin networks describe pictures of the elementary processes that underlie space and time. That is, to describe what is happening at a fundamental level, he postulated that it was not necessary to say where these particles moved in space or when the interactions happen. The picture of discrete interactions instead is to be seen as the primary level of description, from which the geometry of space and time should emerge.

I believe that Penrose thought of his spin networks as a kind of warmup, as a game that had some elements of a true picture of discrete space and time. But what we have found is that exactly this game emerges from applying the principles of quantum theory to general relativity, without any need for further elaboration. Each quantum state of spacetime is a spin-network.

If space is constructed from such knotted networks, why do we have the impression that it is a smooth, featureless continuum? The answer is the same as why skin or cloth appears smooth: the lines and knots of the network are so tiny that they are visible only on the Planck scale, twenty orders of magnitude smaller

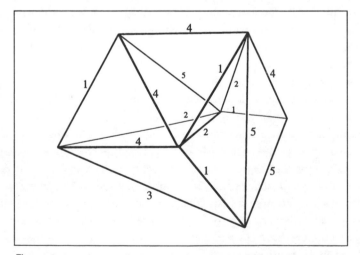

Figure 4 A spin network corresponding to a possible quantum state of the geometry of space. Courtesy of Carlo Rovelli.

than atomic nuclei. The image that we have is that the familiar space around us, smooth and featureless as it seems, is actually woven from an enormous number of fundamental quantum knots.

If this picture is going to work we must ultimately be able to reduce all the geometry that we learned in high school to properties of these networks. This is possible, but it took several years of work to realize. To do it we had to remember what Leibniz, Mach, and Einstein had taught us: to discuss spatial relations meaningfully, we have to ask questions about the relations between real physical things.

Let me illustrate this with an example that turns out to be particularly simple. Suppose we have a surface in space. How do we determine what its area is? The surface will be imbedded in the network, which represents space. It must represent something physically meaningful; for example, it might be the surface of a black hole, or the surface where the electrical potential vanishes. To answer this question it was at first necessary to carry out a rather intricate calculation. Thus, we were surprised when the answer turned out to be very simple. The area comes in discrete units. For each network and surface it is computed by adding up the spins of each of the edges that intersect the surface and multiplying times a fixed unit which is the Planck length squared.

I would like to emphasize that what I have just described is a physical prediction, which has come from finding a way to consistently describe space simultaneously in the language of relativity and quantum theory. The theory makes other predictions as well. It has told us that volume also comes only in certain discrete units, which are valued in units of Planck length cubed. These also come from counting; it turns out that to measure the volume of some region, one must look at all the nodes of the graph that are inside that region. One then adds up certain numbers which describe the routings of the strings through the nodes of the graphs. These calculations were rather more intricate, and indeed we made several mistakes before we discovered the right formulas for these little quantum units of volume.

Unfortunately it is very difficult to do experiments to test these predictions; the Planck scale is just too small. But the mathematics that leads to these results is too beautiful and robust to be accidental, and I am very sure that when a way is found to measure the areas of surfaces or the volumes of regions to a precision of 10^{-33} centimeters, the result will be that the possible answers are discrete, as are the energy levels of an atom.

There is another question that must be answered if we are to take this picture seriously. What if we consider a quantum state that corresponds to a simple network—just a single piece of string tied in a loop, for example? How do we make a correspondence with our classical picture of space? The answer is that we cannot. It can only be done if the network is extremely complicated—sufficiently complicated that on scales thirty and more orders of magnitude larger than that of a single of its knots, it has some average, uniform structure. Thus, the very existence of space around us—that which Newton took as an eternal absolute of the world, is actually a very special and contingent property of the universe. Were the quantum state different there would be nothing that we could describe as space—there would only be a network of relations encoded in some collection of knots.

In the last chapter I discussed a different line of thought that led to the conclusion that the quantum world must be discrete. This was based on the Bekenstein bound that requires that any surface with a finite amount of area must only enclose a finite amount of information. Does this have anything to do with the fact that the area of the surface itself must be discrete? I think that the answer is yes, although I must hasten to add that this line of thought is very new. One clue is that the spin networks appear to encode a deep structure that emerges from several different approaches to quantum gravity; they play a big role also in the topological quantum field theories.

The line of thought I've described actually had its origins in Penrose's spin networks. Beginning in the mid-1960s, Penrose began to elaborate them into a picture of spacetime, which he called twistor theory. The basic idea of twistor theory is that space and time are to be defined from the relationships among processes. In this picture the notion of a process is supposed to be prior to space, time, or the idea of a particle. Twistor theory then led to certain very provocative results in general relativity theory, which suggested that the equations of general relativity might simplify dramatically if we were willing to give up our intuitive notion that the world is symmetric under exchanges of left for right.

This is very surprising, but goes back to one of the seminal discoveries of particle physics. In the 1950s an experimental physicist at Columbia University named Chien Shiung Wu discovered that neutrinos have the unique property of not appearing the same when looked at in a mirror. Just like hands, neutrinos could come in two varieties, right and left handed. This is because they spin around the direction in which they move. Facing the direction in which they travel, they spin either to the left or the right. Most particles in nature come in both varieties. But Wu discovered that only one kind of neutrino exists, which turn out to be the ones that spin left. The possibility that this might be the case had been pointed out by two theorists, Tsung Dao Lee and Chen Ning Yang. But, the discovery was still surprising and, to this day, no one understands why it should be this way.

In Penrose's twistor theory, spacetime is described in terms of what a neutrino sees as it propagates. Because of this its description is intrinsically asymmetric; twistor theory gives a kind of left-handed view of space and time corresponding to the left-handedness of the neutrinos. What is surprising is that Penrose discovered that in certain ways this description, in which the world looks different than its mirror image, seemed simpler than the usual symmetric description.

Later, in the mid-1980s, two Indian physicists working in the United States discovered how to exploit the simplicity Penrose had uncovered. First, Amitaba Sen found that he could reexpress part of the Einstein equations in a form that was much simpler than their usual expression. He did this by following the hint from twistor theory and asking how a neutrino would see the geometry of spacetime. This then inspired Abhay Ashtekar to invent a complete reformulation of Einstein's theory in terms of new variables adapted to describe the proposition of a neutrino through spacetime. In this new language, the equations of Einstein's theory are not only much simpler, they are strikingly similar to the equations of the gauge theories that underlie the standard model.

The reader may recall that the key idea of the gauge theories is to take some particle on a trip and compare how it looks at the end with its state at the beginning. This makes it natural to describe gauge theories in terms of loops that represent the path of a particle. This way of understanding gauge theories has a long history; our work emerged when we applied this description to Ashtekar's new formulation of general relativity. Each loop in our pictures of knots and networks is then a representation of a process in which a neutrino is carried around in space, and then brought back to its beginning point.

Our picture may be summarized by saying that we are describing the world in a language in which the geometry of space arises out of a more fundamental quantum level which is made up of an interwoven network of such processes. And what for me is most beautiful is that this solution to the problem of how to describe space, coming as it does out of attempts to combine general relativity and quantum mechanics, does truly vindicate the visions of Leibniz, Mach, and Einstein and their insistence that in any fundamental theory space must arise as an aspect of relations among physical things.

The translation of Abhay Ashtekar's reformulation of general relativity into the picture of networks has taken so far ten years of work. The basic ideas were set out in work I did with Ted Jacobson and Carlo Rovelli, but many other physicists and mathematicians have helped to bring it to the present point. Still, in spite of successes such as the prediction that area and volume are discrete, I must warn the reader that it cannot be taken as final; it is a step along a possible road to quantum gravity, but we have not yet arrived at the door. For one thing, as we are not able to do experiments at the Planck scale, we cannot be sure that the picture has anything to do with reality.

The picture of quantum geometry in terms of networks is still developing, and there are still big problems yet to solve. Not the least of the open problems is to find a way to combine it with the results of other approaches to quantum gravity, principally string theory. It is possible that these different investigations have uncovered complementary aspects of quantum gravity, which may each play a role in the final synthesis. At any rate I think this possibility is more likely than that one or the other of these approaches is simply right, while the others are simply wrong. There certainly seem to be connections between these different approaches, beginning with the fact that both string theory and the spin-network picture tell us that the quantum gravitational field must be described in terms of one-dimensional objects. A loop in an enormous and complex network could turn out to be a microscopic closeup of the same phenomena that string theory describes as a string moving in a smooth spacetime geometry. Or it could not, only the future will tell.

As I described in Chapter 5, string theory is itself in need of new principles if it is to succeed as a quantum theory of gravity. The principal problem it must solve is how to overcome its reliance on a picture in which the strings move in a fixed background space. We have seen how Leibniz's relational philosophy of space and time made it possible for general relativity to achieve exactly this. String theory must, if it is to succeed, undergo the same transformation; it must be reformulated as a theory of pure relations.

There is thus reason to suppose that the principles that arise out of Leibniz's philosophy may be exactly what string theory needs. Further, graphs and knots are pictures of the possible relations that may hold between strings. It may then be that if string theory succeeds in throwing off the remnants of absolute space and time it will look something like an elaborate version of the picture I have described here.

Beyond the question of the relationship between the different approaches to quantum gravity, no problem ahead is more intimidating than that of extending what is so far a picture of the geometry of space into a true quantum mechanical picture of space *and time*. I will tell what I can of it in the next and final chapter.

TWENTY THREE

THE EVOLUTION
of TIME

If I stretch my imagination, I can just begin to believe in the idea that space is not something fundamental, but emerges only as an approximate way of describing the way things are organized and interrelated. Temperature is such an emergent property; it has no meaning on the atomic level. It is only a measure of the average energy in vast collections of molecules. In the picture I described in the last chapter, space is something like this; there is a fundamental level in which there are only the connections among the nodes and edges of a network. These networks do not exist in space—they simply are. It is their network of interconnections that define, in appropriate circumstances, the geometry of space, just as the jumps and dances of all the atoms in a cubic centimeter of air define its temperature. Perceived at vastly larger scales than the Planck length, the network seems to trace a continuous geometry, just as the cloth of my shirt is woven from a network of threads. Perhaps, just perhaps, this is the way the world is.

But what about time? Could time also be something that emerges from some more fundamental level? Is it possible that at this level there is no time, no change?

Following Leibniz and Einstein, we have so far come to accept that there may be no meaning to time besides change. It has no absolute measure, there is no crystalline clock on the wall of the universe ticking away the UNIVERSAL RIGHT TIME. This is surely right. But this relational time is still a kind of time. It is grounded in change, but change then must be something real.

It is another thing altogether to wonder whether time and change themselves might be constructs—whether there might be some fundamental way of perceiving the world in which they play no role at all. Speaking personally, my imagination quails before a world without change and time. I don't know if there are any real limits to what the human mind can imagine, but thinking about this question brings me closer than I like to the limits of what my own mind has the language or means to conceive.

The problem of time in quantum cosmology is hard exactly because it seems to lead us to confront the possibility that time and change themselves are illusions. This is because it turns out to be hard to extend from general relativity to the quantum world the notion that time is no more than a measure of change. If we cannot do this, we may have to come to terms with a world which, at the most fundamental level, must be described in a language that includes no words for time or change.

Of course, the idea that time and change are illusions is very old and has always held an attraction for certain people. The opposite view, that time, change and novelty are real has an equally long history. The stakes are thus not small; quantum cosmology is the arena in which it will be decided which philosophy of time will pass into science, and which into history.

It is easy to see why there is an impulse to escape time. There is no greater tyrant. Nothing could be taken from us more valuable than what time takes away, which is our past. Nothing could be withheld that we have a greater desire to possess than what time holds forever unattainable—knowledge of the future. Sometime, hopefully not too far into that future, people will understand the answer to the problem of time in quantum gravity, but I'm sure that no one alive does now. So the story I can tell here about time must end unresolved, like an old-fashioned radio play.

The founders of quantum mechanics knew general relativity. Still, when they invented quantum theory, they choose to put to one side everything they had learned from Einstein about time. The theory they built was breathtakingly radical in many aspects of its description of reality. But in its treatment of time it was disappointingly conservative as it took over, unchanged, Newton's absolute notion of time.

In both Newtonian physics and quantum mechanics there is an absolute and universal time, with respect to which change in time is measured. This time is measured by a clock outside of the system being studied, one that is not influenced by anything that happens to the system itself. Just like a Newtonian particle, a quantum system evolves in time in a deterministic fashion. As it does so it traces a trajectory, not in space, but in the infinite-dimensional state space. If we know the initial point in the state space, and the forces among the particles, this trajectory is completely determined for all time, just as is the trajectory of a particle in Newtonian mechanics.

If we are to bring quantum theory and relativity together, we must find a way to replace this notion of time and change with one that is natural to relativity and cosmology. We must find a way to bring into quantum theory a relational notion of time that does not depend on the presence of an outside observer.

This is really the crucial point that makes the problem of time in cosmology so difficult. If the system we are studying is the cosmos as a whole, then just as there is no observer outside the world, there can as well be no clock that is not in the world.

As I emphasized in Part Four, relativity theory tells us that, to speak about time, we must refer to a physical clock that is somewhere in the world. Time can only be spoken of in terms of correlations between things that happen and the readings of that clock.

But, what is a clock? At the most basic level, a clock is a subsystem of the universe that changes in a predictable and regular way, so that its appearance can be used as a label to distinguish different moments, against which the changes of everything else in the world may be measured. Certainly there are no perfect clocks. Once we insist that time is to be defined through its measurement by clocks, we introduce a bit of messiness into the notion. But this is something we can live with; a messy notion of time is still useful. What is much worse is the fact that, given the basic laws as we understand them, we could easily imagine universes that contain nothing that could function like a clock. For example, a universe in thermal equilibrium, in which nothing happened except random motion of the atoms, would have no clock. Nor would a universe so chaotic or disordered that no feature could be identified whose evolution is predictable enough to be used as a clock. Nor could there be a clock in a universe that was too small, or too simple, to divide it into subsystems.

A world with a clock is then one that is organized to some extent; it is a world somewhere on the boundary between chaos and stasis. The world must be sufficiently dynamical that there is no danger of reaching equilibrium, after which it is chaotic at the microscopic level and static on all larger scales. But it must be organized enough that distinct subsystems may be identified that preserve enough order to evolve predictably and simply. Thus, we are led back to the conclusion that any consistent cosmological theory will only be able to describe a world that is complex and out of equilibrium.

This is an important lesson. But it is double edged, for it leads right away to a problem that we do not know how to solve. A universe that is complicated enough to have a clock in it is almost certainly going to contain many different things that could be used as clocks. Certainly, in our world there are a great many processes that could be used to measure time. The problem is that we could use any of them to discuss how the universe is changing in time. If we believe that time is nothing but relations, then each of these must serve equally well as a measure of time.

Einstein's theory of general relativity easily admits such a pluralistic notion of time. This is after all why it is called the theory of *relativity*. One can use any clock one pleases to measure the evolution of the universe. In the end no question about what really happened will depend on which one you choose.

The key question we face now is whether such a pluralistic conception of time can be realized in the quantum theory. So far, no one has found a way to do this. People have, with great effort, managed to make formulations of quantum gravity in which time is defined by one of the clocks in the system. But it turns out that the answers to questions about what really happened all seem to depend on which clock one uses to define time in the quantum theory.

One of the ways this problem arises comes from trying to describe what happens to the quantum state *when* an observer makes a measurement. Quantum theory tells us that we must change the state of the system we are observing at the moment that an observation has been made. But if different observers use different clocks to label when measurements are made, they are going to believe they are talking about different quantum states.

This suggests that a theory that allows all possible clocks must be a pluralistic quantum theory of the kind I described in Chapter 20. Disagreements about the evolution of the different quantum states may in this way become simply an aspect of the fact that different observers hold different information, which is represented in different quantum states. The challenge, as before, is to make sure sense can be made of the relationship between the views held by all the different observers, so that the world retains a sufficient coherence, and that to some approximation it can still be described in the language of classical space and time.

It is too early to tell if a proposal like this is going to succeed. But let me stress that it can work only if the world has the right balance of complexity and order. There must be sufficient complexity to ensure that different observers are completely distinguished from each other by their having different views of the universe. But there must be sufficient order to ensure that the different observers can agree that they are speaking about the same universe. Thus we are led once again to the notion that a quantum universe must be a self-organized world balanced between order and variety.

The consistent-histories formulation may also be able to express such a pluralistic approach to time, as different sets of consistent histories may be chosen that correspond to how the world would seem to evolve according to time as measured by different clocks. This is a strength of this kind of approach. The fact that the interpretation is built around a notion of histories means that time and change are built into the language of the theory. The key issues here again are the balance of complexity and order that the world must have if the pluralistic description of time in general relativity is to be recovered from the quantum theory.

All of these approaches posit that time, as measured by clocks, is fundamental. But if there is a possibility that the theory must describe universes that contain no clocks, this may not be the most basic language to describe the world. One may then ask if there is available a language for the interpretation of quantum cosmology that does not presume that time and change have meaning.

Such an approach to the problem of time has been developed by Julian Barbour, who for the last several years has been arguing that the notion that time is what is measured by a clock cannot work in a quantum theory of gravity. Instead, he has proposed a radical view of quantum cosmology in which time has no fundamental meaning at all. The proposal is simplicity itself. According to it, what exists——the universe——is nothing but a great collection of moments. Each moment is a snapshot of the universe, a simple configuration of things. He calls the collection of all of these moments the heap. The heap contains a great many moments. But there is no sense in which the different moments can be ordered in time. They just simply are. Period. The quantum state of the universe serves only one function, which is to give the probability that any given moment may be found in this collection.

In this conception, all of physical law reduces to one kind of question. Some godlike being reaches into the heap and pulls out a moment. What is the probability that such a randomly selected moment will have some particular characteristic? To describe this, we need no notion of time or change. Nothing is changing in time because there is no time—the whole heap just is, period, and it is all that is.

Why then, do we have an impression that there is time, that we are changing, that you and I were here five minutes ago, and so on? According to Barbour's idea, each moment—say, this one—is an entity in itself, and does not change. We believe in time, because our world is structured in a very special way. Each moment is structured so as to give us the impression that other moments also exist. We have memories and we see all around us evidence that can only be construed as telling us about other moments that we might like to say happened in the past. Barbour likes to say that the world is full of "time capsules"—aspects of the configuration of a moment that speak to us of other moments.

Suppose we are given a box containing a thousand pictures taken at parties. We are asked to try to order them in time. Now it is clear that this will not always be possible. We may have been given one picture each of a thousand different parties. In this case there is no way to order them; we might as well think of them as having occurred simultaneously. This is what Barbour asks us to think is the general situation with the heap. But if the pictures have the right combination of order and diversity, we will be able to order them. They all need not show the same people, partygoers may come and go. But if we are to order them, many must have people in common. If we use this information, together with other clues we get from the pictures—and use a bit of what we know from our theories of the evolution of parties—we may be able to order a great many of the pictures. The order may not be absolute, and there may remain a few we are unable to place. Pictures of someone's trip to Florida may have been mixed up mistakenly with that wild New Year's party we thought we had put behind us. Or, towards the beginning and the end, there may not be enough happening that will allow us to precisely order the sequence. But if there is the right combination of order and variety, we will be more or less able to construct from the pictures a narrative of a single party.

This is what the universe is like according to Barbour. It is a collection of moments, of snapshots, with the right combination or order and variety, and with a sufficient complexity, to enable us to more or less order most of them. By doing so, we recover, as an approximation, a notion of a single universe evolving in time.

That time is not fundamental does not prevent us from having a strong sense of the persistence and continuity of time, for these sensations are parts of the present state of our brain. According to Barbour's theory, the fact that we believe in the existence of time is a contingent property of the particular state of the universe; it reflects the fact that Einstein's equations and Newton's laws give a good approximation to the more fundamental quantum description.

Thus, Barbour's conception can work if the probable universes, according to the quantum state of the universe, are like the box of pictures. We can thereby deduce from our present memories that it is very probable that there is a moment in the heap corresponding to our being here five minutes ago. But we cannot be exactly sure. And when we ask more cosmological questions—such as the probability that there is in the heap a configuration that corresponds to what classical physics might like to call the universe as it would have been thirty billion years ago—the probabilities can become significantly less than one.

Thus, according to this theory, time is an approximate concept that works well enough for ordinary things, but becomes less useful when we consider either the scale of the universe as a whole, or the scale of the very, very small.

This view gets us out of the problems we have been discussing. Barbour can simply say that when the different quantum clocks begin to disagree, all that has happened is that time has become an inappropriate concept. But he can say that only at some cost. Still, the cost is interesting, and worth contemplating.

Besides the obvious cost of trying to do without a concept that we thought of as fundamental, it is very interesting that, like the others, Barbour's proposal can only work in the case that the universe is sufficiently complicated. If there are around us no configurations that speak to us of other times, then, in his proposal, there would be no sense to our speaking of time. For time to be a useful concept, the world must be sufficiently complex to permit each moment to tell stories about other moments.

I have discussed three different kinds of proposals to the problems of quantum cosmology: many worlds and its successor the consistent histories interpretation; the pluralistic conception based on the views of many observers; and Barbour's heap proposal. They are very different from each other, and we certainly do not know which, if any, of these will turn out to be right in the end. It is then remarkable that, despite their differences, each of these proposals leads to the same conclusion: time can only exist in a structured and complex universe, with a sufficient balance of order and variety.

There seems to be no viable approach to the problem of time in quantum cosmology that does not lead to this conclusion. But if it is correct then this must be the final blow to the Newtonian, atomist conception of the world. If time requires a universe that is structured enough to have clocks and time capsules, then we cannot speak of even the simplest possible physical process, such as the motion of a single particle, without at least implicit reference to the configuration of the universe as a whole. The things in the world may very well be built from elementary entities, but it is no longer possible to take seriously a view of the world in which their properties are independent of each other, and of the overall configuration and history of the universe.

It thus seems to be the case that a theory of quantum cosmology cannot be logically consistent if it does not describe a complex universe. But this has a very strong consequence. It means that the theory must somehow explain why it is that the universe is complex. A theory of cosmology must, if it is to be self-consistent, be a theory of the self-organization of the universe.

It might seem completely crazy that a quantum theory of gravity should be also a theory of the self-organization of the universe, but we have already seen at least one example of this in part two. There we saw that two very simple conjectures about what happens in black holes lead directly to a world which structures itself according to the principles of natural selection. These conjectures must, if they are true, be consequences of the quantum theory of gravity. So it is not impossible that mechanisms of self-organization are built into the quantum theory of gravity.

This brings me to the very last idea of this book. A self-organized universe must evolve through a succession of configurations that are distinguished by their increasing organization. This is an observable property; we can define objective measures of organization that can distinguish the different eras from each other. We may then ask if a measure of organization or complexity may give us a measure of change sufficiently robust so that different observers in different parts of the universe will be able to agree about it. If so, such a measure of organization may provide a meaningful measure of universal time. Thus, the idea that the universe is self-organized not only might resolve the problem of time by explaining why the universe is complex enough to have clocks; the very process by which this is achieved might provide a clock.

Seen in this light, the idea of using a principle of self-organization such as natural selection in cosmology seems almost necessary, whatever the fate of the particular proposal I made in Part Two. But, if this is required just for consistency, it cannot be put in after we have developed the quantum theory of cosmology. It must somehow be implicit already in the principles of that theory. Furthermore, there are not very many options besides natural selection available on which to base a theory of self-organization. There may very well be principles of which we are presently unaware, but at least until the present time, the only principle of self-organization that science has studied that has the power to make the extraordinarily improbable likely is evolution through natural selection. Thus, is it not possible that self-organization through processes analogous to natural selection is, indeed, the missing element without which we have so far been unable to construct a quantum theory of cosmology? Is it not possible that in the future, when the history of the great scientific revolution of our era is written, it may be said that the three great minds that showed us the principles by which our universe is ordered were Einstein, Bohr—and Darwin?

Because of this, the arguments I have described concerning space, time, cosmology and the quantum, bring us back to the questions we discussed in the earlier parts of the book. In the first chapters we saw the extent to which the universe is enormously, improbably organized. We have seen in these last chapters that this improbable organization is necessary.

If we put everything together, we see the possibility that such a theory of cosmology may in the end be able to tell us something about what we are doing here in the world. It is possible to argue that a universe complex enough to have clocks, and so describable by a quantum theory of cosmology, might also be a universe that is hospitable to life. One conclusion of Part Four was that a universe complex enough to be described in the relational language of Einstein's general theory of relativity cannot be in thermal equilibrium. But we have seen also that only very special universes will avoid permanently the stasis of the universal heat death. Among the things that are necessary for this are stars and galaxies.

As we saw in earlier chapters, a universe in which the conditions and the parameters have been tuned so that it is full of stars is a universe in which many of the conditions required for life to exist are satisfied. For example, carbon seems to be necessary for copious star formation at all but the earliest eras of the universe. Thus, without any need to postulate the existence of life as a special or necessary condition, we see that it may be that the very requirement that the world have a completely rational explanation, consistent with Leibniz's principle of sufficient reason, may go a long way towards explaining why there is life in the universe. For the conditions that seem necessary simply to guarantee that a theory of cosmology based on relational notions of space and time be consistently formulated are also conditions that we require for our existence.

We have not come to the end of this story, but we have come to the point to which the road has so far been explored. The way ahead is open but in cosmology, as in every other branch of science, we cannot be really sure that we are on the right road until it has been traveled. It is possible that all that I have done here is cobble together a set of false clues that only seem to have something to do with each other. But if this is not the case—if something of this reasoning proves, when a theory of quantum cosmology is finally achieved, to have been reliable—then we will have arrived at a theory of cosmology in which the world is understood to be hospitable to the existence of living beings like ourselves. Not because the world is created in our minds, or for any other mystical or metaphysical reason—and certainly not because we, in particular, are somehow necessary or important for the universe— but only because living systems exist as a byproduct of a much larger pattern of self-organization and self-structuring that must be the history of any world which is completely amenable to rational comprehension.

In the Peter Brook adaptation of the great Hindu saga *The Mahabharata*, the wise king Yudhishthira must, on penalty of the death of his family, answer a god who demands of him to tell what is the greatest marvel in the world. His reply is that, *"Each day death strikes. And yet we live as though we were immortal. This is the greatest marvel."* And, yes, is it not possible that the greatest marvel of all is that we find ourselves in a universe in which everything around us, from the Earth, to the stars, the galaxies, and indeed the whole of what we can see, lives and is bounded by time, while at the same time revealing, through an infinite variety of relations that we are only just beginning to untangle, an order and a harmony that, while perhaps still not immortal, is far older and far richer than anything we have so far let ourselves imagine.

If the goal of modernism in art was to burn the old house down, all
that postmodernism has been doing is playing with the little charred
pieces that are left, which is a pretty puerile thing to be doing
considering that winter is coming.

—*Saint Clair Cemin*

Epilogue/
Evolutions

Whatever else we share, we are all children of the twentieth century—this most surprising, most violent, and most hopeful of times, this time when more humans were killed by violence than in all of previous history, but in which for the first time art, politics, science, popular culture, and commerce became international, and in which, for the first time, it has become possible to meet anywhere, in any city, airport, train compartment, or campground people who view themselves as inhabitants of a planet, and not only a country, region, or city. Surely this uncertain time is a moment of transition, from which humanity, if it emerges, will emerge slowly to a new world; not a utopia, certainly, but perhaps a world infinitely more varied, more interesting, and, yes more hopeful, than that sterile dream could ever have been.

And what will the inhabitants of that world see when they look back on our time? Certainly, the signs of a transition are everywhere, but they perhaps stand out most clearly in the great artistic and intellectual achievements of our time, in the paintings from Piccasso to the surrealists and the abstract expressionists; in the music from Stravinsky to Pert; in the dances of Martha Graham; in Wittgenstein's anti-philosophy; in Gödel's theorem; in modern topology; in literature and theater; in molecular biology and the visions of Margulis and Lovelock. In these inventions of the imagination, and in so much else, we see a great shift of where humans are looking to find and to create coherence and beauty in the world. And I hope I have convinced the reader that the signs of a great transition may be read nowhere more clearly than in the incomplete and unresolved state of our physical and cosmological theories.

It should be apparent by now to the reader that this book aims to say something about more than cosmological science, however inadequately it has satisfied even that ambition. I don't know if the reader will believe me if I plead that it didn't start out that way, and that I was surprised and more than a little puzzled at how many references to God, or asides to artistic or political developments, kept cropping up in my mind as I composed these pages. Finally, it was only the comment of a friend that showed me the ultimate irony of the project in which I was engaged, so that my opening reference to the Copernican revolution introduces a tale of a kid who grew up on dreams of revolutionizing society, sublimated them into dreams of scientific revolution, but then came to believe that the underlying structure of our world is to be found in the logic of evolution.

Indeed, I do think that there are connections between the kinds of ideas that those of us who work on cosmology are thinking about and the ideas that one hears now being spoken about in philosophy, art, theology, and political and social theory. When I walk across my campus and listen to talks by colleagues in postmodern art criticism, critical legal studies, feminist epistemology, psychoanalytic theory, and process theology, or when I listen to friends in art and philosophy as they struggle to define the futures of their ancient practices, I hear arguments whose logic often mirrors those my colleagues and I are wrestling with as we try to invent a sensible quantum theory of cosmology. I know that to draw such parallels is a dangerous business. But, I also know that what these friends and colleagues and I have in common is that we are all trying to make sense of youths lived through the ecstatic utopianism of the 1960s and adulthoods that witnessed the collapse of the very different Marxist utopianism and the revelations of the violence that that dream imposed on its peoples. We are all trying to understand what democracy might mean in a world dominated by consumer capitalism, a growing ecological crisis, a widening gap between rich and poor, and the permanent confrontation of peoples with radically different cultures and expectations about life. We are all waiting to see, as we have been waiting since childhood, whether our world will perish from the unintended consequences of violence and greed or whether we will discover a way for human beings on the whole planet to form a single society based on mutual respect and not violence. Perhaps not surprisingly, we are all, in one way or another, trying to understand what it means to construct a description of a complete universe, from the inside, without reference either to fixed external structures, a single fixed point of view, or absolute imperatives. In the words of the painter Donna Moylan, we are all trying to construct cosmologies of survival.

Without feeling able to make a convincing argument for it, of the kind I know how to make about a technical point in physics, I believe that if we do succeed in forging a community out of humanity it will be at least partly because we are able to envision ways to answer these last questions, in all of the fields in which they arise from politics to art to cosmology itself. I leave it those wiser than myself to explain how, but I believe it cannot be a coincidence that the view of the universe invented by Descartes and Newton resembled to a remarkable degree the ideal society as envisioned by Locke and Hobbes. Atoms moving individually, their properties defined by their relation to a fixed and absolute structure that is identified with God, interacting via absolute and immutable laws that apply equally to all—is it no accident that this describes both the Newtonian universe and the eighteenth- and nineteenth-century European ideal of liberal society? Or that both medieval society and Aristotelian cosmology were based on a hierarchical view of a fixed and finite universe, with different levels composed of different essences governed by different laws, with the earth in the center and the highest level connected directly to God, the prime mover? Cosmology mattered then, and I believe it will continue to matter, even as both it and society transform themselves in unimaginable ways.

At the same time, I want to say—to avoid any misunderstandings—that I am not an intellectual relativist and I do not think that science is freely invented by us or that scientific truth is no more than a consensus among those who are officially called scientists. I believe in nature, in its dominance over us and in its recalcitrance to our fantasies and schemes. Indeed, I believe that what is most dangerous in both contemporary art, social theory, and theoretical physics is the ease to which people seem able to imagine their disciplines to be divorced from contact with nature. Physicists can no more invent the final theory by pure mathematical game playing, without reference to experiment, than artists are free to produce arbitrary artifacts, without regard to the requirements of form, craft and beauty.

More particularly, I believe that while there is no useful sense in which there is a scientific method that leads necessarily to the discovery of truth, and while it is true that, as Einstein said, our theoretical concepts are free inventions of the human mind, science has led to the discovery of truths about nature, and continues to do so. I don't believe that this fact can be explicated by any a priori theory of science that does not take into account certain facts about nature, not the least of which is that we are part of it. Again, this is a topic that I do not feel I now have enough wisdom about to develop here, but I think that science works, in spite of or even because of the fact that our ideas are developed in the milieu of our culture, so that the convergence of scientific ideas with ideas from other fields is something that should only be expected.

Perhaps the reason why science works, in the absence of a fixed method or a fixed set of rules, is that it is based on an ethic which recognizes that while any individual is obligated to champion what they honestly believe, no individual is the arbitrator of the correctness, or even the interest or usefulness of their own ideas. Experience teaches us that no matter how sure of ourselves we may feel, and how clever we may think we are being at certain instants, nature is always smarter, and anyone's individual achievement may only survive to the extent to which it is superseded by the achievements of others.

Perhaps then, this is the most important reason that science does matter to society, because it is in this way a part of the centuries old experiment to discover what democracy is. In its ideal form a science is a network of consensus shared among individuals without propaganda or coercion, as a democratic society is envisioned to be a society of free individuals living with each other without coercion or violence. When I look at the research center I am privileged to be a part of, and see twenty-five individuals from eighteen different countries working together for common purpose, I can only hope that this is a vision of future human society. Thus, whether it is ultimately a useful idea or not, the idea of envisioning the universe as something analogous to a community is one that perhaps had at least to appear and be tried out sometime in the parallel development of the projects of science and democracy.

Having said this I am reminded that I cannot conclude this book without emphasizing one last time that the views I have been presenting here are those of a single individual who happens to be fortunate enough to be a scientist, but are—at least at the present time—neither strongly supported by the evidence nor widely embraced by my colleagues.

In the past few years I have spent several months a year in Italy, in order to work on quantum gravity with my friend Carlo Rovelli. Of course, I have taken the opportunity to look around and try to understand another society, and one of the things I have been most curious about was how the diverse cultures of Europe are combining into the dream of a unified Europe. One of the lessons of this transition, which I think we may hope is an experiment that presages the eventual universal dissolution of nation states, is the extent to which certain aspects of culture merge easily, becoming instantly international, while others stubbornly remain bound to their particular languages. Perhaps not surprisingly, visual art, which needs no translators, moves easily, so that it is often impossible to tell at the international art festivals which countries the artists come from. At the same time, literature, which requires translation, remains national. For philosophy this is especially true, as for most philosophical books the market is insufficient to justify the commissioning of a translation. This is particularly true in Italy, as most philosophers around the world can read English and German, while rather fewer read Italian.

The Italian philosophers have what I think is an interesting way to refer to the transition taking place in the twentieth century in philosophy. They refer to what they call strong theory and weak theory. Strong theory is what philosophers aspired to do before this century, which was to discover by rational reflection and argument the absolute and complete truth about existence and elaborate these truths into complete philosophical systems. Weak theory is what philosophers have been doing since Wittgenstein and Gödel taught us the impossibility of doing this.

I would like to describe the same thing in different terms, borrowed from Milan Kundera. I would like to contrast the heaviness of the old and failed attempts at absolute knowledge with the lightness of the type of philosophy we are now aspiring to develop. Nietzsche also talked about heaviness, the heaviness of life yoked to the eternal return, weighed down by the impossibility of novelty. Nietzsche's darkness and heaviness were exactly reflections of the weight of the heat death of the universe, carrying with it its implication that life has no permanent place in this world, so that any joy—indeed any change—was at best transient. Furthermore, Nietzsche was right to worry about the impossibility of novelty, because on the physics of his time it was indeed impossible to imagine how it might occur.

Even more than this, the old search for the absolute is heavy and it has weighed us down for long enough. It implies that there is a stopping point, a final destination; it reeks, really, of the Aristotelian belief in the meaningfulness of being at rest, of Newton's absolute space, of hierarchy, in knowledge as well as in society, of stasis.

We have also had enough of the weight of utopianism, which comes of the idea that it is possible to arrive at a description of the ideal society by pure thought, indeed by the thought of one or a few solitary individuals. And, we have had enough of the weight of violence, and its justification in terms of any and all systems by which people can be made to believe in their special access to absolute knowledge.

Against this I would like to set the lightness of the new search for knowledge, which is based in the understanding that the world is a network of relations, that what was once thought to be absolute is always subject to evolution and renegotiation, that the complete truth about the world is not graspable as any single point of view, but only resides in the totality of several or many distinct views. We understand now that there is no meaning to being at rest, and hence no sense for stasis; this new understanding of knowledge might be said to be imbued with the freedom of the principle of inertia and grounded not in space but only in relations. And these develop not in absolute time but only in succession, in progression. Finally, this new view of the universe we aspire to will include a cosmology in which life has a proper and meaningful place in the world. That is, in the end the image I want to leave is that life is light, both because what we are is matter energized by the passage of photons through the biosphere and because what is essential in life is without weight, but only pattern, structure, information. And because the logic of life is continual change, continual motion, continual evolution.

Finally, the new view of the universe is light, in all its senses, because what Darwin has given us, and what we may aspire to generalize to the cosmos as a whole, is a way of thinking about the world which is scientific and mechanistic, but in which the occurrence of novelty—indeed, the perpetual birth of novelty—can be understood.

The old image of the Newtonian universe was as a clock: heavy, insistent, static; in this metaphor one feels both the iron hold of determinism and, behind it, the threat of the clock running down. Further, this was always a religious image as a clock requires a clockmaker, who constructed it and set it in motion. Against this, I would like to propose a new metaphor for the universe, also based on something constructed by human beings.

For reasons that I thought were quite irrelevant to its content I was drawn to finish this book here, in the greatest city of the planet, my first home. A few weeks ago I took a walk around, looking for a metaphor with which to end this book, a metaphor of a universe constructed, not by a clockmaker standing outside of it but by its elements in a process of evolution, of perhaps negotiation. All of a sudden I realized what I am doing here for, in its endless diversity and variety, what I love about the city is exactly the way it mirrors the image of the cosmos I have been struggling to bring into focus. The city is the model; it has been all around me, all the time.

Thus the metaphor of the universe we are trying now to imagine, which I would like to set against the picture of the universe as a clock, is an image of the universe as a city, as an endless negotiation, an endless construction of the new out of the old. No one made the city, there is no city-maker, as there is a clockmaker. If a city can make itself, without a maker, why can the same not be true of the universe?

Further, a city is a place where novelty may emerge without violence, where we might imagine a continual process of improvement without revolution, and in which we need respect nothing higher than ourselves, but are continually confronted with each other as the makers of our shared world. We all made it or no one did, we are of it, and to be of it and to be one of its makers is the same thing.

So there never was a God, no pilot who made the world by imposing order on chaos and who remains outside, watching and proscribing. And Nietzsche now also is dead. The eternal return, the eternal heat death, are no longer threats, they will never come, nor will heaven. The world will always be here, and it will always be different, more varied, more interesting, more alive, but still always the world in all its complexity and incompleteness. There is nothing behind it, no absolute or platonic world to transcend to. All there is of Nature is what is around us. All there is of Being is relations among real, sensible things. All we have of natural law is a world that has made itself. All we may expect of human law is what we can negotiate among ourselves, and what we take as our responsibility. All we may gain of knowledge must be drawn from what we can see with our own eyes and what others tell us they have seen with their eyes. All we may expect of justice is compassion. All we may look up to as judges are each other. All that is possible of utopia is what we make with our own hands. Pray let it be enough.

Appendix:
Testing Cosmological Natural
Selection

In this appendix I would like to complete the discussion of cosmological natural selection by considering all the arguments I am aware of that have been offered against the theory. Several scientists have proposed counter-examples in which a parameter of particle physics or cosmology may be varied to produce more black holes. To the best that I have been able to determine, all of these fail, and I need to explain why. After this I will discuss a number of general objections that have been offered against the theory. Finally, I will describe several proposals to modify the basic postulates of cosmological natural selection.

ASTROPHYSICAL TESTS

Along the path which a cloud of gas travels to become a star and then a black hole, there are several forks at which most are diverted to other fates. We may examine each of these in turn to see if there is a way to modify them so that more stars stay on the road to becoming black holes. If we can find a way to do this by changing one or more of the parameters of the standard model, we will have found a counter-example to the prediction of cosmological natural selection. As most stars fail to become black holes, it seems there ought to be a good chance to do this.

The first stage we may examine is the process of star formation itself. As only a small fraction of stars are massive enough to become black holes, a good way to make many more black holes would be to make more massive stars. Unfortunately, we do not understand star formation in enough detail to know exactly what processes are important for determining the distribution of the masses of the stars. Even so, it is not easy to see how any of these process might be modified by changing the parameters, without destroying the whole process of star formation. The problem is that we are not free to vary the parameters in any way that leads to the carbon nuclei being either unstable or less copiously produced in stars. If there is no carbon, there will be no giant molecular clouds out of which the stars can form. This puts strong limits on how much we can vary a number of parameters, including the masses of the proton, neutron, and electron, and the strengths of the electromagnetic and strong interactions.

I have already mentioned that we can vary the strength of the weak interaction in such a way as to turn off supernovas. However, the cost of this seems to be to shut down the processes which drive star formation in the first place. Thus, it seems that this also will lead to a world with fewer rather than more black holes. In any case, if a black hole is going to be produced, we want it to have as little mass as possible, so that as much as possible of the bulk of a massive star is recycled to turn into other stars and black holes. Supernovas accomplish this very well, as they produce black holes with close to the least possible mass, and redistribute the rest.

The next place we might seek to interfere with the process is to try to have as many supernovas as possible end up as black holes. One way to do this is to find a way to lower the value of the upper-mass limit. The reader may recall that this is the amount of mass above which a remnant will become a black hole; thus the lower it is the more black holes will be made. There is, in fact, one parameter whose value may directly affect the value of the upper mass limit, without affecting how much carbon is made. This is a consequence of a theory proposed by the nuclear theorist Gerald Brown together with Hans Bethe, who many years ago was one of the people who figured out how stars work. Their proposal is new and has not yet been completely confirmed observationally. But, even so, it leads to a very pretty test of cosmological natural selection.

Neutron stars are called so because it has been believed that they consist almost entirely of neutrons. The standard theory of how a neutron star is formed is that under the tremendous pressure at the center of a dying star electrons combine with protons to make neutrons and neutrinos. The neutrinos fly away, and only neutrons are left in the central core. Hans Bethe and Gerald Brown propose an alternative scenario. According to them, the electrons do not combine with the protons; instead they transform themselves directly into another kind of particle, called a *kaon*. These kaons are a kind of strongly interacting particle, which are quite unstable to radioactive decay. As a result, they are not found in nature, they are known only because they have been produced in particle accelerators.

Normally a kaon is much heavier than an electron. However, in the very dense environment of a neutron star, kaons can lose much of their mass. Something happens to them in an environment of a certain density very like what happens to electrons in a superconductor. All of a sudden they can move around freely inside the material without encountering any resistance, which means that they become effectively very light. It is possible they may even become lighter than the electrons. If this happens, the electrons will become unstable, as each will be able to decay to a kaon and a neutrino. The neutrinos fly away as before, leaving the kaons in the center of the star, together with the protons and neutrons.

Each of these is a possible scenario for what happens during the collapse of a massive star. The question is which actually happens in nature? Do the electrons convert to kaons, or do they react with the protons to produce neutrons? The answer is not definitely known; but there is some evidence in favor of the Bethe-Brown scenario, which I will describe in a moment. But what is most interesting for our purposes is that which one is chosen by nature turns out to depend on the mass of the kaon. If the kaon is light enough, the electron will prefer to become a kaon and the Bethe-Brown scenario will be the right one. Otherwise the electrons will prefer to react with the protons, leading to the old picture in which the star is composed only of neutrons. What is not so easy to determine is how light is light enough. Of course, we know the mass of the kaons; the problem is that a difficult calculation is required to find out exactly how light it would have to be for the Bethe-Brown scenario to be preferred.

What does this have to do with the problem of making black holes? A lot, because the value of the upper-mass limit, above which the remnant must become a black hole, is very different in the two scenarios. In the new Bethe-Brown scenario the upper mass limit is much lower—about one and a half times the mass of the Sun. In the older scenario it is likely between two and three times higher. This means that if Bethe and Brown are right, many more black holes are made than would be were the older scenario right.

This means that if cosmological natural selection is right the kaon must in fact be so light that all neutron stars choose the Bethe-Brown scenario. This is good, as the mass of the kaon depends on a parameter of the standard model that is independent of all the others that have so far been mentioned. This is because it contains a kind of quark—the strange quark—that is not found in ordinary matter. We may then adjust the mass of the kaon as we like by changing the mass of the strange quark. If we do not make the mass too low, this will not affect most astronomical processes. As a result, nature seems to have the chance to adjust a parameter so as to make the critical mass quite low without side effects, and by so doing to dramatically increase the number of black holes. If nature had this chance, but did not take advantage of it, then it is hard to believe that nature is trying to maximize the number of black holes.

What is the actual situation? There are a few cases where the mass of a neutron star can be measured very accurately, which is when it is one of a pair of neutron stars in orbit around each other. Neutron stars spin very rapidly, and they give off a series of radio pulses which allow us to measure their rotation rates quite accurately. From these pulses we can, using a fancy bit of relativity theory, deduce the masses of the two neutron stars. What is then interesting is that all of the masses which are so far known for neutron stars lie in a narrow range between about 1.3 and 1.45 times the mass of the Sun.

This certainly supports the Bethe-Brown picture, as it puts the upper limit just above this, at about 1.5 solar masses. If the standard scenario is true, it is hard to understand why none of these neutron stars are more massive. Of course, with only a few masses measured we cannot be too sure of this. But we may expect that new pairs of neutron stars will be discovered from time to time. If any of them has a mass more than the amount predicted by Bethe and Brown, it would destroy that theory and also count as evidence against the theory of cosmological natural selection. On the other hand, if say a hundred more are discovered, and all of them are in the narrow range predicted by Bethe and Brown, then this would count as good evidence for Bethe and Brown and, indirectly, for the theory of cosmological natural selection.

However, this is not quite the end of the story. For even if Bethe and Brown are right and the upper mass limit is low, it is not so low that all supernova remnants become black holes. As things are, perhaps half end up instead as neutron stars. We might then ask whether it is not possible to further lower the upper-mass limit so that every supernova becomes a black hole. If it is possible to do this by tuning the mass of the strange quark or some other parameter, without affecting anything else in the star forming process, this would be an argument against cosmological natural selection.

It turns out that this is not a simple problem, because if the mass of the kaon, or strange quark, is too low, they will begin to play a role in ordinary nuclear physics. There is then a danger that lowering the strange-quark mass may cause side effects, such as disrupting the formation of carbon. There also may be competing effects which can tend to raise the upper-mass limit. These are issues that can only be settled by detailed calculations in nuclear physics. There is also a possibility that lowering the value of the strange quark mass may effect processes in the very early universe. This is then another case in which the hypothesis of cosmological natural selection leads to a prediction about the result of a theoretical calculation: if the theory is right it should not be possible to further lower the upper mass limit below its actual value, without somehow disrupting the process of star formation.

Another set of tests comes from tuning parameters in directions opposite to those that disrupt the stability of nuclei. We know that increasing the mass of the electron or the difference in mass between the neutron and proton will lead to a world without nuclei. What about decreasing them? A less massive electron would have many consequences for the physics of stars, from the mechanisms which transfer energy from the core to the surface to the rates of cooling of the clouds. This is another case in which it is difficult to predict the overall effect. This means that the theory of cosmological natural selection makes a prediction: it must be that the overall effect of decreasing the mass of the electron is not to increase the production of black holes.

There is, however, a dramatic effect in the case we decrease the mass difference between the proton and neutron. As this is already small, we can imagine lowering it to the point that it changes sign, leading to a world in which the neutron is lighter than the proton. Such a world would be much changed from our own, for if a proton were more massive than a neutron and a positron put together, it would become unstable and decay to them. As it emerged from the big bang, such a world would consist primarily of a gas of neutrons rather than hydrogen atoms. In such a world clouds of primordial gas would cool much more slowly, for hydrogen in space cools primarily by processes involving its electrons. This would drastically slow down the processes that form galaxies. The result is quite likely to be a world with many fewer galaxies, hence fewer stars, hence fewer black holes. This then seems to be another case that falls on the side of cosmological natural selection.

There are a few more possible changes in parameters whose effects can be discussed. A question that comes up every time I give a talk about cosmological natural selection is whether you could increase the number of black holes by decreasing the mass of each star. A consequence would be that the more stars would be made out of the same given amount of mass. If everything else were the same, so that the proportion of stars that became black holes was unchanged, the result would be more black holes. There is, in fact, a way to accomplish decreasing the mass of each star, which is to increase the gravitational constant. Unfortunately, everything else does not stay the same when this happens. First of all, there are likely to be effects on the processes that form galaxies. Given our ignorance about that subject, however, no firm conclusions can be drawn about the ultimate effect on the number of black holes formed.

Increasing the gravitational constant also changes dramatically the way that stars work. This is because of a curious coincidence pointed out first by Branden Carter, and discussed since by several proponents of the anthropic principle. The way things are in our world, there are two broad categories of stars: those that transfer their energy to the surface by radiation; and those that transfer their energy to the surface by convective processes. With the value of Newton's constant as it is at present, the more massive stars are radiative, while the smaller stars are convective. If Newton's constant is either increased or decreased, the result will be that all stars will be, respectively, radiative or convective. This would have a big effect on processes of stellar evolution, including the critical processes by which massive stars return much of their mass to the medium through evaporation. However, I do not know of a clean argument that leads to a definite conclusion about the overall effect of this on the production of black holes.

Increasing the gravitational constant may also affect the amount of matter returned to the interstellar medium in a supernova. Decreasing the matter recycled from a star may then decrease the total number of black holes that are made. Finally, increasing the gravitational constant will, it turns out, decrease the lifetimes of the stars. This can affect the balance of the various feedback effects we discussed in chapters 9 and 10, as they depend on the fact that the shortest-lived stars—which are the ones that put the most energy into their environments—still live longer than the time scales involved in the processes by which giant molecular clouds form stars.

It thus appears that increasing the gravitational constant, while keeping the other parameters fixed, will have a multitude of effects on the processes that lead from gas to black holes. Some of these tend to increase the numbers of black holes produced, while others apparently have the opposite effect. It is difficult to predict the overall outcome, given the complexity of the interrelations between the processes of the galaxy. For the same reasons, it is difficult to deduce what would be the overall effect of the opposite change in which the gravitational constant is decreased. Thus, these also seem to be cases in which we may predict that if cosmological natural selection is right, the overall result must be that the number of black holes does not increase.

REPLIES TO PUBLISHED CRITICISMS

The development of the idea of cosmological natural selection owes a great deal to a set of criticisms published by Tony Rothman and George Ellis. They present arguments that favor the anthropic principle over cosmological natural selection. They argue that there are certain processes that would lead to the production of more black holes in a world without carbon than in ours. Although several of their arguments are correct, they miss a key point, which is that the majority of black holes in our universe are produced through processes that require the presence of carbon and oxygen, which, as I've emphasized, are necessary for the processes by which massive stars are formed in giant molecular clouds. As result, it is not correct to base estimates of star formation and black hole formation rates on simple models involving only gravitational collapse and fragmentation of clouds, (such as in the 1976 model of Martin Rees) without taking into account the role that carbon and oxygen play in shielding and cooling the star forming clouds. This means that small changes in parameters that lead to a world almost identical to ours, but in which Hoyle's coincidence that leads to resonant formation of carbon in stars fails, would be a universe with both no life and fewer black holes. This means that one must argue rather carefully in order to distinguish the consequences of the anthropic principle from cosmological natural selection. This is why tests, such as the one involving the strange-quark mass and the upper-mass limit, in which the two proposals may be separated, are critical for the discussion.

More specifically, Rothman and Ellis make five criticisms of cosmological natural selection, as it was developed by the time of my first paper on the subject, Smolin (1992a). Three of them are as far as I know, correct. A scenario for a cold big bang discussed in Smolin (1992a) is most likely not viable, given the difficulty of thermalizing the cosmic black body radiation sufficiently. Two cases, of universes dominated either by helium, or by neutrons—corresponding to the neutron proton mass difference being either zero or negative—present cases which are difficult to analyze given present knowledge. As I indicate in the text, if black holes are created in such universes, they are made through processes that have nothing to do with how black holes being produced presently in galaxies. At best the processes of black hole creation in such universes could be compared to hypothetical processes by which an early generation of stars may have been formed shortly after recombination, before any carbon or oxygen had been produced. But we are too ignorant of those processes to reach reliable conclusions about the rate of black hole formation in a universe composed mostly of either helium or neutrons.

In Smolin (1992a), I refer to an argument that the fragmentation of cluster size masses to galaxies depends on efficient radiative cooling summarized in Barrow and Tipler (1986, p. 389). If correct, this would imply that galaxies would not have formed in a neutron universe, leading to less formation of black holes. However, this simple argument ignores the role of dark matter, of whatever origin, in galaxy formation. As such I do not think we know enough to predict the number of black holes formed in a universe made out of neutrons or helium, just as we are unable to predict the initial mass function for stars formed shortly after recombination. As I emphasize in the text, these cases then must be considered to yield predictions. If cosmological natural selection is correct then these universes must produce fewer black holes than our world.

Another possible criticism of cosmological natural selection is found in a recent paper of John Barrow (1996). He considers consequences of increasing Newton's constant based on considerations of black hole entropy. However, the effects on the lifetimes of stars and the formation of galaxies are not taken into account, so the argument about this case remains, I believe, inconclusive.

COSMOLOGICAL TESTS

Still another set of tests of cosmological natural selection comes from cosmology rather than astrophysics. This is because the evolution of the universe as a whole is governed by a set of parameters, which we may imagine varying. Just as in the case of the parameters of the standard model of particle physics, we may ask if the effects are to increase or decrease the numbers of black holes produced. It is possible to get useful answers to this question because what happened very early in the universe had a big influence on its overall structure and constitution, for the same reason that the shape of a vase is formed during the first few minutes of a potter's work.

There are several parameters that come into the description of the overall shape and history of the universe. We may ask whether any of these may be adjusted in such a way that the universe produces more black holes. To approach this question we may begin by summarizing the stages of the cosmological processes which are relevant for the formation of black holes.

The first postulate of our theory is that the universe is created as an explosion in the extraordinarily compressed remnant of a star that has collapsed to a black hole. At this stage the star may be assumed to be compressed to a density given by the Planck scale, as we expect that effects having to do with quantum gravity are responsible for the explosion that begins the expansion. From this point on the universe expands as in the usual Big Bang scenario. Like other origin stories such as those that concern the beginning of life or the evolution of *Homo Sapiens*, there are theories about the early stages of the universe which are plausible, if certainly unproven. There is a standard scenario, some stages of which are reasonably well supported by the evidence, while the description of other stages is much more conjectural. But, in spite of its obvious gaps and difficulties, it seems prudent to stick close to the standard theory, as it stands up better to comparison with the real world than any of the alternatives that have so far been offered.

Close to the beginning of the expansion there may have been a stage of rapid inflation. If so, it determines roughly how large the universe created in the explosion is going to be. After this is a stage during which the strongly interacting particles (such as protons and neutrons), are created. This is called the "era" of baryogenesis. It is during this period that the density of the universe, which is to say the average number of protons or neutrons per cubic centimeter, is determined. Somewhat after this is the era of nucleosynthesis, during which a fraction of the protons and neutrons are bound into helium and a few other light elements. This brings us to a few minutes after the Big Bang.

After this, the universe expands and cools until decoupling, after which the galaxies begin to form. The era of the formation of the galaxies is followed by the present period, during which the spiral galaxies make stars. It is during this period that most black holes made by the universe are produced. After some tens of bil-

lions of years, all the available gas will have been converted to stars, and the galaxies will die. There have been some interesting speculative discussions of the far
future of the universe, some of which do have bearing on the number of black
holes produced, as we will see.

Given this cartoon sketch of the history of the universe, we may ask at what
points a change in the parameters might lead to an increase in the numbers of
black holes produced. We may begin with the inflationary era. Despite some technical difficulties, the basic idea that the universe went through an early period of
inflation remains an attractive hypothesis. According to this idea, the size of the
universe produced from a single explosion is determined by the time the universe
continues its rapid expansion (the size is in fact proportional to the exponential of
the time the inflationary era lasts). As the density of the universe is determined by
later processes, it seems reasonable to conclude that the number of black holes
the universe ultimately makes will be proportional to its size at the end of the
period of inflation. If the hypothesis of cosmological natural selection is true, this
should be as large as possible.

In the standard models of inflation, there is a parameter that determines how
long the period of inflation lasts. It measures the strength by which a certain new
kind of particle hypothesized by the theory interacts. These new particles are
called *inflatons* and the new parameter is the *inflaton charge*. It turns out that the
region that inflates is larger the smaller the inflaton charge is. (In fact the dependence is very strong, the amount the universe inflates is proportional to the exponential of the inverse of the square root of the inflaton charge.) Thus, if cosmological natural selection is to select for as big a universe as possible, it will try to
make the inflaton charge as small as possible.

This is good, because it turns out that the inflaton charge must in any case be
smaller than a certain critical value if inflation is to happen at all. This critical
value depends on the model, but is usually less than one millionth. At the same
time, the inflaton charge cannot be too small. The reason is that it measures the
strength of certain interactions which produce clumping in the early universe.
These clumps are the seeds that, according to the standard inflationary scenario,
later grow into galaxies. The weaker the inflaton charge, the less clumpy the universe will be. If it is too weak, the galaxies will never form. This may lead to a universe that is very large, but with very few black holes.

To give rise to a universe that makes the most black holes, the inflaton charge
must then be within these two limits. It must be large enough to ensure that
galaxies form, but it must be small enough to permit inflation. Within these limits, the charge should be as small as possible; this leads to a universe with the
largest possible volume, and hence the largest number of black holes. Thus, given
the hypothesis of cosmological natural selection, we arrive at a prediction about
the value of the inflaton charge which may be applied in any of the detailed models of inflation. This prediction is that the charge take a value that is as small as

possible, while still giving rise to enough clumpiness that galaxies form. This is an example of how cosmological natural selection can increase the predictability of other hypotheses.

This circumstance is also relevant for the proposal that there are choices of the parameters that lead to universes that produce a great many primordial black holes. As I've already pointed, out, this does not really count as an objection to the theory, as it is unlikely that this could be accomplished with a small change in the parameters. However, this being said, there is also a problem with the idea that the parameters could be tuned to make enormous numbers of primordial black holes. The most obvious way to do this would be to dramatically increase the lumpiness of the early universe. If the universe were initially much more lumpy and inhomogeneous than it was, there might have been many clumps where so much matter was concentrated that they collapsed immediately to black holes. The problem is that the clumpiness of a universe that grows by infla-tion is essentially determined by the strength of the inflaton charge. But, as we have just seen, there is a limit to how large the inflaton charge can be. This means that there may very well be a limit to how lumpy a universe blown up by inflation could be. To go further with this question, we must descend to a much more technical level of discussion, but the point is that it is not necessarily true that there is any choice of parameters that produces a universe that is blown up by inflation, but full of primordial black holes.

After the inflationary era, we come to the era of baryogenesis, which deter-mines how many protons and neutrons are made. There is a rather standard sce-nario of how this happens, which was put forward initially by Andrei Sakharov, who was, in his very interesting life, one of the inventors of the Russian hydrogen bomb, the author of a number of interesting physical hypotheses, and a much-admired advocate of democracy. Sakharov proposed that the density of protons and neutrons is under the control of an independent parameter which measures the extent to which the elementary interactions in nature are changed under a symmetry operation called CP. In this operation one looks in at the world through a mirror and simultaneously replaces all particles by their antiparticles. It turns out that the world is almost, but not quite, unchanged by this operation. Why this is so is a mystery, but it is tempting to believe that it is because this is the way it must be for the world to be full of protons and neutrons.

Cosmological natural selection should predict that the density of protons and neutrons is as high as possible, as this will give rise to more stars, and hence more black holes. We may then ask if there is any limit to how high we may set the den-sity, before there are other effects which may tend to decrease the numbers of black holes. The answer is that there is. In the era of nucleosynthesis, helium is synthesized, as well as a number of other lighter elements. It turns out that the amount of helium that is produced is proportional to the density of protons and neutrons. With the present density, the primordial gas is about one quarter helium. If the density of protons and neutrons were to significantly increase, then there would be more helium. We discussed the case of a world with more helium, but without supernovas in Chapter 8, and concluded that it would be so different from our world that we are unable to predict the outcome. The case here is different, as this is a change that would lead to a world with more helium, but without affecting the processes that cause supernovas. However, as before, there are competing effects that make it difficult to draw a firm conclusion with present knowledge. This is then another case in which the principle of cosmological natural selection makes a prediction, which is that a world with a higher density than ours would in any case have fewer black holes, most likely because of the consequences of increasing the amount of helium.

There is one more set of cosmological parameters that we may consider in our search for ways to contradict the predictions of cosmological natural selection. These control whether the universe will eventually recollapse, and how long it may live before doing so. These provide at least rough tests of the theory.

In the end the universe will either recollapse, or it will expand forever. How long the universe exists before one or another of these happens depends on a competition between the initial speed of expansion and the initial matter density. This is completely analogous to a very familiar example: if I throw a ball into the air, the time it stays up depends on a competition between the speed with which it is thrown upwards and the magnitude of the gravitational forces pulling it downward. In the case of a ball, the gravitational force is proportional to the mass of our planet; for the universe, the gravitational force that seeks to reverse its expansion is proportional to the density of matter. If the initial speed of expansion is too small, it is easily reversed by the gravitational attraction of all the matter, and the universe collapses quickly. On the other hand, if it is too large, the gravitational attraction of all the matter is not enough to reverse the expansion, and it expands forever, eventually becoming fantastically dilute.

To have a universe that lives a long time, without either collapsing or becoming rapidly dilute, it is necessary that the speed of expansion and the matter-density be closely balanced. If they are not, there can only be one natural scale for the lifetime of the universe, which is the Planck time, 10^{-43} of a second. This is because without fine tuning, we expect all the parameters to come out to be reasonable numbers, not too large and not too small. But near the beginning, all scales must be set in Planck units, as they are the only physical units relevant. This is because the dominant force near the beginning must be gravity, and the Planck units are based only on the strength of the gravitational force, the speed of light, and Planck's constant. This means that the lifetime of the universe will also come out to be not too far from one in Planck units. But a Planck unit of time is equal to 10^{-43} of a second.

We live in a universe which has lived so far around ten or twenty billion years, which is about 10^{60} Planck times. During this time it has neither recollapsed, nor has the expansion run so quickly that the density of matter has become negligible. For this to happen it must be that to an incredible precision the initial expansion speed and matter density were balanced quite close to the line between these two possibilities. How close? If we measure everything in the Planck units, then these two things must have been initially in balance by better than one part in 10^{60}. This means that each quantity must be set to a precision of *at least sixty decimal places.*

With this information we see that the most interesting question is not whether the universe eventually collapses or not. It is to understand why it is here at all. If the initial parameters were picked at random, then it is overwhelmingly probable that the universe would either have recollapsed by now or by now be essentially empty. The probability for one or the other of these things not to have happened by now is this same one part in 10^{60}. If we want to have a rational understanding of the universe, this is another fact that needs an explanation.

One way to speak of this situation is in terms of a parameter that is conventionally called Omega. It is defined to be the ratio of the matter density measured now to the density required for the universe to recollapse. If Omega is larger than one, the universe will recollapse. If it is smaller than one, the universe will expand forever, and after a time become extraordinarily dilute. The closer Omega is to one, the longer the universe will exist in something like its present state before it begins either to collapse or to become very dilute. Only if Omega is exactly equal to one will the universe remain forever exactly balanced between the two possibilities of collapse and runaway expansion.

The fact that the universe, at this late stage, has neither collapsed nor taken off in an accelerating expansion means that Omega must be within a few powers of ten of one. It is then natural to ask if Omega might be exactly equal to one. This could only occur if the initial matter density and speed of expansion were precisely balanced. At present we know they are balanced to a precision of at least sixty decimal places, but if Omega is equal exactly to one they must be balanced exactly, which means they match to an infinite number of decimal places.

In the case that the cosmological parameters are governed by some fundamental theory, it is very difficult to imagine that the parameters could be tuned a precision of sixty decimal places without being tuned precisely. This is because it is much easier to construct a mathematical theory that produces the number one than it is to construct one that produces a number which differs from one only by some digit coming in the seventieth, or the thousandth, decimal place.

Indeed, the only theories that have been so far proposed to explain how this balance is achieved do predict that it is achieved to incredible accuracy, which means that Omega must be almost exactly equal to one. Among these is the hypothesis of inflation. According to most models of inflation, Omega should differ from one by a very small amount, like one part in 10^6 Actually, it is possible to have inflationary theories in which Omega is not exactly equal to one. These are described in Bucher, Goldhaber and Turok, (1994) and Bucher and Turok, (1995). But this can only be achieved if some parameter that sets the time that the era of very fast expansion lasts is tuned incredibly precisely.

On the other hand, if the balance is set by the mechanism of natural selection involving the formation of black holes, we may not be surprised to find that the balance has been tuned to allow the universe to live at least as long as it has. But we should be very much surprised if it turns out to be tuned much more precisely than that. For if the balance has been tuned to make the production of a large number of black holes possible, what is required is that the universe live at least as long as a galaxy may maintain a continual process of forming stars, some of which become black holes. But there is no reason that the mechanism of natural selection would lead to the parameters being tuned so that the universe lives much longer than this.

As many galaxies are presently forming stars, the time that the galaxies form stars is at least as long as the lifetime of our present universe. At the same time, it cannot be forever, eventually all the gas is turned into stars, and no more stars can be made. There is some evidence that the rate of the formation of new stars in the universe is slowly decreasing. There is also evidence that there were more galaxies vigorously forming stars earlier in the history than there are at present.

Thus, if the hypothesis of cosmological natural selection is correct, we would expect that Omega would be tuned well enough that the universe lives about as long as galaxies vigorously form stars, before either collapsing or going into a mode of runaway expansion. But for the same reasons that no creature lives many times longer than is necessary for it to reproduce, we should not expect Omega to be tuned any closer to one than that.

This gives us a rough—but I think a rather strong—test to distinguish between the hypothesis that Omega is fixed by a fundamental microscopic theory like inflation and the hypothesis that it is fixed by a statistical mechanism such as I am proposing here.

To measure Omega we must determine the amount of matter in our universe. This can be done in several different ways. First, one can simply add up all the mass coming from things that are directly visible, such as stars, gas and dust. At the present time the evidence is that these constitute between about one and three percent of the amount needed for Omega to be equal to one.

But, there is a second way that matter may be counted, which is by looking for its gravitational effects. If one studies the motions of the stars in a galaxy, or of the galaxies in a cluster of galaxies, one can deduce from the laws of gravity how much matter must be around. These calculations lead to an estimate of between five and ten times more matter than the first method gives. This leads to the conclusion that at least eighty percent of the matter in the universe is composed of things other than the stars, dust, and gas we can see.

As to the crucial question of how much dark matter there is, the best estimates coming from studying both the motion of stars in galaxies and of galaxies in large clusters of galaxies yield enough to get Omega up to between one tenth and one fifth. This is closer to the value of one, but it is not equal to it. The evidence that Omega is less than one is summarized in White, Navarro, Evrard, and Frenk (1993); Coles and Ellis (1994); and Bahcall, Lubin, and Dorman (1995).

Actually, the situation is a bit more complicated, because there is another possible contribution to the density of matter in the universe. This is the cosmological constant, described in Chapter 3 as the parameter that governs the ultimate scale of the universe. It was originally introduced by Einstein in response to his discovery that his theory of general relativity predicts that the universe must be expanding or contracting in time. At that time, in 1916, everyone, including him, imagined the universe to be eternal. To make the theory agree with this expectation, he found a way to modify the equations of general relativity in order to allow them to describe universes that, rather than expanding or contracting, are unchanging and eternal. The cosmological constant measures the magnitude of this modification.

Unfortunately (or fortunately), the modification did not actually serve Einstein's purposes very well, as he quickly realized. With the modification, there are still solutions in which the universe expands or contracts, and those that were static are in fact unstable, so that any little cosmological tickle can send the universe into irreversible expansion or contraction. Thus, in the late 1920s, when Hubble discovered the first evidence of the universal expansion, Einstein proposed eliminating the cosmological constant, calling it the biggest mistake of his life.

Actually, many physicists think the biggest mistake of Einstein's life was not the cosmological constant; it was his opposition to quantum mechanics. One of the ironies of his life is that once quantum mechanics is taken into account, the cosmological constant becomes difficult to kill. This is because quantum effects connected with the uncertainty principle tend to give empty space a non-zero energy density. The effect is as if there were a cosmological constant, even if one were not put in originally. This is disconcerting, as the energy density of empty space is, if not zero, very close to that. Luckily, at the cost of tuning a parameter it is possible to set the energy density of empty space to zero. To achieve this it is necessary to tune the original cosmological constant of Einstein so that it exactly cancels the energy density that comes from quantum mechanics. This can be done, but the cancellation involved is very delicate; it requires that the cosmological constant be tuned accurately to better than one part in 10^{60}.

The mystery of why the cosmological constant coming from Einstein's theory of relativity must be tuned to this accuracy to eliminate certain unwanted effects of quantum mechanics is one of the major problems associated with the question of the relationship between relativity and the quantum theory. In the context of our discussion here, there are again two possibilities to consider. The first is that it is tuned exactly to zero by some fundamental theory. The second is that it is tuned to some very small, but non-zero value, by the mechanism of cosmological natural selection.

In fact there seem to be at least three parameters that can be tuned to affect the overall lifetime of the universe. These are Omega, the cosmological constant, and the mass of the neutrino. The latter is relevant because if the neutrino has a mass in a certain range, it can contribute to the dark matter, and thus to the total density of matter.

If all three are fixed by the mechanism of natural selection, we should expect them to be each tuned about as well as is required so that the universe lives as long as galaxies produce black holes, but not much better. Furthermore, there is no reason why any one of the three should be tuned much more accurately than the other two. This means that we may expect that when all the observations have been sorted out, there will be a small cosmological constant, there will be a neutrino mass, *and* Omega will not be exactly equal to one. Such an outcome would perhaps be the worst nightmare of the cosmologist working from the expectation that the parameters are tuned by a fundamental theory. But, by the same token, it would be good evidence that the parameters of the universe are tuned by some blind and statistical process, as in the hypothesis of cosmological natural selection.

OBJECTIONS OF PRINCIPLE

We now come to a number of objections of principle that have been made against cosmological natural selection. These objections have nothing to do with the details of empirical test; they have instead to do with the logic of the formulation of the theory. First, the theory predicts only that the parameters of our world are near a value that produces a maximum number of black holes. They need not be exactly at a maximum. Does this not make it harder to test? Suppose one change is found to lead to an increase in the number of black holes. Do we then rule the theory out? Probably not, especially if the increase in question is not very much compared to the enormous decreases we expect if what we have been saying here is true. If we are near the summit of a mountain, we can still expect to find a direction that leads to the top. But what if three changes lead to increase, but thirty-seven lead to dramatic decreases? Where do we draw the line?

I don't think this is something that we can decide ahead of time. If a theory like this is going to be right, it is going to work very impressively. And it is also going to be the case that no theory is discovered that does better. If it is to succeed in convincing us of its plausibility, cosmological natural selection must make a number of predictions that are impressively confirmed, and it must continue to do so better and more often than any of its possible rivals. It must also happen that our understanding of physics at the Planck scale eventually improves to the point that the two postulates of the theory may be confirmed or refuted. Thus, what is good about this theory is that as our knowledge of both astrophysics and quantum gravity increases, it becomes increasingly vulnerable to disproof. If some years from now the theory of cosmological natural selection has survived a number of additional tests, and has been found to be consistent with the predictions of a theory of quantum gravity—and if at the same time no unified theory has been invented which predicts the correct value for all the parameters—then I think it will be hard not to take it seriously.

An objection that has sometimes been made against the theory is that we are not really varying the right parameters. The parameters we should be varying are those that decide between the various choices for the physics at the Planck scale, which—according to the basic scenario—is the scale at which the decisions are made about which laws are chosen. We cannot really know how many parameters are free at that level, and how the parameters we have been discussing are related to them.

Of course, this is simply true. The point is only that we have to try to do the best we can with the knowledge that we have. There is nothing that prevents us from improving the theory if at some time we learn about some more fundamental parameters, or we understand that certain of our parameters are actually related to each other.

It should also be said that similar issues arise in biology. What varies under mutation and sexual recombination is the *genotype*, which is what is represented in the actual coding in the DNA. But what matters for natural selection is the *phenotype*, which is the expression of the gene in the actual organism. Not every variation in phenotype that can be imagined can be the expression of a possible DNA sequence. This is perhaps fortunate, as we do not need to worry about being superseded by very smart flying cats who never sleep. At the same time, Darwin, Mendel, and their successors did quite well without knowing anything about how the genetic information was represented. They were able to do this because they were able to make good guesses about how the information was organized at the level of the genotype by studying the patterns of variation of the phenotype.

Of course, we do not have the advantage they did of studying the range of variations of individuals. However, I do not think that this can be taken as an objection against the theory. The reason is that it is hard to see how the empirical adequacy of the theory could be worsened by using a more fundamental set of parameters. For example, it might be that in the future we will learn that several of the parameters are tied together, so they cannot be changed independently. This might enable us to ignore certain changes, which might otherwise have led to an increase in the number of black holes. But the reverse is not going to happen. Suppose there is a way to change several parameters together, that leads to an increase in the numbers of black holes. This will stand against the theory regardless of whether there is a more fundamental theory that requires that they always be changed in this way.

A related issue is the adequacy of the assumption that the variations of the parameters at each formation of a new universe are truly random. Is it not possible that the detailed microphysics allows only certain transitions, or makes the variation of the parameters depend on details of the collapse such as the mass of the black hole? The answer is that, in our ignorance, we should make the weakest possible assumption, which is that the variations are random. If the theory that results is not refuted by the evidence, the situation will not be worsened when we know enough to make a more detailed assumption about the actual variations.

For the same reason, Darwin assumed that variation is random, even if it was clear that in many cases it might not be true. The theory of natural selection was certainly strengthened as our knowledge of genetics on the molecular level increased. But the theory was still testable when nothing was known about the actual microscopic structure of the genes.

Similarly, there is nothing about the present proposal that is incompatible with there being a fundamental theory that describes the Planck scale physics. As long as the fundamental theory has at least one free parameter, the postulates of cosmological natural selection will almost certainly lead to a predictive theory. The only way this could not happen is the unlikely case that the fundamental theory predicted unique values for the masses of the elementary particles and strengths of the forces. Therefore, the possibility of the discovery of a more fundamental theory cannot be used as an argument against the theory.

Another argument made against the theory is that it is not legitimate to consider variations of the parameters that lead to universes that do not contain intelligent life. Given that we know that the universe we live in does contain us, shouldn't we only consider variations of the parameters that result in universes in which we could live? If so, we could only argue that the theory was confirmed observationally if it turned out that more black holes are produced in our universe than in any other universe with intelligent life.

As far as I can tell, this argument is based on a misunderstanding. It is quite possible that most of the small variations of the parameters from their present values, which lead to a world with fewer black holes, lead to a world without life. But if we are interested in testing the theory of cosmological natural selection, the two things must be considered to be independent. It is quite possible that our universe could have the property that it contains life, but fail to have the property that all small variations of the parameters would lead to a universe with fewer black holes. If so, it is quite likely that we will be able to discover that this is the case. This means that the two properties can and, indeed must, be taken to be independent, for the purposes of testing the theory.

Furthermore, if all small variations in the present values of the parameters would lead to universes with fewer black holes, this is a property of our universe alone. The actual ensemble of other universes does not have to exist for us to discuss whether or not our universe has this particular property. This is shown by the fact that it is possible, using only knowledge about the physics of this universe, to discuss whether or not this property is true or not. This means that there cannot be any logical problem with the testability of this hypothesis, as this objection suggests.

But, someone might argue, suppose that after everything were said and done, all changes in the parameters led to dramatic decreases in the numbers of black holes, except one, which led to a modest increase, but at the cost of making life impossible. Might one then use a weak form of the anthropic argument to save the theory? Could we not then argue that we live in the most probable universe that is also consistent with the existence of life?

I imagine that were we in this situation, we might be satisfied to make such an argument. The danger is that if it were applied too soon or too often, it might tend to decrease the falsifiability of the theory. Given the comments I made earlier about such cases, I think the best thing that might be said is that such an argument might be appropriate only if it were already established that almost every change in the parameters led to a decrease in the number of black holes, so that there would be good reason to believe the theory whether or not the argument were made. As something that removes one troublesome case in the face of a large number of confirming cases, it might be possible to consider it. But we are certainly some distance from this situation. At present it is best not to exclude any case from consideration as a possible counter-example to the theory, whatever the implications for life or our own existence.

It is true that, as it is not possible to literally vary the parameters, any test of the theory involves a combination of observation and theory. But, as philosophers of science have been pointing out for many years, any test of a scientific theory involves a mix of theory and observation. There is nothing in principle to rule out the use of theory to refute a prediction about what would happen were the laws of physics slightly different. The test is, of course, only as good as the theories it employs are reliable. The theories we have used to test cosmological natural selection are being continually tested, in experiments in basic physics and in comparison with astrophysical observations. To the extent that these theories become better over time, the testability of the hypothesis of cosmological natural selection will steadily increase.

There is a final technical objection that may be discussed. Black holes can merge by coming close enough to each other that their horizons overlap. This may happen from time to time to real astrophysical black holes. Moreover, if the universe itself collapses, then before the end all the black holes ever created may merge together into one great final crunch. The question is how many universes are created when two black holes merge? One or two?

This is a question that depends on the details of the physics that produces the bounce. While the answer is unknown, it is rather plausible that the answer is two universes. This is especially likely if the first thing that happens after the bounce is a period of inflation.

PROPOSALS FOR MODIFICATION OF THE THEORY

Now I would like to turn to consider several proposals which have been made to modify the postulates I gave in Chapter 7. These lead to alternative theories, whose predictions may, as we will see, differ from those that follow from the theory we have so far been considering.

The first of these is to use the parameter space of string theory, rather than the standard model, as the space of parameters that are varied. If we believe in string theory, then we should assume that the physics before and after the bounce are described by a phase of string theory. During the extreme conditions of the bounce, there might be either a change in the parameters that describe that phase or a transition from one of these phases to another one. (In technical language, a phase corresponds to a classical perturbative string theory, such as might be described by a Calabi-Yau compactification, and the parameters that describe each phase are the "modular parameters.") These transitions might not be completely random, for it may be that the transition is more likely to occur to a phase that is in some sense "nearby."

This hypothesis would certainly be worth investigating, and might even result in a possibility to test string theory. For example, it might be the case that the universe was not at a maximum of black hole production, in terms of the parameters of the standard model, but was when the variations were restricted to those allowed by string theory. If this were the case, it would stand as evidence for string theory.

To pursue this idea, we need a better understanding of the parameter space of string theory. But recent progress in the theory goes exactly in this direction; it suggests the possibility that the whole parameter space may be connected, so that it is possible, by a kind of phase transition, to move between different sectors, in which the extra dimensions are curled up in topologically distinct ways. As string theory continues to develop, it may become possible to investigate this proposal in some detail. By doing so, we might discover that the apparent weakness of that theory—that it describes consistently a very large number of different phases—might be revealed instead as its great strength.

A second modification that may be proposed is that the numbers of new universes which are born from each black hole may differ according to the mass of the black hole. There are in fact calculations, done by Claude Barrabes from Paris and Valerie Frolov, a Russian physicist who now works in Canada, that suggest that a large number of universes may be created inside each black hole. Furthermore, they predict that the number of universes produced may grow as the mass of the black hole increases. These calculations are based on certain approximations, so it is not certain that they reflect accurately what a quantum theory of gravity would predict. But they certainly suggest that this is a possibility that should be considered.

The calculations which Barrabes and Frolov (1995) have done are not reliable enough to predict exactly how the number of universes created depends on the mass of the black hole. Thus, for the present we may make different hypotheses about this. These lead to theories which make predictions that differ from each other as well as from the simple theory we considered up till this point, according to which each black hole leads to one universe, no matter what its mass is. This is because it might easily turn out that our universe is near an extremum for producing black holes only under a particular assumption about how many universes are produced per black hole.

For example, if we assume that the number of universes created inside of each black hole is proportional to the black hole's mass, this can lessen the selective pressure to create black holes with as small mass as possible. If all black holes are counted equally, then two small black holes are certainly better than one large black hole, even if the large black hole has ten times the mass of the small ones. In this case a mechanism that makes black holes that are as small as possible, and redistributes the rest of the matter so that it is available for the formation of more stars and black holes, is clearly favored. However, if the number of universes made grows proportionally to the mass of the black hole, then the situation is different. As most of the matter that is blown off by a supernova does not end up as a black hole, it is certainly better for a large black hole to collapse instantly to form a number of universes. So in this case, mechanisms that make fewer but larger black holes are favored over those that make more small black holes.

Recently, evidence has been reported suggesting that many galaxies have large black holes in their centers, with masses of at least millions of times that of the Sun. In the one world per black hole theory, this discovery is not very important, as each galaxy may create hundreds of millions of small black holes from the collapse of stars. But if the number of universes made in each black hole grows rapidly with the black hole's mass, then the one huge black hole in the center of a galaxy may become as significant as the many small ones. In this case there could be selective pressures to create very large black holes that are not present in the theory I've been considering.

In fact, Barrabes and Frolov suggest that the number of universes produced could grow very quickly as the mass of the black hole increases, by perhaps as much as the fourth power. In this case, even if it were the case that every star became a black hole, a single million solar mass black hole at the center of a galaxy would be much more important. Such a theory would in fact greatly favor changes of the parameters in which whole galaxies collapse directly to form enormous black holes, without any of the complex phenomena associated with the formation of stars. It may then be the case that this is a theory that could be ruled out by present knowledge.

Thus, even with current astrophysical knowledge it is possible to distinguish between these different versions of cosmological natural selection. This is good, it means that cosmological natural selection is a scientific theory, whose assumptions are subject to test. Even more, as we may hope that a quantum theory of gravity will ultimately predict whether new universes are created by black holes, and how many are created, we may look forward to the possibility of testing jointly cosmological natural selection together with various versions of quantum theories of gravity.

Another kind of alternative arises if we come back to the question I opened the book with: Why are the laws of nature compatible with the existence of life? Certainly cosmological natural selection goes a long way towards a possible answer, as many of the ingredients that life requires, such as stars, or carbon and oxygen, also play a role in increasing the production of black holes. But what if the conditions that made for the formation of the most black holes astrophysically were not, in the end, quite the same as the conditions required for life? It is quite conceivable that there may be small changes in the parameters from their present values that would lead to an increase in the numbers of black holes produced, but at the cost of making conditions inhospitable for life. A candidate for such a possibility might be a change that resulted in the formation of only very massive, but short lived, stars. If there were a change in the parameters that had this effect, this might very possibly be a world with more black holes than ours, but without life.

In fact, I do not know if there is any change in the parameters that could accomplish this. But if there were, we might ask whether this would disprove the hypothesis of cosmological natural selection. As pointed out by the mathematician Louis Crane (1994b), it would not, if it could be established that there were a selective advantage for universes that create life. This may seem unlikely, but it is not impossible. For example, Crane suggests that we consider the possibility that very far in the future, after all the stars have burned out, the continued survival of life will depend on the discovery of new ways to keep warm. One way that might be available to them is to create small black holes. According to the great discovery of Steven Hawking, black holes radiate heat through quantum effects. The temperature of a black hole varies inversely to its mass, so what one wants is a black hole small enough to give off a substantial amount of heat. This may sound outlandish, but who is to say what intelligent living things will be able to do in twenty or thirty billion years?

Very far in the future, after all the stars have burnt out and all matter has been fused to heavy elements, it may not be so easy to find ways to keep warm. Perhaps it is not impossible that in this situation the universe will be populated by creatures that have learned to make small black holes. If there are enough of them that the black holes they make outnumber those made by stars, this could provide the selective advantage needed for life.

Surely this seems more fiction than science, but let me note that it is still a hypothesis that can be tested by observation. For were there only one change in the parameters that did not strongly decrease the number of black holes, and were the result both an increase in the number of black holes and a world without life, one possible explanation consistent with cosmological natural selection would be that it is probable that sometime in the future living things make lots of black holes.

Another reason intelligent creatures might want to make black holes was pointed out by Edward Harrison (1994). If sometime in the future cosmological natural selection became an established theory, people might understand that one way to increase the probability that future universes would contain intelligent life would be for them to make as many black holes as possible. This would increase the number of universes with parameters hospitable to intelligent life.

I must say that I hope that we can understand the open problems in cosmology without having to resort to such kinds of ideas. Perhaps the best argument against such a proposal comes from natural selection itself: given the tremendous resources that black hole production might absorb, any civilization that became fixated on making new universes would go the way of those societies, from the Aztecs to the Stalinists, that allowed obsession with an imagined future to blind them to the possibilities of making life better for those alive now.

If I may come back to the original proposal of cosmological natural selection, I hope in this appendix to have at the very least convinced the reader that it is a legitimate scientific theory, and that the fact that it has so far withstood the challenges raised against it is not trivial. Given the implausibility of there being any connection between the parameters of physics and cosmology and the number of black holes formed in the universe it might easily have turned out the other way. I think we may be impressed by the fact that we have eight changes in the parameters that lead to worlds with fewer black holes than our own, but none that clearly have the opposite effect. To recount them, they are the original five that unbind atomic nuclei; the change in the strength of the weak interaction that leads to a universe without supernovas, but still made of hydrogen, making the neutron lighter than the proton; and the increase in the strange-quark mass, which raises the upper mass limit. Several more changes seem to lead to a combination of competing effects, so that it is not possible to conclude with present knowledge whether the result is to increase or decrease the number of black holes. Examples of these are increasing or decreasing the gravitational constant, decreasing the mass of the electron, the change in the strength of the weak interaction, which produces a world made mostly of helium, and the effect of lowering the mass of the strange quark still further. Each of these remains a test that the theory will have to pass as astrophysical knowledge improves.

It is also good news that alternatives to the theory may be proposed, and that ultimately observation may decide between them. This is something that must happen to all good scientific theories. We may look forward to the possibility that progress in quantum gravity and string theory may lead to the situation in which different hypotheses about quantum effects in black holes or the parameter space may lead to different versions of cosmological natural selection that may be distinguished by their astrophysical predictions.

Cosmological natural selection is then a theory that competes not with a unified theory, such as string theory, but only with the hypothesis that there will be a unique unified theory that unambiguously predicts the values of all the parameters. Given the speed at which our knowledge of the astrophysical processes involved in the formation of galaxies, stars, and black holes is increasing, we may be fairly certain that if the theory is wrong, it will be possible to definitely disprove it in the near future.

Notes and Acknowledgments

In these notes are collected a variety of comments, references, and clarifications. They are followed by a few final words of acknowledgment.

The reader will also find suggestions for further reading in the bibliography, which consists mainly of books that are accessible to the general reader. A more complete list of references is also available on the internet at http://www.phys.psu.edu/SMOLIN/book. At that site the reader will also find information about developments relevant to the book that have occurred since the publication.

PART ONE

To avoid confusion, let me stress that I use the word *Newtonian* as an adjective to refer not only to what Newton invented, but to everything that followed, including electromagnetism, until the break with the basic principle that a system is described completely by a deterministic dynamical law in an absolute space and time early in the twentieth century. Specialists often use the adjective *classical* in the same sense.

CHAPTER 3

Most of the arguments I employ for the specialness of the parameters were invented by advocates of the anthropic principle, particularly Brandon Carter, Bernard Carr, and Martin Rees. Concern about the smallness of several of the parameters began with works by Herman Weyl, Paul Dirac and Robert Dicke. A good summary of all of these arguments, with complete references, can be found in the book of Barrow and Tipler (1986), which also contains very interesting discussions of the historical and philosophical roots of the anthropic principle. The best treatment, of which I am aware, of the history of elementary particle physics is Abraham Pais's *Inward Bound* (1986). It gives a perspective weighed towards the experimental developments, which is complementary to the emphasis here of the role of philosophical principles. The best arguments for the hope of their being a unique "final theory" are in the book of Weinberg (1992).

I now sketch the calculation of the probability that a universe with randomly chosen values of the parameters of the standard model will have stars that live for billions of years. We may begin with the six masses involved-the Planck mass, the masses of the four stable particles, and the mass associated with the cosmological constant. To ask how probable they are we must consider their ratios. To do this we will consider the largest, which is the Planck mass, to be fixed, and express the others in terms of units of that mass. Each of the others is then some number between zero and one. Let us assume God created the universe by throwing dice, and so chose these numbers randomly.

The lifetime of stars depends on the ratio of the proton mass to the Planck mass. That they live more than a billion years, requires that this ratio be less than 10^{-19}. The probability for this to occur randomly is one part in 10^{19}. We saw that for there to be many nuclei, the neutron must have about the same mass as the proton, while the electron must have a mass on the order of a thousand times smaller. The accuracy to which the neutrons mass must approximate that of the proton is a few electron masses. This means that the masses of the electron and neutron must come out to within an accuracy of about 10^{-22} in Planck units. The same is true also for the neutrinos. The probability for this to happen in three roles of the dice is about 10^{-22} cubed, which equals 10^{-66}. If we put this together with the probability that the proton mass comes out as no larger than it is, we get a probability of one part in 10^{85}.

We have to take into account also the cosmological constant, for if it is too large the universe will not live long enough for stars to form. In order for the universe to live at least until the time of the formation of the galaxies, the cosmological constant must be less than 10^{-60}. The probability to get this number randomly is then one part in 10^{60}. Putting this together with the previous results, we now have a probability of one part in 10^{145}.

We have yet to mention the nongravitational interactions. If we start with their strengths, then we may again compute the probabilities by taking ratios. Taking again the strongest, which is the strong nuclear interaction, as the measure, the weak and electromagnetic interaction are each about one part in 100. This multiplies the above probability by 10^{-4}, which gives us one part in 10^{149}. Finally, we have to take into account the ranges of the forces. The largest range is that of electricity, which is at least the radius of the universe. The ratio of the radius of the nuclei, which is the range of the two nuclear interactions, to the radius of the universe, is at most 10^{-40}. The probability to get two such tiny ratios randomly is one part in 10^{80}. If we combine this with our previous result, we reach the conclusion that the probability for the world to have turned out as ours, with stars lasting billions of years, and thus with nuclear and atomic physics more or less like ours— were the parameters of the standard model picked randomly—is at most one part in 10^{229}.

The reader may wonder why I don't consider variations of the constants G, c, and ℏ. The version is that there must be a set of fixed constants that define the units in which other quantities are measured. It is most convenient to take these to be G, c, and ℏ. Then all other quantities are measured in Planck units.

CHAPTER 4

A point of clarification regarding the relationship between Leibniz's thought and modern physics is in order. Certainly, I make no pretense to saying anything comprehensive or original about Leibniz as a philosopher. What I am claiming is that several aspects of Leibniz's philosophy provide philosophical motivation for some of the key concepts in twentieth-century physics. These are primarily his relational conception of space and time, as well as the principles of sufficient reason and the identity of the indiscernible. However, there is also much in Leibniz's philosophy that seems irrelevant to modern physics, such as his idealism and his denial that his monads have any real interaction. To take parts of Leibniz as inspiration for physics does not mean I have any commitment to these other aspects of his philosophy.

I must also emphasize that while one can introduce and justify the gauge principle through Leibniz's philosophy, as I have done here, I do not mean to give the impression that this was an argument that was used, or even would have mattered, to most of the theoretical physicists who developed the idea. Herman Weyl, like Einstein, was likely influenced by the philosophical tradition begun by Leibniz, but once gauge invariance and general relativity were invented, they were adopted because of what they were able to explain, not because of philosophical arguments.

I would also be remiss if I did not I also tell the reader that there is a long tradition, starting with Kant, that holds that Leibniz was wrong about his philosophy of space and time. For example, in his *Critique of Pure Reason*, Kant claims to disprove Leibniz's principle of sufficient reason, which is the basis of his relational philosophy about space and time. It is certainly true that scholars may disagree about the extent to which one philosopher may or may not have established their points in their arguments with each other. I can only speak as a scientist whose aim is to use the writings of philosophers as a source of ideas and inspiration. As such, I, personally, find what Kant has to say about both space and time and Leibniz unconvincing and uninspiring. Some of the arguments Kant uses seem to assume the existence of distinctions that Leibniz would deny, such as there being an absolute meaning to where something is, or an absolute distinction between left- and right-handedness.

If the proof of a philosophical argument is its eventual relevance for science, then certainly Leibniz seems more interesting than Kant. His thought led eventually to both general relativity and to profound developments in mathematics, such as topology and symbolic logic, while Kant's writings on space and time claim to establish the logical necessity of both Newtonian mechanics and Euclidean geometry. It is not hard for me to choose which writer to ponder over when I am in need of inspiration.

Perhaps this is a sign of philosophical naiveté, on the other hand, I find Kant's presumption to define a domain of discourse in which final conclusions could be reached, which limits once and for all what we might perceive about the world, highly implausible (even leaving out the fact that they led Kant to conclusions we now know are false). Beyond this, there seems something amiss in the whole presentation: the writing is ponderous and inelegant, while the style is that of one who takes it for granted that words are sufficient to solve puzzles whose resolutions in fact required conceptual and mathematical discoveries that had not yet been made. What is most bothersome in Kant is that one gets a feeling he leaves no room for the future, he writes as if everything one would need to know to settle all the deep issues of science and philosophy was known then. In order to deal with what he doesn't know, there is a proliferation of categories, and verbal distinctions, and there is an insistence on setting up dialectical oppositions, when a mere confession of confusion and ignorance, which is no shame for an honest scientist, would have sufficed.

An example of this is the antinomies in Kant's *Critique of Pure Reason*, in each of which two arguments are presented. One leads to a conclusion such as, time or space are infinite, the other to the opposite conclusion. For Kant these arguments show ultimate limits to the power of rationality. However, in several of the cases, such as that of the finiteness or infiniteness of space, we know now that the dilemma can be resolved by the invention of a new idea, which was not known at the time. Indeed, I use arguments very similar to the antinomies in several places such as the introduction to Chapter 7. However, I would ask the reader to note the way they are employed, which is to either introduce new conceptual ideas or to argue for the need of such ideas.

CHAPTER 5

During 1995 and 1996, just before this book was finished, several dramatic developments took place in string theory. These suggest that some of the string theories that were originally thought to be distinct are in fact just the same theories described in terms of different variables, thus reducing the number of different theories. It has also been discovered that there may be transitions between the phases described by the different perturbative string theories. Another intriguing result was that it became possible to use string theory to describe certain black holes quantum mechanically (a development that makes it very difficult to believe that string theory is not at least part of the truth.)

While this work continues, it is of course not possible to predict the outcome. The results so far make it possible to do some calculations in string theory non-perturbatively, at the very least this provides strong evidence that the theory exists beyond perturbation theory, which makes more urgent the problem of finding a non-perturbative formulation. (Perturbation theory is the name of an approximation that may be applied to cases in which small disturbances travel on a stable and uniform background. A non-perturbative formalism is one that applies to all cases.) As to whether a non-perturbative formulation will resolve the problems I discuss here—and so lead to unique predictions for the dimensionality of space and the properties of the elementary particles—opinion is presently split. This is possible because these string theories have a symmetry, not realized directly in nature, which is supersymmetry. To describe the world supersymmetry must be spontaneously broken, through some process similar to the process involving the Higgs mechanism that breaks the symmetry of the weak interactions. It seems clear that as long as supersymmetry is unbroken there are many consistent realizations of string theory, which are labeled by a set of parameters that are not fixed by the theory. Thus, the theory at this level certainly has a problem of free parameters, just as does the standard model. If this problem is to be solved by the theory, it must be in some way that is connected with how the supersymmetry is broken. At present a number of different proposals about this are being studied, but it is fair to say that there is no one that seems preferred, nor has there yet emerged a natural reason why supersymmetry must be broken. (I mean a reason internal to the dynamics of the theory, it must be broken if the theory is to describe nature.) It is reasonable to speculate that the symmetry breaking will reduce or eliminate the freedom in the parameters, and as a result many string theorists believe that the theory will eventually predict a unique set of parameters for the standard model. At the same time, it is also possible that the problem of parameters will not be resolved by the supersymmetry breaking, so that there will be many consistent descriptions of string theory even after supersymmetry is broken. In this case some cosmological mechanism, such as the one I propose in Part Two, will become necessary if string theory is to lead to predictions about the real world. Thus, at the very least, I would suggest that until it is proved unnecessary, it is reasonable to study such mechanisms, especially if they lead to predictions about the real world that can be falsified, as is the case with the proposal I made in Part Two.

Even if string theory does not make unique predictions, it may still be true, and useful. The theory of quantum electrodynamics (QED) describes the interactions of charged particles, so it in principle predicts the properties of all metals and semiconductors. But it does not tell us what material will be found in a particular region of space, or what phase that material may be found in. To know that we must know something about the history of the world. It seems to me quite possible that string theory will describe a large number of different possible realizations or phases of the world, any of which may be realized in some region of space and time, depending on its history. In this case, just as in the case of QED, whether the theory is right does not depend on its making unique predictions. It does, of course, depend on it making some unambiguous predictions that can eventually be tested, but there is no reason to believe such tests might not eventually be done.

The key problem in string theory then remains the necessity of constructing a non-perturbative formulation, which does not rely on the approximation that there exists a fixed classical spacetime background. My own view is that this will require incorporation of discoveries made in other approaches to quantum gravity such as the holographic hypothesis and the discreteness of spatial geometry as expressed by the spin-networks.

It is of course likely that, whatever else happens, principles will be discovered which reduce the number of parameters in elementary particle physics. These may come from string theory or from a grand unified theory. There is in fact an interesting proposal of a supersymmetric grand unified theory due to Dimopoulos, Hall, and Raby that reduces the number of parameters in the standard model by 6. One may argue about the extent to which this particular ansatz requires justification by some still undiscovered principle, but it remains true that it is unlikely that all the parameters of the standard model are independent. But, unless the number of free parameters is reduced to zero, a mechanism such a cosmological natural selection will still be necessary to explain our world.

PART TWO
CHAPTER 6

The reader may wonder why I claim that quantum effects eliminate may eliminate singularities in black holes, but I say nothing about the possibility that those effects may eliminate the horizons that prevent light from escaping from a black hole. The reason is that the horizons for astrophysical black holes occur in regions where the gravitational field is not terribly strong, so that there is no reason to expect that quantum effects may play an important role. Having said this, there is a possibility that quantum effects may over a very long period of time modify or eliminate the horizon. This time is at least 10^{57} times the present lifetime of the universe, for astrophysical black holes. This is because of the Hawking process, by which black holes radiate light as if they were hot bodies at a very low temperature. Left in isolation, any black hole is expected to radiate away its mass by this process. What happens then is a fascinating question, which has resisted solution in spite of a lot of work devoted to it in the last twenty years. It is possible that light trapped inside the black hole will be able finally to escape it. One possibility is that after such long times, the different regions of the universe created inside the different black holes may become accessible to each other. Thus, if these hypotheses are true, and we wait 10^{67} years, we may be able to see light coming from parts of the universe created from inside a black hole in our universe (and they may be able to see us, although we will appear to be 10^{67} light years away from them.) On the other hand, it may be that the black hole disappears after 10^{67} years, and we are completely disconnected from the world inside it. We may only hope that the theoretical problem of what happens then will be solved by this time.

Why no material that satisfies reasonable physical conditions can be opaque to gravitational waves is discussed in Smolin (1985).

CHAPTER 7

Cosmological natural selection was proposed in Smolin (1992a) and discussed in Smolin (1992b,1994b,1995a) and (1995b). The hypothesis is also discussed in books by Davies (1991); Dennett (1995) and Halperin (1995); and in papers of Ellis (1993); Rothman and Ellis (1993); Smith and Szathmáry (1996); and Barrow (1996). The reader may find other introductions to the problems of singularities and quantum gravity in Penrose (1989). Inflation is described in Linde (1994); and Kolb and Turner (1990).

Since making the proposal, I discovered that I was not the first to imagine that the laws of nature may have evolved by a process akin to natural selection. To my knowledge the earliest such proposal is by Peirce (1891). The particle theorist Yochiro Nambu (1985) suggested that the different generations might be fossils of a process by which the elementary particles evolved. Andrei Linde (1994) has often written of inflationary models in which many universes evolve, but he employs the weak anthropic principle rather than a mechanism of natural selection to explain the parameters of our universe.

The hypothesis that quantum effects might eliminate singularities and cause collapsing universes or stars to bounce is very old, and goes back to the early twentieth-century cosmologists. Modern discussions of this proposal is found in Frolov, Markov, and Mukhanov (1989,1990); and Barrabes and Frolov (1995). Two scenarios according to which string theory may predict that singularities bounce have been proposed by Emil Martinec.

The pioneering ideas of John Wheeler, who proposed that the parameters of particle physics may change at the birth of each new universe are described in Wheeler (1974).

The principles of natural selection are described in masterly books of Dawkins (1986) and Dennett (1995). Reading Dawkins together with Lovelock (1979,1988), and Margulis and Sagan (1986), while confronting the failure of string theory to lead to unique predictions for the properties of the elementary particles, was the immediate inspiration for the invention of the model described in Smolin (1992a). Descriptions of the use of fitness landscapes are found in Dawkins (1986) and in Kauffman (1995).

CHAPTERS 9 AND 10

The view of galaxies, star formation and the interstellar medium I present is grounded in a number of articles by Elmegreen (1992a, 1992b). More references are found in Smolin (1995b, 1996), which also develops this view of spiral galaxies in more detail. The models of Gerola, Schulman and Seiden; and Elmegreen and Thomasson are described in articles referenced under their name. A good view of the present situation in astrophysics and cosmology can be gotten from the papers collected in Bahcall and Ostriker (1996).

PART THREE

I want to emphasize again the extent to which my understanding of the philosophical issues raised in this part of the book is due to the insights that I have gained in conversations with friends and colleagues. Above all, Julian Barbour's view of the centrality of Leibniz's relational conception of space, time, and property for general relativity and quantum gravity underlies much of my thought in the whole book. I would like also to thank Stanley Rosen and Kay Picard for enlightening me about the importance of Plato's myth of the reversing cosmos, as well as about Boscovich's influence on Nietzsche. Conversations with them and with Paola Brancaleoni and Drucilla Cornell have given me whatever small understanding of the line of thought from Nietzsche through Derrida that I have.

CHAPTER 11

The view of life presented here owes a great deal to the influence of Harold Morowitz (1986,1987,1992). A meeting with him as an undergraduate greatly affected my interest in this subject; he also gave me then an invaluable piece of advice, which was that a good scientist focuses on questions he or she would like to understand, and is willing to learn any technique to accomplish that, rather than looking for problems to solve with the techniques he or she already knows. As many others, I was also inspired to worry about *What is Life?* by the book of Schroedinger (1945) of that title. Recent discussions with Stuart Kauffman have also had an influence on the ideas presented here.

The Gaia hypothesis is presented in Lovelock (1979,1988); and Margulis and Sagan (1986). The work on theoretical models of natural selection, which led to the discovery of the importance of collective effects is due to Stuart Kauffman (1993,1995); Bak and Sneppen; Bak and Paczuski (1994); and Paczuski, Maslov, and Bak (1994). The role of symbiosis in evolution was proposed and demonstrated by Margulis (1981). A provocative window into the debate among evolutionary theorists is found in Brockman (1995).

It is possible to imagine definitions of life different from that presented here that do not rely on the existence of stored information. It might be that a deeper understanding of the statistical physics of far from equilibrium systems would enable us to see the evolution of mechanisms of stored information as a consequence of some deeper principles of self-organization.

CHAPTER 12

I am not aware of any popular account of critical systems or the renormalization group, although these ideas have had a greater impact on the development of physics than ideas such as chaos that have been much written about. Self-organized criticality was introduced by Bak, Tang, and Wiesenfeld, and is described in a recent book by Bak (1986). Articles of Holgar Nielsen and collaborators on random dynamics may be found in Frossatt and Nielsen (1991). It seems to me that even if one believes that the answer in the end will be a string theory or grand unified theory, these papers are full of provocative hints and suggestions that are worth mulling over. More discussion about the possibility that the universe as a whole may be conceived as a critical system in Smolin (1995b) .

CHAPTERS 13 AND 14

The ideas about the foundations of mathematics given here have been inspired partly by the work of the mathematician Louis Kauffman, who by imbedding logic in topology has shown how the paradoxes of self-reference can, in non-standard logics, lead to the description of structure. More particularly, what he has done is invented a notation for a logic that can be read in drawings of knots. Related ideas have also been expressed by Gregory Chaitin and Gregory Bateson. I am also indebted to John Baez and Louis Crane for discussions on the relationship between foundational problems in mathematics and the categorical foundations of the mathematics used in quantum gravity and string theory.

I want to stress that here and in the following chapters I am not attacking religion, nor am I attacking Christianity. I am trying to understand the metaphysical presumptions that I believe are presently hampering the progress of science, and part of this involves locating their historical sources. It would be very strange if the metaphysical presumptions of the sciences that grew up in Europe during the last four centuries did not reflect, both overtly and in subtle ways, the religion which formed that civilization. Thus, I have nothing against religious persons, of the Christian or of any other faith. Nor do I have anything to say against religious faith, per se. What I am against is only the claim that modern science is somehow free of metaphysical presuppositions, for this leads to a confusing situation in which any questioning of those presuppositions is felt as an attack on science itself.

Of course there is nothing wrong with a scientist letting his or her religious or philosophical beliefs guide their choice of questions or research problems. This happens commonly all the time, and it probably helps the progress of science as much as it hinders it. Indeed, as Feyerabend (1975) argues, it is probably necessary that people be guided by larger ideas in order to thereby gain the energy and the courage needed to do good science. Given that this is the case all that can be asked is that people be as open about their views as they can be. In the end it is in any case not us, but the whole community, that will decide if something is useful or true.

The quotation at the beginning of Chapter 13 is from Roberto Mangabeira Unger (1987), page 180. The quote comes at the end of a very interesting discussion of the implications of the Big Bang cosmological model for philosophy and social theory. In that discussion he says, "You can trace the properties of the present universe back to properties it must have had at its beginning. But you cannot show that these are the only properties that any universe might have had. . . . Earlier or later universes might have had entirely different laws." Discussing the possibility of a Phoenix universe he goes on to say, "Within this cyclical extension of the standard model, the universe has a history. To state the laws of nature is not to describe or to explain all possible histories of all possible universes. Only a relative distinction exists between lawlike explanation and the narration of a one-time historical sequence." The fact that a social and legal theorist draws these kinds of conclusions from cosmology suggests the existence of common themes of the sort that I point to in the epilogue.

Another common theme between contemporary social theory and science is the investigation of the idea that society is a an "autopoietic," or self-organized, system by social theorists such as Niklas Luhmann (1987) and Drucilla Cornell (1992). According to Cornell "The central thesis of autopoiesis . . . is that legal propositions or norms must be understood within a self-generated system of communication which both defines the relations with the outside environment and provides itself with its own mechanisms of justification. Autopoiesis conserves law as an autonomous system that achieves full normative closure through epistemological constructivism. . . . Law maintains the consistency of legal reality through the very recursiveness of its system of communication." (p. 122). A few pages later she understands time as arising from the differentiation of a subsystem from the environment, "The very distinction between the system and its environment means that there is an inevitable temporalization of the system." (p.125). Perhaps I am reading in, but I hear in such discussions analogies to the issues that arise in attempts to understand what is physically observable in quantum gravity, such as I discuss in Parts Four and Five.

CHAPTER 15

The relationship between theology and the anthropic principle is discussed in the book of Ellis and in Barrow and Tipler (1986). The nonexistence of other intelligent life in the galaxy is argued by Barrow and Tipler and in the papers in the collection edited by Zuckerman and Hart (1995). The estimate of the timescale for galactic exploration comes from the article of Eric Jones in Zuckerman and Hart (1995) and from Barrow and Tipler (1986). The argument for the non-existence of intelligent life in the galaxy from its absence here is one of the most curious I have ever encountered; it seems a bit like a ten-year-old child deciding that sex is a myth because he has yet to encounter it.

PART FOUR

The best treatment of the history of ideas of space and time I am aware of is the book of Julian Barbour (1989). The view that general relativity precisely realizes Leibniz's relational conception of space and time has been gaining preeminence over earlier substantive views, at least among practitioners of relativity if not philosophers of science. This is due largely to the influence of the historical and philosophical work of Julian Barbour and John Stachel (1986,1989). Interestingly enough, an important part was played by a shift in the notation used by relativists from the original coordinate based notations to the coordinate free, geometrical notations, as epitomized in the influential textbooks of Misner, Thorne, Wheeler, and Wald.

The question has been muddied by confusion about the meaning of Mach's principle, and its applicability in general relativity. It did not help that what Einstein called Mach's principle is apparently different than what actually appears in the writings of Mach (see, for example, Feyerabend (1987) and Barbour (1990).) It is also true that general relativity does not realize Mach's principle when it is applied to the study of isolated systems defined against a background of flat empty spacetime. Fair enough, but what is relevant for the discussion of cosmology is only the question of whether general relativity fully realizes Mach's principle in the cosmological context, which it does. There is, of course, no substitute for reading Einstein himself on these questions.

The key question for the interpretation of general relativity is what constitutes an observable quantity. This problem may be to some extent avoided if one studies only special, highly symmetric solutions, but it becomes unavoidable when one confronts the quantization problem. The relational view presented here is, as far as I know, the only one that has a chance to lead to an acceptable quantum theory. Having said that, I must also mention that there are contrary views. A masterful summary of them, together with an attempt to reconcile the debate between absolute and relationalists, is in Earman's book (1989). The proposal of variety is described in Barbour and Smolin (1996); Barbour (1989); and Smolin (1991, 1992b, 1995b).

The reader may be disturbed that I have described general relativity while making no mention of geometry, curved spacetime and so on. This is because, while the analogy between spacetime, as Einstein's theory describes it, and geometry is both beautiful and useful, it is only an analogy. It has little physical content apart from the ideas I have already described here. Further, the analogy is in an important sense not completely true. It is not true that to a spacetime there corresponds a geometry, defined by a Riemannian manifold. What is true is that to the physical spacetime there correspond an infinite class of manifolds and metrics, which are related one to the other by certain transformations, which are called diffeomorphisms. These transformations preserve only those relations among the fields that describe physically observable properties of space and time.

The fact that in general relativity the physical world is represented not by geometries, but only by those relationships inherent in geometries which are preserved by these transformations, is often missed in discussions of the theory. This is a serious mistake; it is one that is easy to make and Einstein himself struggled with it for many years before coming to understand it himself. The argument that led to his understanding is called the "hole argument" and on a technical level it is the key to the interpretation of the theory. I refer the reader to discussions of it by Stachel and Barbour, which have settled the matter for most physicists I know who have considered the matter.

There are many books available on the problem of the interpretation of quantum theory, from which the reader can gain an impression of the key problems. Any serious student of this subject must not neglect a close study of the originators of the theory, including Heisenberg (1989), but especially Bohr (1958). The approach to quantum mechanics in chapter 19 is strongly influenced by Rovelli (1996).

For purposes of historical accuracy I should also point out that I have somewhat simplified the story in Chapter 18 for pedagogical purposes. In particular, Galileo believed in a principle of inertia different from the modern one, which was that things left to themselves move in perfect circles around the Earth, rather than straight lines. This is a perfect example of how key new ideas often first arise mixed up with vestiges of an old conceptual framework.

PART 5

A lot of attention has been given to the problems of the interpretation of quantum theory by people in quantum gravity, primarily because of the problem of extending quantum theory to cosmology. The students of this question will find invaluable the discussions of the many worlds interpretation in the book of DeWitt and Graham (1973), but they should also consider the contrary views presented in Penrose (1989, 1994), Shimony (1986), and Smolin (1983, 1984b). One approach to non-local hidden valuables theory is in Smolin (1983, 1986). In recent years the proposals for consistent histories interpretations have been much debated. The criticisms of Dowker and Kent seem formidable, unless one is willing to contemplate the possibility that natural selection is responsible for the fact that the world we see obeys Newtonian physics on the right scales, as proposed by Gell-Mann and Hartle. The view that quantum theory must be modified to incorporate quantum gravity is put forward most forcefully by Penrose (1989, 1994). I must emphasize that this is a field on which experts disagree strongly.

The pluralistic interpretation of quantum cosmology described here has been developed in papers by Louis Crane (1994, 1995), Carlo Rovelli (1995), and the author. Louis Crane is responsible for the proposal that category theory is relevant for quantum gravity and cosmology. Related mathematical developments are found in papers by Louis Crane in collaboration with Igor Frenkel and David Yetter, as well as in papers by John Baez, John Barrett, and their collaborators. The holographic hypothesis is due to Gerard 't Hooft and Leonard Susskind, and was inspired by results and conjectures of Jacob Bekenstein.

The discovery that Einstein speculated about a discrete physics is due to historical researches of John Stachel. The developments I describe here began with the formalism invented by Abhay Ashtekar, which has been the basis on which much important work in quantum gravity has been since built. The loop representation was invented independently for QCD by Rodolfo Gambini and Antony Trias and for quantum gravity by Carlo Rovelli and the author, based on earlier results I obtained with Ted Jacobson and Paul Renteln. The role of loops was inspired by the works of Alexander Migdal, Alexander Polyakov, and Kenneth Wilson, who applied them first to QCD. The picture of discrete quantum geometry described here was invented by Carlo Rovelli and the author. Criticisms and suggestions by Renata Loll and Michael Reisenberger have also played an important role. A related set of rigorous mathematical results, which were inspired by and confirm this picture, were developed by Rayner, Abhay Ashtekar and Chris Isham, John Baez and Abhay Ashtekar, Jurek Lewandowski, Don Marolf, Jose Morao and Thomas Thiemann. A somewhat different, but closely related development of the loop picture has been developed by Rodolfo Gambini, Jorge Pullin and their collaborators. Others whose contributions, suggestions or criticisms have been invaluable include Roumen Borissov, Berndt Bruegmann, Riccardo Capovilla, John Dell, Viqar Husain, Hideo Kodama, Juniche Iwasaki, Karel Kuchar, Seth Major, Fotini Markopoulou, Lionell Mason, Vince Moncrief, Ted Newman, Roger Penrose, Jorge Pullin, Chopin Soo, Edward Witten, and Joost Zegwaard.

Finally, it goes without saying that there are many interesting developments in quantum gravity not mentioned here. Among the most interesting of these are the non-communicative geometry of Alain Connes, David Finkelstein's spacetime code, Roger Penrose's Twistor theory, the null structure approach of Frittelli, Kozemeh, and Newman, and the causal sets of Rafael Sorkin and collaborators. It would require a book to do justice to the variety of ideas and technical developments underway in this field.

ACKNOWLEDGMENTS

It is impossible to give sufficient thanks to the people who gave help and advice (both taken and ignored) during the writing of this book. Friendship, collaborative work and arguments, stretching over many years with Julian Barbour, Louis Crane, and Carlo Rovelli have first of all provided the framework within which my thinking has developed. During the writing of the book, my ideas were often shaped in conversations with Saint Clair Cemin, and Donna Moylan, whose insights and encouragments were as invaluable as the examples of their own work. I am also indebted to a number of people for critical comments and expert advice on drafts of all or part of the manuscript. These include, in astronomy, Jane Charlton, Bruce Elmegreen, Eric Feigenbaum, Piet Hut, Martin Rees and Larry Schullman; in biology, Stuart Kauffman and Harold Morowitz; in mathematics and physics, Louis Kauffman, Chris Isham, Ted Jacobson, Fotini Markopoulou, Silvia Onofrie, Roger Penrose, Andrew Strominger, and John Stachel; and in philosophy, Paola Brancaleoni, Drucilla Cornell, Simon Saunders, and Abner Shimony. Conversations with several others were critical in helping me form my point of view on several subjects, including Abhay Ashtekar, John Baez, Per Bak, John Barrow, Gerald Brown, Mauro Carfora, Greg Chaitin, Paola Ceseri, Shyamoli Chaudhuri, Elizabeth Curtiss, John Dell, Freeman Dyson, Valerie Frolov, James Hartle, Gary Horowitz, Anna Jagren, Alejandra Kandus, Karel Kuchar, Peter Meszaros, Holgar Nielson, T. Padmanhaban, James Peebles, Anna Pigozzo, Jorge Pullin, Amelia Rechel-Cohn, Stanley Rosen, Alan Sokal, Steven Shenker, Frank Shu, David Smolin, Leonard Susskind, Gerard 't Hooft, and John Archibald Wheeler. Useful advice and feedback was also provided by Ioanna Makrakas, Julius Perel, and Madeline Schwartzman. Several anonomous readers also provided useful criticisms, which I have tried to answer in the final draft. Most of all my mother and father, Pauline and Michael Smolin, provided invaluable advice about style, in all its aspects.

I owe a great debt to John Brockman and Katinka Matson, for guiding this book through the vicissitudes of the publishing world, as well as for encouragement and friendship. My first editor, Peter Guzzardi, provided guidance at the early stages, and I am indebted to my editor at Oxford, Kirk Jensen, for many suggestions and much encouragement. During the period of the writing the book I have also benefited from the stimulating atmospheres for research and reflection provided by The Institute for Advanced Study in Princeton, The Newton Institute of Cambridge University, The Rockefeller University, and SISSA in Trieste. The National Science Foundation has supported my research for many years, and through them I owe a debt both to the community of relativists and to the American people at large for support that makes this kind of work possible. Finally, it is a pleasure also to thank Angela and Franco Rovelli, Alejandra Kandus, Engelbert Shucking, and Alan Sokal for providing havens in which much of the writing was done.

1990–1996
Syracuse; Verona; Buenos Aires;
State College; New York City.

Selected Bibliography

The following is a representative, but far from complete, list of references for the subjects mentioned in this book. In the case of books, I have limited myself to those that are either written for lay people or to a large extent can be followed by them Following this is a selective listing of articles in which ideas proposed here are discussed For a more complete list of references, please consult http://www.phys.psu.edu/SMOLIN/book on the internet.

J. Bahcall and J. P. Ostriker. *Unsolved Problems in Astrophysics.* Princeton: Princeton University Press, 1996.

P. Bak. *How Nature Works.* Springer-Verlag, New York, 1996.

J. B. Barbour. *Absolute or Relative Motion? A Study from the Machian Point of View of the Discovery and Structure of Dynamical Theories.* Cambridge: Cambridge University Press, 1989.

John Barrow and Frank Tipler. *The Anthropic Cosmological Principle.* New York: Oxford University Press, 1986.

Niels Bohr. *Atomic Physics and Human Knowledge.* New York: John Wiley and Sons, 1958.

John Brockman. *The Third Culture.* New York: Simon and Schuster, 1995

Drucilla Cornell. *The Philosophy of the Limit.* New York: Routledge, 1992

Charles Darwin. *The Origin of the Species by Means of Natural Selection or the Preservation of Favored Races in the Struggle for Life.* London: Penguin Classics, 1986.

Charles Darwin. *The Variation of Animals and Plants under Domestication.* New York: Organe Judd, 1868.

Paul Davies. *The Mind of God* New York: Simon and Schuster, 1991.

Paul Davies. *The Accidental Universe*. Cambridge: Cambridge University Press, 1982.

Paul Davies and John Gribbin. *The Matter Myth* New York: Simon and Schuster, 1992.

Richard Dawkins. *The Extended Phenotype*. San Francisco: W. H. Freeman, 1982.

Richard Dawkins. *The Blind Watchmaker* . London: Longman, 1986.

Daniel Dennett. *Darwin's Dangerous Idea*. New York: Simon and Schuster, 1995.

Bryce S. DeWitt and R. Graham (eds.) *The Many Worlds Interpretation of Quantum Mechanics*. Princeton: Princeton University Press, 1973.

John Earman. *World Enough and Space-Time*. Cambridge MA: MIT Press, 1989.

George Ellis. *Before the Beginning: Cosmology Explained*. London: Borealean Press, 1993.

Albert Einstein. "Autobiographical Notes," in *Albert Einstein: Philosopher Scientist*. ed. P. A. Schilpp. La Salle: Open Court, 1949.

Albert Einstein. *Relativity: The Special and the General Theory*. New York: Bonanza Books, 1961.

Paul Feyerabend. *Against Method: Outline of an Anarchistic Theory of Knowledge*. London and New York: Verso, 1975.

Paul Feyerabend. *Farewell to Reason*. London and New York: Verso, 1987.

C. D. Frossatt and H. B. Nielsen. *Origin of Symmetries*. Singapore: World Scientific, 1991.

Murray Gell-Mann. *The Quark and the Jaguar: Adventures in the Simple and the Complex*. New York: Freeman, 1994.

G. Greenstein. *The Symbiotic Universe*. New York: Morrow, 1988.

John Gribbin. *In the Beginning: After COBE and Before the Big Bang*. New York: Little Brown, 1993.

Paul Halperin. *The Cyclical Serpent*. New York: Plenum, 1995.

S. Hawking. *A Brief History of Time*. New York: Bantam, 1988.

W. Heisenberg. *Physics and Philosophy* . London: Penguin, 1989.

John Holland. *Hidden Order,*. Reading: Helix Books, 1995.

Fred Hoyle. *Galaxies, Nuclei and Quasarss*. London: Heinemann, 1965.

Stuart Kauffman. *The Origins of Order: Self-organization and Selection in Evolution*. New York: Oxford University Press, 1993.

Stuart Kauffman. *At Home in the Universe*. New York: Oxford University Press, 1995.

E. W. Kolb and M. S. Turner. *The Early Universe*. New York: Addison-Wesley, 1990.

Leibniz. *The Monadology* and *The Leibniz Clark-Leibniz correspondence* in *Leibniz, Philosophical Writings* . ed. G.H.R. Parkinson, trans. M. Morris and G.H.R. Parkinson. London: Dent, 1973.

James Lovelock. *Gaia: A New Look at Life on Earth*. New York: Oxford University Press, 1979.

James Lovelock. *Ages of Gaia*. New York: W.W. Norton, 1988.

Benoit Mandelbrout. *The Fractal Geometry of the Universe*. San Francisco: Freeman, 1982.

Lynn Margulis. *Symbiosis in Cell Evolution*. San Francisco: Freeman, 1981.

Lynn Margulis and Dorion Sagan. *Microcosmos*. New York: Simon and Schuster, 1986.

Harold Morowitz. *Energy Flow in Biology*. New York: Academic Press, 1968.

Harold Morowitz. *Cosmic Joy and Local Pain*. New York: Scribners, 1987.

Harold Morowitz. *Beginnings of Cellular Life*. New Haven: Yale University Press, 1992.

Isaac Newton. *Mathematical Principles of Natural Philosophy*. Trans. A. Otte and F. Cajori. Berkeley: University of California Press, 1962.

Abraham Pais. *Inward Bound*. Clarendon Press, 1986.

J. Peebles. *Principles of Physical Cosmology* Princeton: Princeton University Press, 1993.

C. S. Peirce. "The Architecture of Theories" in *The Monist*, reprinted in *Philosophical Writings of Peirce*, ed. J. Buchler. New York: Dover, 1955.

Roger Penrose. *The Emperor's New Mind*. London: Oxford University Press, 1989.

Roger Penrose. *Shadows of the Mind*. London: Oxford University Press, 1994.

Ilya Prigogine. *From being to becoming*. San Francisco: Freeman, 1980.

Erwin Schroedinger. *What Is Life?* Cambridge: Cambridge University Press, 1945.

Roberto Mangabeira Unger. *Social Theory: Its Situation and Its Task.*. Cambridge: Cambridge University Press, 1987.

Steven Weinberg. *The First Three Minutes*. Glasgow: Collins, 1977.

Steven Weinberg. *Dreams of a Final Theory*. New York: Pantheon Books, 1992.

Ben Zuckerman and Michael Hart, eds. *Extraterrestrials, Where Are They?*, second ed. Cambridge: Cambridge University Press, 1995.

SELECTED ARTICLES

J. B. Barbour, 1993. "The emergence of time and its arrow from timelessness," in *Physical Origin of Time Asymmetry* eds. J. Halliwell, J. Perez-Mercades and W. Zurek. Cambridge University Press.

J. B. Barbour and L. Smolin, 1996. "Complexity, variety and cosmology," Penn State Preprint.

C. Barrabes and V. P. Frolov, 1995. "How many worlds are inside a black hole?" hep-th/9511136, Phys. Rev. D53 (1996) 3215.

John D. Barrow, 1996. "The gravitational selection of universes by black hole production: Some comments." Preprint.

N. Bohr. "Discussions with Einstein on epistemological problems in atomic physics" in *Albert Einstein: Philosopher Scientist*. ed. P. A. Schilpp. La Salle: Open Court, 1949.

T. Breuer, "The impossibility of accurate state self-measurements." Philosophy of Science, 3056.

B. J. Carr and M. J. Rees, 1979. Nature 278, 605.

B. Carter, 1967. "The significance of numerical coincidences in nature," unpublished preprint, Cambridge University.

B. Carter, 1974. in *Confrontation of Cosmological Theories with Observational Data*, IAU Symposium No. 63, ed. M. Longair Dordrecht: Reidel. p. 291.

L. Crane, 1994. "Possible implications of the quantum theory of gravity" preprint hep-th/9402104.

L. Crane, 1995. "Clocks and Categories: is quantum gravity algebraic?" J. Math. Phys. 36, 6180.

F. Dowker and A. Kent, 1996. J. Stat. Phys. 82, 1575.

G.F.R. Ellis, 1993. "The physics and geometry of the universe: changing viewpoints." Q. J. R. astr. Soc. 34, 315–330.

B. G. Elmegreen, 1992a. "Triggered Star Formation," in the Proceedings of the III Canary Islands Winter School, 1991, eds. G. Tenorio-Tagle, M. Prieto and F. Sanchez. Cambridge University Press,.

B. G. Elmegreen, 1992b. "Large Scale Dynamics of the Interstellar Medium", in *Interstellar*

Medium: Processes in the Galactic Diffuse Matter , ed. D. Pfenniger and P. Bartholdi, Springer-Verlag.

B. G. Elmegreen and M. Thomasson, 1993. "Grand design and flocculent spiral structure in computer simulations with star formation and gas heating." Astron. and Astrophys., 272, 37.

V. P. Frolov, M. A. Markov and M. A. Mukhanov, 1989. "Through a black hole into a new universe?" Phys. Lett. B216 272–276.

A. Lawrence and E. Martinec, 1996. "String field theory in curved space-time and the resolution of spacelike singularities," Class. and Quant. Grav. 13, 63; hep-th/9509149.

A. Linde, 1994. "The self-reproducing inflationary universe," Scientific American 5, 32–39.

E. Martinec, 1995. "Spacelike singularities in string theory." Class. and Quant. Grav. 12, 941-950. hep-th/9412074

Y. Nambu, 1985. Prog. Theor. Phys. Suppl. 85,104.

N. Luhmann, 1987, "Closure and openness: On reality in the world of law", in *Autopoietic Law: A New Approach to Law and Society,* ed. Gunther Teubner, trans. Ian Fraser, Berlin: de Gruyter.

M. Paczuski, S. Maslov and P. Bak, 1994. Europhys. Lett. 27, 97; 28, 295.

R. Penrose, 1971a. in *Quantum Theory and Beyond.* ed. T. Bastin. Cambridge: Cambridge University Press, 1971.

R. Penrose, 1979b. "Singularities and Time Asymmetries" in *General Relativity An Einstein Centenary* . ed. S. W. Hawking and W. Israel. Cambridge: Cambridge University Press, 1979.

M. Rees, MNRAS 176 (1976) 483;

T. Rothman and G.F.R. Ellis, 1993. "Smolin's natural selection hypothesis," Q. J. R. astr. Soc. 34, 201–212.

C. Rovelli, 1995. *Relational Quantum Mechanics.* International Journal of Theoretical Physics 55, 1637.

C. Rovelli, 1997. "Half way through the woods," in *The Cosmos of Science,* J. Earman and J. D. Norton, eds., University of Pittsburgh Press.

C. Rovelli and L. Smolin:, 1988. Phys. Rev. Lett. 61, 1155;

C. Rovelli and L. Smolin, 1990. Nucl. Phys. B133, 80.

P. E. Seiden and L. S. Schulman, 1990. "Percolation model of galactic structure," Advances in Physics, 39, 1–54.

Shimony, Abner, 1986. In *Quantum Concepts in Space and Time,* R. Penrose and C. J. Isham, eds., Oxford University Press.

J. M. Smith and E. Szathmáry, 1996. Nature 384, 107.

L. Smolin, 1983. "A Derivation of quantum mechanics from a deterministic, non-local hidden variables theory I. The two dimensional theory," Institute for Advanced Study preprint.

L. Smolin, 1984b. "Quantum gravity and the many worlds interpretation of quantum mechanics," in *Quantum Theory of Gravity: Essays in Honor of the 60th Birthday of Bryce S. DeWitt,* S.M. Christensen, ed. Bristol: Adam Hilger.

L. Smolin, 1985. Gen. Rel. and Grav., 17, 417.

L. Smolin, 1976. In *Quantum Concepts in Space and Time,* op. cit.

L. Smolin, 1991. "Space and Time in the Quantum Universe" in *Conceptual Problems of Quantum Gravity* ed. by A. Ashtekar and J. Stachel, Boston: Birkhauser.

L. Smolin, 1992a. "Did the Universe Evolve?" Class. Quantum Grav. 9, 173–191.

L. Smolin, 1992b. "Time, structure and evolution," Syracuse Preprint (1992); to appear (in Italian translation) in the Proceedings of a conference on *Time in Science and Philosophy* held at the Istituto Suor Orsola Benincasa, in Naples (1992) ed. E. Agazzi.

L. Smolin, 1993. "Recent developments in nonperturbative quantum gravity," in *Quantum Gravity and Cosmology*. Singapore: World Scientific.

L. Smolin, 1994. "On the fate of black hole singularities and the parameters of the standard model" gr-qc/9404011/

L. Smolin, 1995a. "Cosmology as a problem in critical phenomena," in the proceedings of the Guanauato Conference on Self-organization and binary networks, Springer.

L. Smolin, 1995b. J. Math. Phys. 36, 6417.

L. Smolin, 1996. Galactic disks as reaction diffusion systems, astro-ph/9612033.

John Stachel, 1986 "What a physicist can learn from the discovery of general relativity", in *Proceedings of the Fourth Marcel Grossmann Meeting on Recent Developments in General Relativity*, ed. R. R Ruffini. Amsterdam: North-Holand.

J. Stachel, 1989. "Einstein's search for general covariance 1912-1915," in *Einstein and the History of General Relativity* ed. by D. Howard and J. Stachel, Einstein Studies, Volume 1 Boston: Birkhauser

John Stachel, 1993. *The Other Einstein*, Context 6, 275.

L. Susskind, 1995. "The world as a hologram," J. Math. Phys. 36, 6377. hep-th/9409089.

G. 't Hooft, "Dimensional reduction in quantum gravity" Utrecht preprint THU-93/26; gr-qc/9310006.

J. A. Wheeler, 1974. "Beyond the end of time," in *Black Holes, Gravitational Waves and Cosmology: An Introduction to Current Research*. eds. Martin Rees, Remo Ruffini and John Archibald Wheeler. New York: Gordon and Breach.

Glossary of Scientific and Philosophical Terms

Note: *Italicized* words used in a definition are defined here as well.

absolute space and time The philosophy which underlies *Newtonian physics*, according to which space and time exist eternally and are fixed and unchanging, independent of whatever is in the universe.

acceleration The rate of change of the speed and direction (also called velocity) of motion.

black hole A region of space and time where the gravitational field is so strong that nothing can escape, because the velocity required to do so is larger than that of light, which nothing can travel faster than. They are believed to form often from the collapse of a massive star, when it runs out of nuclear fuel and can no longer support its own weight.

category theory is a general framework for mathematics, in which any mathematical structure is described in terms of a set of objects and the relationships between them.

critical system A macroscopic system in which structure can be observed at every length scale from the atomic scale to the size of the system itself. A related property is that every part is correlated with every other part, in the sense that if any one part is perturbed the influence can be felt in any other part of the system.

diffeomorphism invariance The mathematical expression of the principle that space and time are an expression only of the relationships between events in the world.

dynamical Something is dynamical if it changes in time as a result of interactions with other things in the world, subject to the laws of motion.

Einstein's Equation is the precise expression, in the general theory of relativity of the dynamics that relate the geometry of space and time to matter. Einstein's Equation is not $E=mc^2$; it is another, much more complicated equation.

entropy A measure of the disorganization of a system. The lower the entropy the more information can be gotten from observing the configuration of the system. The higher the entropy, the more the configuration is random and uniform, with no significant structure or features.

equilibrium A state of a macroscopic system in which the *entropy* is as high as possible. This means that the configuration is completely uniform. This often means that every part of the system has the same density and temperature.

feedback mechanism Allows a system to act back on itself. There are two kinds; negative feedback acts to restore the system to some preferred state, like a thermometer, so that it acts to reverse changes away from that state; positive feedback is the opposite—it acts to amplify departures from some given state. All the cases I discuss are negative feedback mechanisms.

fitness landscape An abstract representation of the concept of the fitness of a gene, used in evolutionary theory. One imagines an abstract space where the different points correspond to different DNA sequences. The fitness landscape assigns to each point a number, which is analogous to the altitude of a landscape, which is the average number of progeny a creature with those genes would have.

frame of reference consists of an observer, a clock, and a coordinate grid, which makes it possible to describe the motion of objects with respect to the observer.

Gaia hypothesis The conjecture that the climate, as well as the composition of the atmosphere and ocean of the Earth are controlled by feedback effects involving living organisms.

gauge principle The basic principle behind the *standard model,* which expresses the idea that all physical properties of elementary particles are defined in terms of interactions between them. A modern expression of Leibniz's *principle of the identity of the indiscernible.*

general relativity Einstein's theory of gravity, according to which the geometry of space and time is dynamical and contingent.

giant molecular cloud Very cold clouds of gas and dust found in spiral galaxies, within which stars form. The name refers to the fact that the gas contains molecules rather than single atoms, including many organic compounds.

heat death of the universe The name given by nineteenth century science to the belief that the whole universe is tending to equilibrium, after which all life, change and structure would disappear.

Higgs boson A kind of particle, so far not observed, whose existence is required by the standard model of elementary particle physics. Their existence is associated with the mechanism of spontaneous symmetry breaking.

inertial frame A *frame of reference,* with respect to which objects moving freely, without external influence, move along straight lines at a constant speed.

interstellar medium Something like a galactic atmosphere consisting of clouds of gas and dust. In the disks of spiral galaxies they have a density ranging from about one atom for every ten-thousand cubic centimeters to thousands of atoms per cubic centimeter.

manifold A continuous space such as a real number line.

the metric of spacetime A mathematical description of the geometry of space and time, which gives the size of distance and time intervals and determines which *frames of reference* are *inertial*.

neutron star A dead star, consisting of matter the density of an atomic nucleus, with a mass of about one and a half times that of the sun compacted in a diameter of a few kilometers. One of the common remnants of a *supernova* explosion.

neutrino A fundamental particle with no charge that interacts with matter only through the *weak nuclear interaction*. It's mass is not presently known, but it is much less than that of an electron.

Newtonian physics The theory of space, time and motion, published by Isaac Newton in 1687 and the basis of physical science until the beginning of the twentieth century.

newtonian Any theory based on the basic principles of space, time and motion used by Newton, which encompasses all physical theories developed in the eighteenth and nineteenth centuries.

non-equilibrium system A system whose configuration is very different than *equilibrium*.

non-equilibrium thermodynamics The study of non-equilibrium systems, among them self-organized systems.

non-linear equation or process is any one in which the response or output is not directly proportional to the input. Its significance is that all linear equations can be solved, whereas, most non-linear equations are understood only through approximations. Any *feedback mechanism*, or process of *self-organization* is non-linear.

parameters are numbers that come into the description of any physical system that can be freely adjusted. Examples, in the case of elementary particle physics, are the masses of particles and the strengths of the interactions.

perturbation A small disturbance of a physical system

phase One of several possible different configurations of some material, such as ice, water and vapor, in the case of water.

phase transition A change from one phase to another, such as melting or evaporation.

Planck Length The basic unit of length in a quantum theory of gravity, equal to approximately 10^{-33} centimeters.

Planck Mass The basic unit of mass in a quantum theory of gravity, equal to about 10^{19} times the mass of the proton.

Platonism The philosophy that mathematical concepts refer to entities that have some real and eternal existence that transcends the world we perceive directly.

principle of sufficient reason Leibniz's principle according to which any question that can be asked about why the world is one way rather than another must have a rational answer. Used to eliminate theories, such as Newton's *absolute space and time*, that allow one to ask questions that don't have such answers.

principle of the identity of the indiscernible A corollary of the principle of sufficient reason, according to which there can be no two things in the world that are distinct, but which share all their properties.

quantum chromodynamics Part of the *standard model*, that explains how the quarks make up protons, neutrons and other strongly interacting particles.

quantum gravity or **quantum theory of gravity** The hoped for theory that would unify in one framework both quantum and gravitational physics.

quantum mechanics Presently the basic theory of matter and motion, developed during the 1920s.

quark One of the fundamental particles which make up protons, neutrons and other particles that feel the strong interaction.

radical atomism The philosophy according to which the properties of the elementary particles are given eternally, for all time, independently of the history of the universe or its present state.

reductionism A methodological principle according to which the properties of any system made of parts can be understood by knowing what it is made of and how those parts interact with each other.

relational philosophy of space and time The philosophy that space and time are no more than aspects of the relationships among events.

shock wave A pressure wave, like a sonic boom, which carries energy and material through a medium, which here is the *interstellar medium.*

self-organized system A system in which a high degree of structure and organization has arisen as result of processes internal to the system itself. A **self-organized critical system** is one that is, additionally, a *critical system*, in the sense that it has structure on all scales.

singularity An event in which a fundamental physical quantity such as the density of matter of the strength of the gravitational field becomes infinite. This may occur in a region of space in which the gravitational attraction of a star has caused it to collapse to infinite density.

spontaneous symmetry breaking A process by which a symmetry present in the laws of nature is not realized by a particular physical system.

standard model of elementary particle physics The basic theory of elementary particle physics, which unifies the electromagnetic, weak and strong interactions, but not gravity. Based on the *gauge principle* and *Yang-Mills theory.*

statistical mechanics A set of concepts for dealing with large numbers of physical systems through statistical reasoning, such as the molecules that make up gases or materials.

string theory A unified theory of the interactions of elementary particles, that successfully includes gravity and *Yang-Mills fields*. Under intensive development since the early 1980's, it is based on the postulate that the fundamental entities in the world have a one dimensional rather than a pointlike character.

strong interaction The force that binds protons, neutrons, and other particles together into atomic nuclei.

supernova The explosion that takes place in most stars significantly more massive than the sun, after it has burned all its nuclear fuel. The result is to return most of the material of the star to the *interstellar medium*.

thermodynamics The general study of energy, *entropy* heat, and their relationship with matter, developed in the nineteenth century, now largely superseded by *statistical mechanics*.

thermodynamic equilibrium See *equilibrium*.

topology The study of the connectivity of a set, without regard to numerical measures such as distance.

upper mass limit The largest possible mass a *neutron star* may be without collapsing under its own self-gravity to a *black hole*.

velocity The speed and direction of motion.

weak interaction The least known of the four basic interactions, which is responsible for radiative decay and the interactions of neutrinos with matter.

Yang-Mills theory The basic theoretical structure underlying the *standard model*, based on the *gauge principle*, developed in the 1950s.

Index